Quantum Thermodynamic Processes

Quantum Thermodynamic Processes

Energy and Information Flow at the Nanoscale

Günter Mahler

Pan Stanford Publishing

Published by

Pan Stanford Publishing Pte. Ltd.
Penthouse Level, Suntec Tower 3
8 Temasek Boulevard
Singapore 038988

Email: editorial@panstanford.com
Web: www.panstanford.com

British Library Cataloguing-in-Publication Data
A catalogue record for this book is available from the British Library.

Quantum Thermodynamic Processes: Energy and Information Flow at the Nanoscale

ISBN 978-981-4463-73-7 (Hardcover)
ISBN 978-981-4463-74-4 (eBook)

Printed in the USA

To
Anne, Anja, Maike, and Nils

Contents

Preface

The point of departure of this book is a triad of themes: information theory, thermodynamics, and quantum mechanics. These are related: thermodynamics and quantum mechanics form the basis of quantum thermodynamics; information and quantum mechanics underly, inter alia, the notorious quantum measurement problem; information and thermodynamics have much to say about control limits in the tension between micro- and macro-descriptions.

Virtually all theories are "contextual" in the sense that specific properties of a system cannot be identified without reference to contingent contexts external to the system. It is being appreciated only recently that such features—in various disguises—are in fact deeply rooted in all three basic themes mentioned above.

Why does the world around us typically look thermal—from cosmology down to individual embedded spins? Do informational measures constitute additional (independent) parameters beyond physical ones? Is the transition between mechanical and thermal systems gradual or discontinuous? Properties, resources, control limits of the observational tools become decisive.

Pertinent examples can be found in various processes implemented on small quantum systems. Particularly attractive are model systems which can be treated thermodynamically, but—to some extent—also exactly, that is, based on pure quantum dynamics. This possibility may be seen as a special feature of nano-thermodynamics.

On a given level of description such processes can typically be associated with certain "functions". In this sense the book aims at a modern perspective of nanoscale applications, defined here as the potential realization of such functions as constrained by given resources.

The style is not primarily technical; while the "quantum engineer" may locate interesting challenges for further investigations, the prime target is to help establish this field as an interesting subject of its own, for teaching and research alike. This is why considerable room is given to conceptional and fundamental aspects; these are expected to change less in the near future than the body of examples, to which they can be applied. An in-depth discussion of selected case studies is provided. The field of thermodynamic processes is still in progress, completeness is not attempted.

Günter Mahler
Institut für Theoretische Physik 1
Universität Stuttgart
November 2014

Acknowledgments

I thank Armen Allahverdyan (Yerevan), Harald Atmanspacher (Freiburg), George Ellis (Capetown), Hermann Haken (Stuttgart), Thomas Jahnke (Stuttgart), and Dominik Janzing (Tübingen) for valuable discussions.

Chapter 1

Introduction

Jonas Burgert: Head against stillness, 2006, based on photograph by Lepkowski Studios [Reprinted with permission from Schreiber and Wipplinger (2011)].

Quantum Thermodynamic Processes: Energy and Information Flow at the Nanoscale
Günter Mahler

ISBN 978-981-4463-73-7 (Hardcover), 978-981-4463-74-4 (eBook)
www.panstanford.com

Anaximander (610–546 BC) appears to have envisioned the universe as an eternal, continuous process. He would be among the first thinkers to design a cosmology entirely in physical terms, that is, based on observation and rational thinking [Rovelli (2011)]. For the first time, the scientific method emerged—made possible by a society, which succeeded to suppress dogmatism and religious prejudice. Anaximander became the teacher of Pythagoras, who extended his approach by applying mathematical concepts to natural phenomena.

Today we do not know any ultimate scientific reason for the universe to exist. Its creation *ex nihilo* seems implausible.[a] As there is no external control left for the universe as a whole, its development can only be a kind of self-ordering process. Nevertheless, its description, being based on a network of interdependent subsystems, typically calls for an artificial partition into "given constraints" (like the expansion of the universe) and affected physical variables.[b] In fact, a large part of cosmology uses the language of thermodynamics (see, e.g., Ref. [Cheng (2005)]). This is an example of an effective theoretical description—defining what should count as phenomena on this large scale. There is even room for novelty: "It is the evolution itself that constructs the subspace of states that is needed to describe the possible futures of any given state. And by doing so the construction gives us an intrinsic notion of time" [Kauffman (1997)]. Some sort of circularity appears to be unavoidable in the process of understanding.

1.1 Effective Theories

Large parts of fundamental physics are concerned with the search for "ultimate constituents," which derive their appeal from a basic type of reductionism. It was P. W. Anderson [Anderson (1972)], among others, who challenged the view that it should be possible

[a] Recently there have been some speculations to the contrary; but these violate the observation that there is no way to proof something from nothing.

[b] This feature may be seen in an even broader context: The so-called Duhem–Quine thesis states that no scientific hypothesis can be tested in isolation; there are always auxiliary assumptions to be taken for granted.

to reconstruct the universe from such fundamental entities or laws: He argued that phenomena of emergence would imply an intrinsic decoupling between different levels in the hierarchy of theoretical descriptions. As formulated by S. S. Schweber: "High-energy physics and condensed matter physics have become essentially decoupled in the sense that the existence of a top quark, or any heavy particle discovered at CERN or elsewhere, is irrelevant to the concerns of condensed matter physicists" [Schweber (1993)]. "Fundamental" aspects should thus show up independently on each level (domain).

In response to this kind of criticism, the notion of "effective theories" has gained popularity. "An effective theory (ET) is a theory which 'effectively' captures what is physically relevant in a given domain" [Castellani (2000)]. This idea is in accord with the observation that physics is not scale invariant. What happens at a smaller scale may become irrelevant and unobservable at a larger scale. Noteworthy examples are effective field theories (EFT). In fact, the so-called Standard Model is now understood as such an EFT, namely the low energy limit of a more fine-grained theory still to be found.

By construction effective theories are contextual, that is, their validity depends on the context, for which they have been designed. As a consequence the respective basic concepts also refer to this domain of applicability.

Effective approaches abound in quantum modeling: Systems with two discrete (relevant) states, for example, are all equivalent and can be mapped onto the spin algebra, that is, to systems with a fundamental two-level structure or the so-called qubits (cf. [Nielsen (2000)]). This fact underlies the possibility to consider physically different implementations of any given quantum model ("quantum simulation").

Likewise, electrons in the conduction band of a semiconductor can often (for small energy) be described as free particles with an effective mass. The idea of elementary excitations of solid-state many-particles systems is based on an effective description.

Effective theories are "closed," that is, they do not require the explicit input of more detailed descriptions, except for the fit of some phenomenological parameters. This "closure" is quite remarkable and further underlines the usefulness of approaches on different hierarchical levels.[a]

1.2 Partitions

But effective theories need not be related to length or energy-scales only: One may also consider effective theories to result from partitions imposed on given composite quantum systems. Any such partition can be used to introduce a relevance concept: Parts being considered as the system of interest, other parts as the environment. Given certain properties of the environment, the system of interest may exhibit some kind of time-dependent process. Appropriate questions are of the type: Given such constraints, conditions, controls, and contexts (due to the environment), what will happen in the embedded system (on some level of description)? These are phenomena in a strict sense; they are contextual. Examples abound in the physical, chemical, and biological domain. Often they do not so much differ on scales but on the type of partitioning. Certain partitions may introduce preferred scales, though.

Many models of classical mechanics are based on a partition: An external potential, whether constant or time dependent, derives from the interaction with other physical subsystems, which are thus reduced to some (approximate) effect on the system considered.

Thermodynamics with its emphasis on macrovariables (such as internal energy, entropy, volume) is often considered to be reducible to an underlying microscopic level. Originally, this underlying level has been taken to be classical mechanics. This program has been successful provided one accepts certain additional assumptions like the ergodic hypothesis. Quantum thermodynamics (see, e.g., Ref. [Gemmer (2009)] and Chapter 5) aims at a derivation of thermodynamics directly from quantum mechanics. This has been achieved by embedding the object into a sufficiently large quantum environment under the condition of weak coupling. Local thermal equilibrium can thus be considered as an effective property

[a]This closure has nothing to do with the philosophical problem of "physical closure," that is, whether or not the only causes of physical events are other physical events. This question seems to be undecidable [Bishop (2006)].

emerging from the respective partition. Unless the total (closed) system has accidentally been prepared in a stationary or thermal state, it will show eternal unitary motion (coherence) in its huge state space. Nevertheless a small subsystem will typically show stationary thermal properties; this surprising feature will eventually disappear as the size of the respective subsystem is made a larger and larger fraction of the whole.

Different partitions may be said to represent different reference frames. However, the relation between different partitions is not like that for special relativity theory, which connects the same type of properties as seen from different reference systems. Instead, different partitions are often mutually exclusive, thus generating different "realities." Phenomena are not "just out there"—they are in part constructed by the observational process (thus defining a frame).

1.3 Operational Aspects

A version of operationalism has been popularized by P. W. Bridgman almost a century ago [Bridgman (1927)]; he had been influenced by Ernst Mach and his rejection of absolute space (as operationally inaccessible). While Bridgman's extreme position, "the concept is synonymous with the corresponding set of operations," is considered somewhat outmoded today, the reference to operations still proves very useful in various fields of physics, in particular, when mathematical definitions tend to obscure the physics behind. The meaning of a concept can often be clarified once it is supplemented by a method of measurement. One may argue that any theory should eventually be "linked to procedures for obtaining knowledge" [Breuer (1995)]. The related notion of intervention is basic to the idea of causation [Ducheyne (2006)]: If this or that is done, this or that will happen.[a]

---

[a]Operational constraints have been introduced even for the universe as a whole: "Every observable in a theory of cosmology should be measurable by some observer inside the universe, and all mathematical constructions necessary to the formulation of the theory should be realizable in a finite time by a computer that fits inside the universe" [Smolin (2001)].

Information, quantum dynamics, and thermodynamics are the basic input for the study of quantum thermodynamic processes. Remarkably, these three fields are particularly suitable for an operational approach. In fact, the concept of information can hardly be introduced without reference to communicating agents. Observational quantum dynamics with its focus on measurement and statistical features differs substantially from "abstract" (isolated) quantum mechanics. Thermodynamics can be seen as a control theory, its main laws as statements about what can be done by agents equipped with limited resources.

Indeed, the expenditure of resources (cost) is a dominating (though often overlooked) theme of the operational approach. The system in question has to be prepared, designed, and controlled. The respective investments have to be considered even explicitly, when the efficiency of a process is at stake. The respective process would have to be classified as "inefficient" (intractable) once the total cost tends to diverge. On the other hand, hidden costs would undermine any realistic assessment.

Operational aspects[a] will thus constitute a recurrent theme of this book. Sometimes these aspects show up as driving forces or boundary conditions given from the outside. Often they can be "brought on stage" in the form of a specific model of an observer/agent. In either case, the operational features result from an appropriate partition of the total system.

1.4 Processes

A physical process describes a gradual change through a series of states. Obviously such processes may come in various forms, which can be characterized by the pertinent state space, the means of external control (control space), and the type of dynamics (discrete, continuous, deterministic, stochastic). Processes are primarily concerned with effective descriptions (cf. Section 1.1), much less so

[a]Scientific statements conditioned by operational aspects have been termed "weakly objective" [d'Espagnat (2003)].

with fundamental dynamics[a]. Finite processes can be considered to map some input state into an output state.

In the quantum domain, one may distinguish two limiting types of processes: the discrete (externally controlled) unitary process describing quantum computation (via the sequential application of quantum gates [Nielsen (2000)]) and the continuous incoherent process of (quantum-) thermodynamics.

The typical approach to identify quantum thermodynamic processes is via specific models. Models refer to the system as such as well as to the environment. Specifications for the environment may even have to involve models of the observer, of the agent (i.e., of explicitly operational aspects). Many implicit assumptions are, in fact, related to his abilities to isolate the system, his limits to resolve alternative states or processes, his ways to respond and to control. While it goes without saying that his interventions are subject to physical laws, it is often not so clear, what this means in actual situations. "Unattainability" eventually refers to diverging costs, rendering the corresponding physical process intractable. An example is the third law of thermodynamics, that is, the impossibility to reach zero temperature [Allahverdyan (2011b)].

Relaxation toward local equilibrium will typically be accompanied by various fluxes between system and environment. These constitute transient processes. In general, however, quantum thermodynamic processes require more complex embeddings. Typically, such embeddings depend on special design. It is a problem of its own to decide whether the respective process can be described in terms of conventional thermal currents like heat or work.

The main characteristics of a certain process (or a combination thereof) can often be summarized in terms of its "function". A function is here not meant in the mathematical sense. Rather it has to do with "purpose" (or goal)—which is problematic within physics due to its vicinity to teleological reasoning. In the context of technology and of biological systems, however, the notion of function clearly does play a major role. Indeed, merely saying that

[a]It has been argued that "every physical system corresponds to a computational process, and that every computational process, if applicable, has to be physically and operationally feasible in some concrete realization" [Leitsch (2008)].

something, X, has a specific function Z is to offer a kind of important explanation of X. For this to be the case, Z must not be accidental.

A formal definition is this [Wright (1973)]:

The function of X is Z means:

1. X is there because it does Z (i.e., "X is designed to do Z");
2. Z is a consequence of X being there (the reverse does not hold).

Typically, X refers to a "machine" or a device, i.e., a system designed for a special purpose: X = heart, Z = pumping blood. X = heat engine, Z = convert heat into work. X = mechanical filter (sieve), Z = separate small from large grains.

The above examples exhibit decreasing complexity down to the point, where one might think of a mechanical filter having been assembled "naturally," say in a river bed. This would, however, violate the second condition in the above definition of function; the filtering would be an effect rather than a function.

Also the so-called "games" are characterized by a goal (function). Occasionally, specific games will be invoked to serve as tools for describing the effective system dynamics in terms of rules and information.

Functions call for a physical explanation; but a function is usually something useful in a given context only. It often appears as the goal of the intervention undertaken by some agent. It thus transcends a mere physical description: It is part of the epistemic cut (cf. Section 2.2.1); it relates to the pragmatic aspects of information, cf. Section 2.1.

1.5 The "Leitmotif"

Basic ingredients for the study of quantum thermodynamic processes are information, quantum dynamics, and thermodynamics, see Fig. 1.1. Their overlap is partial only; tensions prevail, last not least, due to conflicting conditions of validity. Typically, these input fields are supplemented by a number of further implicit assumptions.

Quantum thermodynamic processes are based on effective descriptions adapted to and constrained by control and observation.

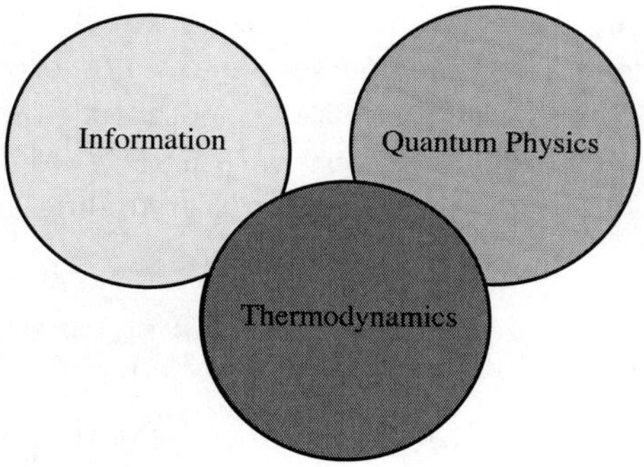

Figure 1.1 Basic input fields.

The modeling of system and observer/agent is essential on the quantum level [Goyal (2012)]. The resulting physical properties ("thermal phenomena") are neither in the system nor in the observer but are a consequence of the subtle interplay between the two.[a] Typically, the phenomena of interest are emergent rather than "fundamental." They can thus be taken as completely worked-out examples of how we can see and manipulate the world around us.

In this enterprise, information plays an important though somewhat unconventional role. A paradigmatic model of information dynamics (computation) has been the Turing machine as a formalization of a human calculation process. But this machine does not only require some implementation, it needs further embeddings: "The Turing machine model essentially presupposes a human as a part of the system—the human is the one who provides material resources, poses the questions, and interprets the answers" [Crnkovic (2011)].

For the processes considered here, information and information dynamics appear as natural (fixed) ingredients of the underlying description (as used by the observer). Effective descriptions allow for the appearance of new "rules" of behavior (beyond mere physical laws).

[a]This is also reminiscent of some modern theories of art: Art is neither in the viewer nor in the artwork but comes into being in the "esthetic process" [Bertram (2011)].

Incompleteness is a common thread in any of these approaches.[a] But effective descriptions and models represent more than unavoidable simplifications: If well selected they capture the essentials of our intuitive understanding—disregarding less important details. They typically underlie the ubiquitous "story telling" in science.[b]

---

[a]Dreams to the contrary pop up now and then: As theories of everything (TOE) or, more recently, as simulations of everything in the form of the Switzerland-based "living earth simulator (LES)" [Moyer (2012)].

[b]"Humans are story-telling species" [Myhrvold (2012)]. This does not mean anything goes: In science, at least, the stories have to pass consistency tests over and over again—as imposed by the scientific method. A never-ending iterative process by itself.

Part I

Basic Input

Chapter 2

Information

Jonas Burgert: Game, 2008, based on photograph by Lepkowski Studios [Reprinted with permission from Schreiber and Wipplinger (2011)].

Quantum Thermodynamic Processes: Energy and Information Flow at the Nanoscale
Günter Mahler

ISBN 978-981-4463-73-7 (Hardcover), 978-981-4463-74-4 (eBook)
www.panstanford.com

Summary: Information is anything but a simple concept; ongoing debates are still characterized by rather conflicting opinions. In what follows information will be considered a basic resource for any effective theory, whether in physics, biology, or other fields of research. Designed to provide a powerful description of certain aspects of reality, effective theories implement the so-called Peircean triad: object–sign–interpretant. The sign becomes information about the object for the interpretant. Information is inherently contextual.

Information should also allow for novelty: But new information cannot deterministically follow from previously obtained information. If, at least, the universe of all possibilities is known in advance, the concept of probability will serve as a means to quantify novelty.

Eventually, information is represented by physical states. It may be interpreted as a kind of "higher-level currency" to be traded in for a set of basic physical or chemical entities. As a result, the underlying (dynamical) laws of physics also apply to such an encoded information. This is the basis of computation/simulation.

However, the respective machine function cannot be seen in isolation; its meaning is derived from its embedding. Effective dynamical equations (rules) are obtained, which are much more flexible than the underlying physical laws.

There is a strange tension between physics underlying information and information within physics. As far as the latter is concerned: Lack of information appears as an efficient way to account for incomplete descriptions. In fact, almost all physical theories, in particular statistical physics and thermodynamics, let alone quantum mechanics (cf. Section 2.2.4) have to deal with uncertainty; this is a fundamental constraint rather than an independent resource located outside physics.

2.1 Historical Remarks

The earliest attempt to record information on a piece of rock is reported to have been found in the Blombos cave in South Africa near the Indian Ocean and is dated back more than 70 thousand years (see [History of Information (2011)]). At this time

Homo sapiens appears to have roamed through this area already for more than 150 thousand years, as indicated by finds of a cranium near Florisbad (north of Bloemfontein).

This early start in human history, besides being of interest in its own right, presents a remarkable challenge to our understanding of the concept of information. In a first step the objects in question have to qualify as artifacts rather than natural structures. Then, assumed to be artifacts the respective pattern have to be identified to represent information rather than mere decoration. This is hard to decide, but the systematic organization has been taken as evidence for that pattern to represent information. Of course, we cannot know what the information stood for and for whom it was. (A similar problem occurs in the context of SETI, the search of extraterrestrial intelligence: Here the goal is to convincingly identify radio signals that could be traced back to alien intelligence [SETI], cf. also Example 2.5.)

Counting and writing as a process (algorithm) may have surfaced around 10 thousand years ago in the form of neolithic tokens replacing the much older paleolithic tally sticks; the first evidence for a mathematical table has been found in Sumer, dated around 2600 bc. [History of Information (2011)].

Here is not the place to further follow the time line of ever increasing sophistication in the use and handling of information. Instead we jump right into the 20th century, where we find the notion of information introduced in physics in the late 1920s. L. Szilard [Szilard (1929)] attempted to exorcise Maxwell's demon, a creature, which could apparently beat the second law of thermodynamics. Szilard argued that the demon would need something for his actions to be operational, which he called information. But it was not before the end of the second world war, when information eventually became a key concept, thus marking the beginning of the information age.

Information theory in a mathematically codified framework started with the seminal work of Claude E. Shannon in 1948 [Shannon (1948)]. Shannon's statistical model of communication is based on three assumptions: (1) A message can be represented by a finite set of signs, the so-called alphabet, which occur with a given probability. (2) Communication is based on the selection of signs. (3) This selection is a stochastic process (performed by the source). He then proved the existence of a lower bound: If each sign of a text is coded in a binary way, then the minimal number of bits (yes-no-decisions) per sign, averaged over all signs of the alphabet, is given by S, the so-called entropy. Any coding of a message of n signs with length larger than $n \times S$ is called redundant. Because of this remarkable minimal property it was believed that S would characterize the information contained in the message. This alleged objective measurability of information has been highly welcomed within many fields of research, ranging from cybernetics to information esthetics and information psychology.

Shannon's approach has been a considerable step forward, but uncritical reception also contributed to widespread misunderstandings about the nature of information [Rechenberg (2003)]. Entirely missing are the semantic (meaning) and pragmatic (action oriented) aspects of information. Meaning and action are contextual, that is, relate to the receiver of the message and thus cannot be an absolute property of the message as such. Two pertinent examples for these limitations have been mentioned above: the tentative interpretation of historical artifacts and the SETI problem. A way to deal with this situation is, for example, the so-called semiotics as developed by Peirce (see, e.g., [Herrmann-Pillath (2010)]). Peirce distinguishes between the object, the sign, and the interpretant, the so-called "Peircean triad."[a] This triad constitutes a special "partition" of the scenario under consideration. It should be clear that the partition as such "is not just out there": It is an element of our (effective) description.

The description of an object (rather than its mere distinction from a given set of objects, as suggested by Shannon) is the target of the so-called Kolmogorov complexity. The recipient of the description (message) should, in principle, be able to reconstruct the object. This ability might be termed the "meaning" of the message—in quite close analogy to the Peircean ideas above. Unfortunately, this does not imply that everything had now been made mathematically clear and explicit. A severe obstacle has been identified in the form

[a] Peirce generally likes to arrange his arguments in the form of triads—reminiscent of Hegel's thesis-antithesis-synthesis structure.

of "unspeakable information," which calls for pre-defined common reference frames (cf. Section 2.3), that is, a context that has to be agreed on by the communicating parties before getting started.

2.2 The Big Questions

What are some of the most pressing questions? The following are my subjective favorites; they should make up a kind of leisurely walk through the "backyard" of information theory.

2.2.1 *Is Information Physical?*

As emphasized by Rolf Landauer [Landauer (1967)], information is physical, that is, needs a physical carrier or representation.

But right from the start an observer/agent has to set the scene: Representations are based on a convenient *selection* of states—in a specific form of description. The choice of such states is due to convention and fairly ambiguous. For the description, the observer/agent might use macro- or micro-variables, complete or entirely incomplete parameter sets. It could be the (coarse-grained) position of a particle, in polar or Cartesian coordinates. It could be anything from the presence or absence of a token to the intensity of a light field.

By definition, the representation of an information as such cannot answer the question, what it represents, that is, what it should stand for (cf. Section 2.1). While this "meaning" cannot be hard-wired, it has obviously to be correlated with its representation as a state. One might argue that the *correlation* between representation and represented should be physical, too, not only the representation of information. This is what C. Rovelli [Rovelli (1996a)] and others have in mind, when going beyond the Landauer dictum.

Indeed, for multipartite quantum systems, say, the state description could be decomposed into local parameters (referring to a single subsystem only) and to parameters describing clusters of two or more subsystems. Whether classical or quantum mechanical, these correlations are, at least, two-subsystem properties, and part of a complete state description—on the one hand. On the other hand, those correlations between local states (of multipartite quantum systems) could be interpreted in this alternative way: Any physical system contains information about its partner-system, to which it is correlated [Rovelli (1996a)]; it becomes a kind of "aboutness." Referencing would thus require a physical interaction that changes the state of the subsystems involved—it is not "free of charge."[a]

Information—even information resulting from physical measurements—is about something *for someone*. In a naturalistic framework, the human observer or agent is just a physical system that interacts with the observed physical system. However, macro-objects can never be directly observed, only via certain aspects = "signs." Even the simplest observable, such as the position of a particle, is not a label imprinted on that particle but refers to an external reference, requiring a relevance concept about what should count as the position (symbol). At the end of the day, there is a pointer reading, which has to correlate with the property in question.

In the Peircean triad [Herrmann-Pillath (2010)], the sign does not carry information: information is created by the interpretant. In physics, the sign is needed to bridge the gap between ambiguous micro-state parameters and some characteristic measurement concept. The "sign" often functions as a kind of physical reference frame: selection, correlation, and interpretation. Without this so-called epistemic cut [Abel (2008)], no information can exist.

The referent of information is physical. But all this is not just "out there," it does not follow from physics alone. One can hardly consider information to be "nothing but physics." We will return to this problem in Section 2.2.6 and in the context of the measurement problem and relational quantum mechanics (cf. Sections 3.2.3 and 3.2.8).

---

[a]Such kind of physical correlations cannot be fixed, though. Otherwise the choice of a "code" would have to be predetermined—an unacceptable situation. In fact, the correlations are subject to dynamical evolution of the total system. And the underlying partition depends on the choice of some external agent to begin with.

2.2.2 *Is There a Conservation Law for Information?*

Representations constitute a connection between information and physics. Only because of this connection can one expect physical properties to carry over to informational properties.

Invariance properties such as conservation laws may result from an analysis of the underlying state dynamics. In this context, though, one would hardly contemplate the behavior of macro-representations such as a printed text in a book. Little could be said except that the lifetime of this information carrier will be finite. Actual lifetimes for modern information storage devices tend to be even shorter than those of old-fashioned books. In this practical realm, no conservation law applies.

Instead, let us turn to the full-fledged micro-states of a closed system. Classically such states can be visualized as points in phase space. Now, Liouville's theorem implies the conservation of phase space volume: This may be thought of as a conservation of information (assumed to be represented in such states). In the quantum domain, the same follows as a consequence of the unitarity of the dynamical evolution, see Section 3.3.7.

The conservation of information can be seen from the perspective of deterministic evolution laws: The basic laws of dynamics in physics allow us to predict the future as well as the past states from the present state. In that sense it is reasonable to claim that the information about what is going on does not change as a function of time.

Such a conservation law may be criticized on the grounds that it is nothing but a property (or interpretation) of the underlying equation of motion for a closed system, an idealization. It has little if anything to say about information processing on a standard computer.

But occasionally, this metaphorical language helps to clarify certain problems in fundamental physics. A well-known example is the potential loss of information about objects falling into a black hole, which has aroused much concern. With great relief, the physics community has embraced ideas according to which that so-called information paradox can be avoided. One way out [Susskind (2008)] is to acknowledge that information and observation necessarily have to refer to appropriate coordinate systems, here to two observers, say, one outside and the other inside the event horizon of the black hole. If the viewpoint of the observer falling into the black hole is taken, all the information will stay with him in the black hole. If the viewpoint of the outside observer is taken, the constituent particles of the falling observer will be scattered across the event horizon. Both these contradictory views, to be sure, cannot be taken together (black hole complementarity). They provide an interesting case for a very fundamental observer-dependence.

2.2.3 *What is Information Processing?*

R. Feynman [Feynman (1982)] pointed to the strange interrelation of information processing and physical dynamics. Computers are physical systems that, inter alia, are built to simulate physics. Efficient simulation is thus, at least, constrained by the physics underlying that very simulation. On the one hand, this means that physical behavior not amenable to simulation (either by computer or by our brain) is not accessible to us. On the other hand, these bounds have created much hope with respect to quantum computation, as different physics could pave the way toward new modes of calculation [Nielsen (2000)].

It has been noted [Toffoli (1982)] that any realistic computer actually uses only a minute part of its state space for actually carrying out a given task. This is for stability reasons, control, and read-out. But if left alone, what would the system do? In this context, an extreme position is taken by computationalism [Lloyd (2006)]: As a whole it would just "calculate itself" by following its trajectory in full state space.

If this picture was true, our actual computers might appear to be nothing but an attempt "to hitch a ride on this huge ongoing computation of our universe, and try to discover which parts of it happen to go near where we want" [Toffoli (1982)].

The interpretation of the universe as a computer [Fredkin (1990); Lloyd (2006)] rests upon the idea that information is primary. The physical laws are taken not to preexist as ideal mathematical entities in some Platonic world but rather to derive from information, in particular, to be constrained by limits imposed by information theory.

Ironically, though, all this cannot be put to work without a context, that is, without an external observer: A kind of Peircean triad [Herrmann-Pillath (2010)] has to be introduced—without admitting explicit reference to this concept. The assumed properties of this observer/agent will be decisive for what can be said about the informational aspects of the universe.

For the evolution of a quantum object in its Hilbert space (cf. Chapter 3), the observer is taken to use the pertinent energy eigenstates as his standard reference frame (accepting that these states constitute abstract concepts and are part of his tools to describe nature). In addition, he artificially decomposes the system dynamics into a sequence of (discrete) "steps": As he could distinguish at most between orthogonal states, he defines an effective computational step to cover the transition from one state to some orthogonal one. Now, there exists a minimum time for that to happen, which is known to depend inversely on the average energy the system has above its ground state [Margolous (1998)]. Based on this simple recipe and the age and energy content of the universe, Lloyd has estimated its computational capacity [Lloyd (2002)]; he found that the universe so far should have performed approximately 10^{120} operations.

There are serious doubts, though. The main motivation for using a computer is its programmability. The universe is certainly not programmable.[a] To say the universe computes itself runs the risk of confusing the description (in terms of our limited theories, cf. Chapter 1) with the described. It remains entirely unclear what "calculation" could mean: numbers, units, etc., have no absolute internal meaning, and change cannot be defined without reference to a clock. Even switching devices have continuous properties; they merely can be run in a mode where they essentially show an all-or-

[a] As argued by [Conrad (1988)], there should be a trade-off between programmability, computational efficiency, and evolutionary adaptability.

none response, that is, behave digital.[a] So, why should the universe operate in a digital mode to begin with?

And finally, the universe as a whole should certainly contain all potential observers. But an observer cannot have a full dynamical description of the interaction of the observed system with himself [Breuer (1995)]; non-self-predictability is inevitable. (It is remarkable that quantum mechanics obeys such a non-predictability with respect to quantum measurements, cf. Section 3.2.3.)

2.2.4 *Wheeler's Participatory Universe: A Feedback Loop?*

What if physical descriptions are inherently incomplete? How then could the missing information be fed into the system? The origin of Wheeler's well-known phrase—it from bit–can be traced back to a generalization of the quantum mechanical idea that observation creates reality [Wheeler (1998)]: The questions we ask and the answers we get (in "bit") determine what is (the "it"). As the observer is itself part of the observed universe, this notion is based on a kind of feedback loop, that is, becomes inherently self-referential.

This picture of a feedback loop tries to put information and physical reality on equal footing, in fact, this is his participatory or agent-oriented universe (cf. Section 1.3). At the same time, it removes (or at least weakens) the functional asymmetry between observer and observed. The idea is much more revolutionary than simply replacing the priority of real things by the priority of (abstract) information. In this latter sense, Wheeler's metaphor has often been misunderstood: J. Barbour [Barbour (2011)] warns that qubits would not float around ready for use and therefore suggests to return to "bit from it." And Wheeler's statements have even been interpreted as implying the universe to be a map of a computation [Pattee (1995)]. But this is most likely what he did not have in mind.

Unfortunately, that feedback loop has not been worked out—neither by him nor by his followers. But similar ideas are being contemplated in biology: There we "have a computer that re-wires

---

[a] As first noted by C. Mead present-day digital devices can be run in an analogue mode, then resembling the behavior of neuronal ion channels [Poon (2011)].

itself in a hardware-software interplay: the hardware defines the software and the software defines new hardware. Then the circle starts again" [Kampis (1991)]. Whether or not this cycle should be based on a digital representation is left open, though.

But one can hardly doubt that Wheeler's loop calls for a digital appearance of the world, in which (observed) reality would somehow emerge via a sequence of yes-no decisions. Along similar lines, Weizsäcker proposed his so-called Ur-theory [Weizsäcker (1980)]. Such kind of propositions seem to imply that reality might not be entirely physical. Plato's ideas about a separate world of ideas lurk behind the horizon.

2.2.5 *Does Information Derive from History?*

"Without history there would be no information, and, of course, no informational constraints would be needed to understand the world" [Salthe (2011)]. Is this a strong statement or rather a triviality?

Fundamental evolution equations tend to come as differential equations; initial and boundary conditions are thus needed in order to select unique solutions. But there is no theory for these conditions. The equations and laws define structural relations and formalize constraints, but without further input data they are empty and mute about the actual reality, cf. [Primas (2007)].

These equations are deterministic:[a] On this level of description, there would be "nothing new"— neither in the future nor in the past (cf. Section 2.2.2). In this sense, time was irrelevant, even if existing as a formal parameter (cf. Section 3.2.7). "Time is, that is to say, our ignorance about the full microstate" [Rovelli (2009)]. All that counts is the complete present state.

Obviously, such a complete picture is not accessible to us.[b] Instead, we have to settle for severely restricted views and effective equations of motion, which are incomplete and often non-

[a]cf. Einstein's famous dictum: God does not play dice.

[b]This may even be seen in a quite general context: "Is it possible that there is no finite procedure by means of which the configuration space of general relativity or some other cosmological theory may be constructed?" [Smolin (2001)]. If yes, incompleteness is fundamental, time cannot be eliminated.

deterministic. Additional information about the past (records) are required to improve the predictability of the future. In physics such recording would refer to measurement results. Biological information encoded in the DNA could, at least in part, also be seen as a document of past evolution, cf. Section 2.2.6.

Everyday situations abound, in which such conditions show up simply as the result of previous developments. Rather than trying to trace back the respective evolution, leading to an "infinite regress," the past is considered to be encoded in the present. The information about the past exhibits itself to the observer willing to accept the respective level of description. For more ambitious observers, part of the constraints ("historical facts") may become accessible to full dynamical modeling. As a consequence, the relevant state space increases, the nature of the boundary conditions changes. These are alternative effective theories.

Constraints may appear also as examples for the so-called top-down causation [Auletta (2008)]: The flow of a river creates a channel (high-level description), which, in turn, constrains the subsequent motion of the water molecules (low-level description). At the same time, the form of the channel documents certain properties about the past history of flows (on the high level of description). As long as the whole theoretical approach is consistent, bottom-up and top-down causation cannot be inconsistent. In any case, causation defines a direction of time from cause to effect, though. In this sense, the corresponding information has to relate to the past.

2.2.6 *Might Information Apply to Biology Only?*

Information is contextual, it cannot be found "inside" a given system. In particular, the categories of meaning (semantics) and function (pragmatics) do not apply to physical and chemical systems as such [Roederer (2005)]. They are not contained in the physical laws.

The question of information and living systems can be given at least two different readings:

(i) The outside view: Does the observer/agent (as an indispensable referent of information) necessarily has to be a living creature?[a]

(ii) The inside view: Are living systems exempted from the condition that there is no "internal" information?

Agency has to be included in any meaningful model of the observer: His abilities and limits to observe, to initiate processes (free will?) are of prime interest. This seems to indicate that, indeed, life may be a prerequisite, though not necessarily full-fledged brains. Even "single-celled organisms seem to make true choices (e.g., approach/avoidance to food sources and noxious stimuli) though they lack physical brains and formal minds" [Abel (2008)].

"Life is guided by information and inorganic processes are not" [Yockey (2005)]. Information can be considered the difference between life and matter. The genetic code must not be understood metaphorically: It is a set of rules (cf. Section 2.2.7) by which the information encoded in the DNA translates into proteins, functionally similar to a computer code (or the US postal code). "Information begins when and where life begins" [Roederer (2004)]. The genetic code is one of the requirements for biological adaption (Darwin). Living systems seem to be in command of the "information tool," without the need for an outside observer.

But is this the final word? Clearly, when we do research on biological systems, we find ourselves just in the role of external observers. So, could it be that the very efficient description of those biological processes, like of any other more or less complex behavior, is just an example of an effective theory? That a demon insisting, instead, on a complete micro-description of that very scenario could do well with an extremely demanding state-to-state dynamics without any reference to meaning and function?

There is good reason to challenge the (apparent) special role of information in the living versus the non-living. On the one hand, information (and top-down causation) is also needed as input for the various effective theories in the entirely physical regime (cf. Section 1.1, Section 2.2.5, Section 6.4, Example 7.4.3). On the other

[a]Additional support for the special role of life might derive from the so-called anthropic principle, which argues that our universe appears to be fine-tuned to allow for the appearance of life, the universe is "bio-friendly" [Davies (2004)].

hand, the effective descriptions in the biological realm—while on a much higher level of complexity—need not be qualitatively different.

And last not least: Information does have physical effects. If information was non-physical, we would thus have to admit that non-physical causes enter physical equations of motion; causal closure of physics would explicitly be violated, not just left undecidable, cf. [Bishop (2006)].

We thus tend to prefer the outside view: Information is as "physical" as other physics-based concepts like entropy—a tool within an effective theory.

2.2.7 *Effective Dynamics Revisited: From Laws to Rules?*

Examples for effective dynamics abound not only in the technical but also in the biological domain. Indeed, many tasks benefit from the informational language, which improves one's abilities to predict the expected behavior of some object and then to react in a most adaptive way. Tasks thus include the pragmatic aspect of information: A model of how information might be applied to produce an effect, that is, how it can become a resource for subsequent behavior. Tasks are based on purposeful action (agency). This means to progress "from description to prescription" [Abel (2008)].

A primitive example for information as an input for an effective description is the so-called "volume": Encoded in the structure of the container, it functions as a mechanical control parameter in the ideal gas law, say. This equation of state is not "fundamental"; and the container as a complicated many-body system has many physical properties beyond serving as a global constraint for the extension of a gas or fluid. But it is this constraint that only characterizes the "information-driven interaction" between container and gas.

An interesting model for "programmable" constraints has been suggested by Judson and Rabitz [Judson (1992)]. In this model, the interaction between light pulses and molecules is modified by means of a feedback loop. The goal is to adjust the pulse shape such that the molecules—immediately after excitation—prefer to land on a certain target state. For this purpose, each excitation process is divided into various substeps: excite, evaluate, and adjust. The actual excitation pattern within the molecules is compared with the target pattern, the difference being minimized by iteratively re-adjusting the input pulse shape for the next step. This strategy is incorporated in a computer code (genetic algorithm). The experiment is run until convergence to the optimum has been reached. In this way, the respective effective light–matter interaction has become an information-controlled rule and thus most flexible (it could be adjusted to various requirements).

Interpreted in the Peircean triad [Herrmann-Pillath (2010)], the "object" is the molecular state, and the "sign" the digital representation of the excitation test result, which is "assigned" to that very state. The "interpretant" is the computer with its built-in code for evaluating the excitation quality and performing the intended response (improving the pulse for the next iteration step).

Effective descriptions are able to generate an unexpected wealth and flexibility of behavioral modes, which transcend the behavior according to physical laws: That a single-celled organism tries to avoid noxious stimuli does not directly follow from physical law, but constitutes a rule based on information. Man-made computer controlled machines or robots translate such rules into the technical domain.

But what is the relationship between such rules and the underlying physical laws? On the one hand, there should be no conflict between the direct physical and the information-driven interactions [Roederer (2005)]. This would be the case if the different information states were physically indistinguishable. But then the world of "pure" information and the physical world would remain separated for ever, rendering information processing and pragmatic effects of information impossible.[a]

Fortunately, such a strict purity is not needed. It suffices to require that in an information-driven scenario the direct physical interactions are negligible. Stopping in front of a red traffic light, for example, is information-controlled; the slightly different radiation pressure exerted on the car by the green versus the red light is clearly irrelevant in this context.

---

[a]This tension is reminiscent of the mind–body problem.

On the other hand, the implementation of those rules, while dependent on special design, cannot do without physics.

2.3 The Formalism

We now turn to the "mathematical language" needed not only to talk about information, but also to come up with quantitative results. At the same time, these tools serve to reconstruct essential features of information theory.

Information is intimately related to uncertainty. The exposition of indispensable tools thus starts with the axiomatic probability theory, followed by a discussion on the "interpretation" of probability. While deductive reasoning can strictly be based on logical rules, inductive reasoning has to deal with incomplete knowledge and thus uncertainty. This is a first application of probability—and a way to infer probability models. A central link is then provided by the so-called entropies: They serve not only as measures of uncertainty, but also as measures of information. Information processing defines the task of computation. Computation, in turn, underlies both computational and Kolmogorov complexity. The laws of physics underlie and constrain the way information can be "processed."

Physics is not only needed to represent information, but information is relevant also within physics—provided it can be translated into concrete action (control). And vice versa, the gain of information is not for free, as the underlying physical processes requires input of energy, etc., in order to run.

2.3.1 *Probability Theory*

Probability theory is a way to cope with incomplete knowledge; as such it presents, *inter alia*, a specific tool for the description of physical reality. Main ingredients of a probability theory consist of

- axioms (rules),
- interpretation,
- models.

The rules define the general mathematical structure, the interpretations specify types of applications, and the models suggest concrete probability distributions.

Axioms (cf. [Ballentine (1998)]): The formulation of probability theory in terms of basic axioms goes back to A. N. Kolmogorov (1933). We consider events taken from some set Ω (*the sample space*). Let *A, B, and C be elements of this set.*

Definition 2.1 Composite events

$$\begin{aligned} \neg A &= \textit{not A, negation}\,. && (2.1)\\ A \cap B &= \textit{A and B, conjunction}\,. && (2.2)\\ A \cup B &= \textit{A or B, disjunction}\,. && (2.3) \end{aligned}$$

Definition 2.2 Probability *$P(A|C)$ is called probability of A under the condition C, if*

$$\begin{aligned} 0 \leq P(A|C) &\leq 1\,, && (2.4)\\ P(A|A) &= 1\,, && (2.5)\\ P(\neg A|C) &= 1 - P(A|C)\,, && (2.6)\\ P(A \cap B|C) &= P(A|C)P(B|A \cap C)\,. && (2.7) \end{aligned}$$

These axioms suffice to prove the following propositions.

We first note that from the axioms it readily follows that

$$P(A \cup B|C) = P(A|C) + P(B|C) - P(A \cap B|C)\,. \tag{2.8}$$

Definition 2.3 Exclusivity *A and B are mutually exclusive under condition C, if*

$$P(A \cup B|C) = 0\,. \tag{2.9}$$

Proposition 2.1 Rule of addition *Let A and B be exclusive. Then, according to Eq. (2.8),*

$$\boxed{P(A \cup B|C) = P(A|C) + P(B|C)\,.} \tag{2.10}$$

Proposition 2.2 Bayes' theorem

$$P(B|A \cap C) = \frac{P(A|B \cap C)P(B|C)}{P(A|C)}\,. \tag{2.11}$$

Putting aside the "background" condition, C, this theorem allows to relate $P(B|A)$ with $P(A|B)$. The theorem follows from $P(A \cap B|C) = P(B \cap A|C)$ and applying Eq. (2.7) on both sides.

Definition 2.4 Independence *B is called independent of A, if*

$$P(B|A \cap C) = P(B|C)\,. \tag{2.12}$$

This is a mutual relationship.

Proposition 2.3 Rule of multiplication *If A and B are independent, it holds that*

$$\boxed{P(A \cap B|C) = P(A|C)P(B|C)\,.} \tag{2.13}$$

This rule is an immediate consequence of Eq. (2.7).

Definition 2.5 Probability density *In physics one often encounters a situation in which an event E consists of a continuous variable x to be found within a certain finite region: The values x are mutually exclusive, the probability $P(E)$ thus additive in this space, that is,*

$$P(E|C) = \int_E dx\; p(x|C)\,. \tag{2.14}$$

$p(x|C)$ (of dimension $1/x$) is the respective probability density normalized by the requirement

$$P(\Omega|C) = \int_\Omega dx\; p(x|C) = 1\,. \tag{2.15}$$

Definition 2.6 Moments

$$\overline{x^m} = \int_\Omega dx\, x^m\, p(x|C)\,. \tag{2.16}$$

For $m = 1$ this is the average value or expectation value of x.

Definition 2.7 Variance

$$\mathrm{var}(x) = \sigma_x^2 = \int_\Omega dx (x - \overline{x})^2\, p(x|C) = \overline{x^2} - (\overline{x})^2 \geq 0\,. \tag{2.17}$$

This is a good measure for statistical fluctuations; it is also routinely applied in quantum mechanics, cf. Proposition 3.19. σ_x is the so-called standard deviation. Basic properties are ($a =$ const.)

$$\mathrm{var}(ax) = a^2\mathrm{var}(x)\,, \tag{2.18}$$

$$\mathrm{var}(a + x) = \mathrm{var}(x)\,. \tag{2.19}$$

Definition 2.8 Two-dimensional probability density *Let an event E be specified by a point $p(x, y)$ to be found within some two-dimensional region characterized by $p(x, y|C)$*

$$P(E_x \cap E_y) = \int_{E_x} \int_{E_y} dxdy\, p(x, y|C)\,. \tag{2.20}$$

If x and y are independent, that is, $p(x, y|C) = p(x|C)p(y|C)$, it holds that

$$P(E_x \cap E_y) = P(E_x)P(E_y)\,, \tag{2.21}$$

(rule of multiplication). In particular, $\overline{xy} = (\overline{x})(\overline{y})$.

Definition 2.9 Covariance

$$\text{cov}(x, y) = \overline{(x - \overline{x})(y - \overline{y})} = \overline{xy} - (\overline{x})(\overline{y})\,. \tag{2.22}$$

For $x = y$, this expression coincides with σ_x^2. The covariance is zero, if x and y are independent.

Definition 2.10 Correlation coefficient

$$\text{corr}(x, y) = \frac{\text{cov}(x, y)}{\sigma_x \sigma_y}\,. \tag{2.23}$$

This measure of linear dependence of the two random variables x and y is normalized and symmetric. Its absolute value is less or equal to 1: +1 means strict correlation, −1 strict anti-correlation. Correlation does not imply any causal relationship.

Remark 2.1 All probabilities are conditioned *It is obvious that the choice of the experimental setting itself cannot be covered by the statistical model. Such different choices are taken to be disconnected and define different points of departure. The specification of the experimental setting thus constitutes a condition in a strict sense, it should never be considered an event. (An event may function as an additional condition, though.)*

This condition in a strict sense is reminiscent of the boundary conditions needed to select solutions of a differential equation: One should refrain from combining solutions with different boundary conditions. Likewise it is meaningless (inconsistent) to combine probabilities referring to different settings.[a]

---

[a] This restriction becomes especially relevant in the context of quantum mechanics.

Interpretations of P: The abstract rules underlying P have to be supplemented by some interpretation. The most prominent examples are:

- P as relative frequency. In this case P relates to a set of concrete experimental observations. Under given condition C, one observes event E or $\neg E$. If there are n (independent) repetitions of such an experiment, and n_E is the number of outcomes E, then

$$P(E|C) = \lim_{n\to\infty} \frac{n_E}{n} . \tag{2.24}$$

  The use of this "classical" approach is widespread, despite the fact that the existence of this limit cannot be guaranteed, and the accuracy is unclear. (The underlying "law of large numbers" is not a law, but an expectation, which may or may not be fulfilled.)
- P as generated by a model process. A direct way to define a parametrized probability law is to assume that the distribution in question can be thought to result from some well-defined stochastic process. In this case, one only has to fit the few model parameters to concrete observations.
- P as propensity. In this case, P is not the property of an actual sequence of observed events; rather, it is directly connected with condition C. It is understood to be a built-in tendency toward a specific behavior. Such a definition is often used in quantum mechanics to deal with "objective" uncertainty. The argument goes as follows:
  Let F be the measurement value of some quantum observable $\hat{F}$. C denotes the state, which the quantum system under consideration has been prepared before measurement. Then there are fixed rules (Born rule, cf. Definition 3.80) by which to calculate $P(F|C)$ for any observable and any state. The quantum case is thus very special in the sense that the appropriate model for $P(F|C)$ follows from the underlying theory in a direct way. However, the conjunction rule, Eq. (2.7), does no longer apply without further qualification: Given $P(F|C)$ and $P(G|C)$, it does not follow that $P(F \cap G|C)$ would exist also. Formally, the

conjunction has to be replaced by a time-ordered "quantum history $\hat{h}$" of measurement projections. Only under rather special (compatibility-) conditions can this history by re-written as a simple conjunction. Quantum theory may be said to differ from classical probability theory by such constraints on the "event-space."
The propensity interpretation has been suggested to apply even outside physics: confer, for example, quantum cognition, Section 9.2.5.

- P as a measure of subjective "believe." In this case, we are concerned with hypotheses H_i (e.g., about the occurrence of event E_i), the conditions C are newly acquired data D. Bayes' theorem can then be used to describe an iterative process of reasoning, leading from a so-called "prior" $P(H_i)$ to a "posterior" $P(H_i|D)$:

$$\boxed{P(H_i|D) = \frac{P(D|H_i)P(H_i)}{P(D)}\,.} \tag{2.25}$$

We have suppressed here any context C needed to specify the underlying experimental setting, cf. Remark 2.1. $P(D|H_i)$ is the conditional probability of observing the data D, when hypothesis H_i is true. Learning from data is the task of inference. The probability for H_i taken to be true (under the new data D) is proportional to the prior believe about the probability of H_i multiplied by the conditioned probability for D to result under condition H_i. For a debate on Bayesian statistics and subjective probability, see, for example, [d'Agostini (2011)].

Models for P—stochastic processes: There are several well-known examples, including:

Example 2.1 Bernoulli process *Let us consider an event E with $E \cap \neg E = \Omega$, so that*

$$P(E|C) = q\,, \tag{2.26}$$

$$P(\neg E|C) = 1 - q\,. \tag{2.27}$$

The finite process consists of n repetitions of an experiment, which leads in n_E cases to the outcome E. Based on the rule of multiplication *and the rule of addition one finds the binomial distribution*

$$P(n_E|C) = \binom{n}{n_E} q^{n_E}(1-q)^{(n-n_E)}, \tag{2.28}$$

$$\binom{n}{n_E} \equiv \frac{n!}{n_E!(n-n_E)!}. \tag{2.29}$$

The expectation value for n_E is $\overline{n_E} = qn$, the variance $\sigma^2 = nq(1-q)$.

Example 2.2 Gaussian process *For large n, the binomial distribution approaches a Gaussian with the same expectation value and the same variance,*

$$P(n_E|C) \approx \frac{1}{\sqrt{2\pi\sigma^2}} \exp\left(-\frac{(n_E - \overline{n_E})^2}{2\sigma^2}\right). \tag{2.30}$$

Example 2.3 Poisson process *The Poisson process counts the number of independent events, n_E, within some time interval τ assuming that the waiting time between consecutive events is exponentially distributed with parameter λ. The result is*

$$P(n_E|C) = \frac{\exp(-\lambda\tau)(\lambda\tau)^{n_E}}{n_E!}. \tag{2.31}$$

The corresponding expectation value is $\overline{n_E} = \lambda\tau$.

But what, if P cannot directly be constructed, because the quantum state or the underlying stochastic process is not known?

In such cases, there are but a few general methods available to find probability laws from finite datasets. Some of these are summarized as follows:

Proposition 2.4 Methods for inductive reasoning *While Bayesian statistics is still considered outlandish by many, Bayesian thinking is, in fact, present in a number of important statistical methods.*

- *Iteration of Bayes' rule: As indicated above, Bayes' rule can systematically be used as an optimization in hypothesis space. Based on the accumulation of more and more data, one can improve on the estimation of probabilities P or probability densities p. Indeed, as the number of iteration becomes large, one can show [Li (1993)] that (at least for well-behaved examples) the posterior probability concentrates around the true hypothesis, independent of the original prior. But in this limit, other estimation methods are often preferable. One has to accept that for finite data the success of Bayes' method depends on a good choice of the prior. The following methods may be considered alternative routes to just this purpose.*

- *Minimum description length principle (MDL): This strategy has been proposed [Li (1993)] as a unifying background principle for inductive reasoning. According to Bayes' rule, Eq. (2.25), we must find the hypothesis H_i such that $P(H_i|D)$ is maximized. Taking the negative logarithm of both sides, one obtains*

$$-\ln P(H_i|D) = -\ln P(D|H_i) - \ln P(H_i) + \ln P(D)\,. \quad (2.32)$$

The last term, $\ln P(D)$, is fixed and can thus be ignored. The maximization task now means to minimize

$$\boxed{M \equiv -\ln P(D|H_i) - \ln P(H_i) \stackrel{!}{=} \min.} \quad (2.33)$$

One then shows that this M can be estimated as

$$M \approx K(D|H_i) + K(H_i)\,, \quad (2.34)$$

where K denotes the so-called algorithmic complexity (cf. Definitions 2.27 and 2.23). So, minimizing M gives the MDL principle: The best theory to explain a set of data D is the one, which minimizes the length (in bits) of data when encoded with the help of the theory, together with the length (in bits) of the description of the theory. This result is in accord with "Occam's razor," which asserts that the simplest explanation is the most reliable [Li (1993)]. The following two methods are special variants of this MDL principle:

- *Maximum likelihood method (ML): In this procedure the last term of Eq. (2.33) is considered fixed, typically by assuming equipartition. One then has to maximize $P(D|H_i)$, or, equivalently, minimize $-\ln P(D|H_i)$ that is, one selects the hypothesis H_i, which leads to the highest probability to observe the data D. This procedure is equivalent to minimizing the so-called Fisher entropy.*
- *Maximum entropy principle (MEP): According to E. T. Jaynes, one requires the prior $P(H_i)$ to maximize the Shannon entropy, $S_I(P)$, subject to given constraints (prior knowledge). As one needs*

nS_I bits to record the outcome of n trials (cf. Proposition 2.7), the MEP implies minimization of $-\ln P(D|H_i)$ and thus, indeed, the fulfillment of MDL. However, two sources of potential failure have been identified: Missing important constraints, choosing the "wrong variables" (the wrong metric) [Caticha (2000)].

Example 2.4 Application of MEP: Macro-constraints *Given $j = 1, 2, \ldots n$ possible events. We know beforehand that the distribution P_j must be such that the following m macro-constraints are fulfilled:*

$$\overline{G^{(\nu)}} = \sum_j P_j G^{(\nu)}(j) \quad \nu = 1, 2, \ldots m. \tag{2.35}$$

Here $G^{(\nu)}$ is a given function of the events, j. Maximizing the Shannon entropy

$$S_I(P) = -\sum_j P_j \ln P_j\,, \tag{2.36}$$

that is, requiring $\delta S_I(P) = 0$, leads to

$$P_j = \frac{1}{Z} \exp\left(-\sum_{\nu=1}^{m} \beta_\nu G^{(\nu)}(j)\right). \tag{2.37}$$

The β_ν are the respective Lagrange multipliers, the factor $Z = \sum_j P_j$ guarantees normalization; it is a function of the parameters β_ν. Because

$$\overline{G^{(\nu)}} = -\left(\frac{\partial \ln Z}{\partial \beta_\nu}\right)_{\beta_\mu \neq \beta_\nu}, \tag{2.38}$$

the m Lagrange multipliers are implicitly determined by the m macro-constraints.

Example 2.5 Challenges for Bayesian reasoning *In the following, we address some hypotheses (questions) that are very hard to assess in terms of "yes" or "no." A way out is to put aside their actual "truth-value." Instead, consider the respective statement as a (plausible?) consequence of some preconditions, whose probability of occurrence appears to be more easy to establish. As it turns out, this reference to probability is rather problematic in each case. Bayesian reasoning may no longer be applicable at all [Aaronson (2013)].*

(1) *"Why is there something rather than nothing"? (This is known as the "primordial existence question," going back to the philosopher G. W. Leibniz.) Nowadays the appeal of this question arises from the unfounded expectation that "nothingness" would be more probable than the existence of "something" [Holt (2012)]. Quite to the contrary, one may say: There is only one "null set," while indefinitely many non-empty sets.*

(2) *"Are we living in a computer simulation?" An argument favoring the answer "yes" goes back to Nick Bostrom [Bostrom (2003)]. In a nutshell, it would go as follows: If an advanced (extraterrestrial) society is typically running a large number of simulated universes (the so-called ancestor simulations) that are so accurate that the "inhabitants" would think they are real, then we ourselves might, in fact, also be simulations. Now, if there are more simulations than real variants, we would likely be in a simulation.*[a]

(3) *"Are we alone in the universe?" One intriguing way to address this question is via the so-called "probability barriers" [Bostrom (2008)]. Such a barrier would consist of an improbable evolutionary step for intelligent civilizations to get started. There is evidence that on earth life occurred only once and it took very many trials. Probability thus appears rather low. In an infinite universe, the likelihood for extraterrestrial intelligence would nevertheless be very high. The observable universe (only this is causally connected to us) is finite, though. Whether or not intelligence was to be expected even in this case: if advanced civilizations tend to undertake space colonization, we should have already come in contact with some aliens, which is not the case (often referred to as the "Fermi paradox"). So it appears likely that we are alone in the observable world.*

(4) *"Is 'doomsday' near?" Consider the total number of people ever living on earth to be finite, N. Our position n is taken to be anywhere in the interval* $[0, N]$ *with the same probability. Thus, our* $f = n/N$ *is with probability* $P(f) = 0.95$ *in the interval* $(0.05, 1]$. *This P is the probability for us to be in the last* 95% *of all humans ever existing. As a consequence* $N < 20 \times n$.

---

[a]The simulated inhabitant must also run simulations; otherwise he could not even contemplate the possibility of being simulated himself.

Taking for $n = 6 \times 10^7$ (the number of people having lived up to now, our rank n) one finds $N = 1.2 \times 10^9$. This finite value is necessarily reached within a finite time Δt from now; with details depending on further (technical) assumptions Δt will be no longer than a few thousand years, a short period compared to our past [Bostrom (2002)]. "Doomsday is near."
This argument depends on the so-called "self-sampling assumption," which has repeatedly been criticized.

Proposition 2.5 Bayesian brain *Kolmogorov probability theory, including Bayes' theorem, may be more than just a somewhat ambiguous definition, which could as well have been chosen otherwise. In fact, it has recently been suggested that perceptual inference (and not just our rational inference as discussed above) could best be understood with our brain working as a kind of probability machine (Helmholtz machine): This machine would constantly make predictions about the world and then update them based on what it senses; the brain would be "Bayes optimal" [Friston (2007)]. The optimization calls for minimization of what has been termed a (generalized) free energy.*

2.3.2 *Entropies*

Entropy measures try to quantify certain aspects of randomness, diversity, or uncertainty contained in a given probability distribution $\{P_j; j = 1 \ldots n\}$. The index j specifies the event E_j in question, P is the "name" of the distribution.

There are various forms of entropy; Tsallis mentions more than 20 forms of entropy [Tsallis (2000)]. Which one to prefer thus appears highly problematic, unless a full set of requirements has been formulated. A pertinent set is as follows:

Definition 2.11 Fedeev's postulates *We require $P_j \geq 0$ and $\sum P_j = 1$. Then the (yet unknown) entropy measure $S(P_1, P_2, \ldots)$ is taken to be characterized by [Renyi (1961)]*

- *$S(P_1, P_2, \ldots)$ is a symmetric function of its variables P_j,*
- *$S(P_1, 1 - P_1)$ is a continuous function of P_1 for $0 \leq P_1 \leq 1$,*
- *$S(1/2, 1/2) = \ln 2 \equiv k_2^{-1}$,*

- $S(tP_1, (1-t)P_1, P_2, \ldots) = S(P_1, P_2, \ldots) + P_1 S(t, 1-t)$ *for* $0 \le t \le 1$ *and any distribution* $\{P_j\}$.

Here, k_2 *is a scale factor defined by*

$$\log_2 x \equiv k_2 \ln x\,. \tag{2.39}$$

These four postulates uniquely specify $S(P) \equiv S_I(P)$ as the Shannon entropy:

Definition 2.12 Shannon entropy

$$S_I(P_1, P_2, \ldots P_n) = S_I(P) = -\sum_{j=1}^{n} P_j \ln P_j\,. \tag{2.40}$$

This is a functional characterizing the probability distribution P_j *by a single real positive number. The first identity defines our notation.*

The Shannon entropy can be seen as a special case of a Renyi entropy:

Definition 2.13 Renyi entropies *Let the last of the Fedeev's postulates be weakened, that is, replaced by the additivity rule for the joint entropy [Renyi (1961)],*

$$S(P\,Q) = S(P) + S(Q)\,. \tag{2.41}$$

P and Q are taken to be distributions characterizing independent *experiments, that is, they refer to different events. Then the set of postulates are fulfilled by the Renyi entropies of order* $\alpha > 0$,

$$S_\alpha(P_1, P_2, \ldots) = S_\alpha(P) = \frac{1}{1-\alpha} \ln\left(\sum_i P_i^\alpha\right). \tag{2.42}$$

This holds correspondingly for $S_\alpha(Q)$.

Proposition 2.6 Special Renyi entropies *Some pertinent properties are as follows [Zyczkowski (2003)]:*

$$S_\alpha \ge S_\beta \quad \textit{for } \beta > \alpha\,, \tag{2.43}$$

$$S_0 = \ln m \quad m = \textit{number of positive components } P_i\,, \tag{2.44}$$

$$S_{1/2} = 2\ln\left(\sum_i \sqrt{P_i)}\right), \tag{2.45}$$

$$S_1(P) = S_I(P) \quad = \textit{Shannon entropy}, \tag{2.46}$$

$$\exp(-S_2(P)) = \sum_i^n P_i^2 \equiv r \quad \textit{"index of coincidence"}, \tag{2.47}$$

$$S_\infty(P) = -\ln P_{max} \quad P_{max} = \textit{largest component of}\,\{P_i\}. \tag{2.48}$$

Contrary to S_2 itself the index of coincidence, r, is non-extensive (non-additive); it is known in quantum mechanics as the so-called "purity", cf. Eq. (3.147). Its inverse, $R = 1/r$, is the "effective number of different events." It should be clear that the Renyi entropies of different order tend to have vastly different meaning and thus different potential applications. But how to select the appropriate entropy measure? An example of paramount importance is communication.

Definition 2.14 Communication task *Communication is a directed interaction between two subsystems: the transfer of a message between sender and receiver counts as an event. The message can be represented as a sequence of symbols. Shannon based his reasoning on the following assumptions:*

- *The universe of possible messages is known to sender and receiver.*
- *There is a finite set of basis messages (symbols)* $a_i, i = 1, 2, \ldots, n$. *These symbols form the so-called alphabet (a).*
- *Each symbol* a_i *is associated with a probability* $0 \leq P(a_i) \leq 1$, *which is also known from the start.*
- *Communication is defined as a stochastic process. In this process, symbols are freely selected from the alphabet to form a message. The* a_i *are thus random variables.*

In these statements, the meaning of the message is entirely ignored.

Definition 2.15 Binary codes *Each symbol* a_i *can be mapped on a string of binary numbers,* $\{0, 1\}$, *such that different symbols correspond to different strings. The number of digits within such a string is its length,* $L(a_i)$, *called "bits."*

The original meaning of the Shannon entropy is related to the following observation:

Proposition 2.7 Minimal property of a code *If each symbol a_i is encoded in binary form, the minimum number of bits per symbol, averaged over all symbols of the alphabet, is given by*

$$\overline{L}_{min} = k_2 S_I \, . \tag{2.49}$$

k_2 is defined in Eq. (2.39). Any message of m symbols, when encoded in binary form, has a minimum length of $\overline{L}_{min}(m) = k_2 S_I m$ bits. Encodings with $L > \overline{L}_{min}(m)$ are called redundant.[a] *Because of the above optimal coding property S_I is taken to represent the quantitative aspect of "information" with respect to the process of communication.*

Example 2.6 Morse code *The Morse code of old-fashioned telegraphy is a concrete example of a binary code. Each symbol a_i is represented as a sequence of "dots" and "dashes." To increase coding efficiency, the code length of each symbol is taken to be approximately inverse to its frequency of occurrence in a typical English text. This gives a concrete example for how to define and find the probabilities $P(a_i)$ assumed to be given in Shannon's theory. The $P(a_i)$ reflects prior knowledge; for different languages and/or restricting to specialized texts the actual $P(a_i)$ could, of course, be different.*

Proposition 2.8 Subjective ignorance *By assumption, the potential receiver is aware of all the possible messages (a). $P(a)$ can thus be said to represent his ignorance before he receives a message. After he has received a message consisting of m symbols, his information gain is on average given by the Shannon entropy $k_2 S_I(P(a))m$ (in bits), cf. Definition 2.7 and 2.16. Note that this information gain is not an internal property of the message (or the object, enumerated by this message). Rather it relates to the ensemble of possible messages and answers the question: Which one?*

Subjective ignorance is highly contextual; it lacks universality.

[a]In turn, the Shannon entropy can be considered the preferable measure for establishing this remarkable property.

Definition 2.16 Shannon information *"The Shannon entropy of a probability distribution can be interpreted not only as a measure of uncertainty, but also as a measure of information: In fact, the amount of information we get when we observe the result of an experiment (depending on chance) can be taken numerically equal to the amount of uncertainty concerning the outcome of the experiment before carrying it out" [Renyi (1961)].*

Because of the above optimal coding property S_I is taken to represent the quantitative aspect of "information" with respect to the process of communication.

We finally extend our considerations to higher-dimensional spaces of random variables.

Definition 2.17 Joint entropy *For a pair of random variables, $\{a_i, b_j\}$ associated with the distribution $P(a, b)$ the Shannon entropy is defined as [Nielsen (2000)]*

$$S_I(P(a, b)) = -\sum_{i,j} P(a_i, b_j) \ln P(a_i, b_j). \qquad (2.50)$$

Proposition 2.9 Subadditivity

$$S_I(P(a, b)) \leq S_I(P(a)) + S_I(P(b)). \qquad (2.51)$$

The equality sign only holds, if (a) and (b) are independent.

Definition 2.18 Conditional entropy *Suppose we have received the message b_j, so we have (on average) acquired $k_2 S_I(P(b))$ bits of information about the pair (a, b). The remaining entropy with respect to (a) conditioned on knowing (b) is thus given by [Nielsen (2000)]*

$$S_I(P(a)|P(b)) \equiv S_I(P(a, b)) - S_I(P(b)). \qquad (2.52)$$

Definition 2.19 Mutual information *The mutual information specifies, how much information the random variables (a) and (b) have in common. It is defined by [Nielsen (2000)]*

$$\begin{aligned} S_I(P(a) : P(b)) &\equiv S_I(P(a)) + S_I(P(b)) - S_I(P(a, b)) \\ &= S_I(P(b) : P(a)) \geq 0. \end{aligned} \qquad (2.53)$$

The mutual information is zero, if the two random variables (a), (b) are independent, that is, according to Eq. (2.13), if $P(a, b) = P(a)P(b)$.

Using Eq. (2.52), the mutual entropy can alternatively be written as

$$S_I(P(a) : P(b)) = S_I(P(a)) - S_I(P(a)|P(b)). \qquad (2.54)$$

Definition 2.20 Relative entropy *The relative entropy defines a distance between two different probability functions, $P(a)$ and $Q(a)$, with respect to the same set of events (a):*

$$S_I(P(a)||Q(a)) \equiv \sum_i P(a_i) \ln \left(\frac{P(a_i)}{Q(a_i)} \right). \qquad (2.55)$$

This distance measure is always positive; it is zero only for $P(a) = Q(a)$. Note that we have to require $Q(a_i) > 0$ for all a_i. The mutual information can be considered a special case of relative entropy with $P(a) \to P(a, b)$; $Q(a) \to P(a)P(b)$.

2.3.3 *Links between Computation and Physics*

What is computation? A convenient operational definition is based on a "machine" working on symbols:

Definition 2.21 Turing machine *Any algorithm can be defined as a process. A generic representation of such a process is the Turing machine [Nielsen (2000)]. In a way, this machine tries to simulate a person doing a calculation on his desk step by step. The basic functional elements of this machine consist of a tape to carry symbols on discretized local areas (specified by number n), a read/write head, and a control unit including the so-called Turing table (i.e., a finite list of instructions). The machine has a finite number of internal states $\{q_j\}$. A calculational step starts from a reading of the symbol $a_i(n)$ appearing under the Turing head and a threefold action conditioned by $a_i(n)$ and momentary q_j:*

(i) *overwrite a_i by a specific symbol,*
(ii) *go to $n + 1$ along the tape, or to $n - 1$, or stay, and*
(iii) *change to a new internal state, q_k.*

The actual process consists of a sequence of such steps and is a combined effect of the information stored on the tape and the Turing table. A universal Turing machine can simulate an arbitrary Turing machine on arbitrary input.

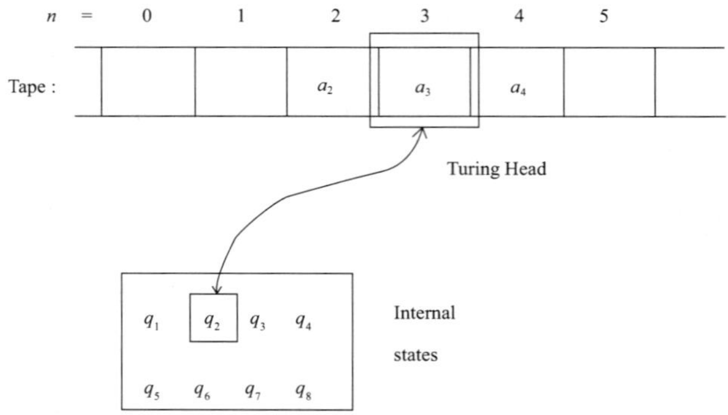

Figure 2.1 Turing machine.

Definition 2.22 Partial recursive function *A function ϕ is defined recursively, if ϕ takes other functions g as input. The respective equations may not define ϕ for any conceivable input and thus provide a partial definition only. The set of recursive functions is equivalent to the functions that can be calculated by (universal) Turing machines.*

Remark 2.2 Computational process *A Turing machine gives a rather intuitive idea of what it means to carry out a computation. However, some of its inherent preconditions should be made explicit: A computational process is implemented as a physical process. But a physical process as such should not count as a computation—even though a theoretical description of this process might be simulated by a computer. To prevent computation and computer from becoming an empty conception, we require programmability and a symbolic interpretation of the underlying dynamics.*

The requirement of symbolic interpretation cannot be fulfilled without explicit reference to a "user," who selects and defines the interpretation. A closed system can never be a computer, cf. Section 2.2.1 and Section 2.2.3.

Remark 2.3 Feynman's missing machinery *R. Feynman [Feynman (1963), Chapter 7-7] complains that physical laws are abstract and do not come with a machinery: "All we have done is to describe how the earth moves around the sun, but we have not said what makes it go." Similar questions abound: How does* *a many-particle system manage to obey the law of conservation of total energy? How does the formation of quasi-crystals observe global symmetry requirements? How are the abstract laws of quantum mechanics put to work? Physical laws are extremely efficacious but somehow "immaterial."*

Excessive metaphorical language offers a superficial answer: "for the Earth to follow in its orbit, it must constantly receive information about the ratio mass/distance … of all other gravitating masses" [Harmuth (1992)]. Computationalism, cf. Section 2.2.4 proposes a somewhat similar "explanation": The system "calculates itself." For this bewildering assertion to make any sense, information and algorithms would have to play a prime role. However, the tension between the abstract law and concrete (executable) sequences of operations cannot be made to disappear. Physical phenomena might easily be equivalent to the solutions of computationally hard problems (high computational complexity) but, nevertheless, appear to be realized routinely: A modern example is based on photon sampling [Crespi (2012)]—celebrated as "beating classical computing without a quantum computer [Franson (2013)]. A somewhat misleading statement, though. The problem being "solved" is nothing but our mathematical description of the basic physical properties of the respective system; its programmability is very limited. This tension is reminiscent of the dichotomy between digital (symbolic) and analogue computing.

Remark 2.4 Brain as computer? *Serious concerns also come from biology. There is a wonderful cartoon in the article by M. A. Boden on animal perception [Hokway (1984)]. It shows "a kingfisher sitting on a willow-branch, staring at a fish in the river below, and thinking to itself,* $\mu = \sin\phi/\sin\theta$ *(*ϕ *and* θ *are the diffraction angles)." But do we really think the bird applies Snell's law in order to correct for the apparent position of the fish under the water surface? Do we really think any such kind of computation is involved—despite our attempts to account for the birds successful actions in terms of performing measurements and calculations? We should not mistake a metaphor for reality, cf. Section 9.2.4.*

Computation is often seen to relate to the notion of complexity. Various and largely different concepts have been suggested. Among those two stick out: the computational complexity and the algorithmic complexity. They serve different purposes.

Computational complexity is concerned with the computational process as such and the resources needed to carry it out, preferably in terms of time and memory. These investigations have led to the introduction of certain complexity classes: They give an idea of what it means to have an "easy" or a "hard" problem.

Algorithmic (or Kolmogorov) complexity will be briefly introduced below. It contains deep and sophisticated mathematics, which is clearly beyond the scope of this book; details can be found in [Li (1993)].

Definition 2.23 Algorithmic or Kolmogorov complexity *As an alternative to Shannon's concept of information, one may try to quantify the information* within *some object, Z. This may be done by a description of the object [Li (1993)]. If complete, the description should allow the recipient of the description to reconstruct the object (given the material resources).*[a] *As a measure for the amount of information (i.e., its complexity C) we could thus take the number of bits needed to encode the description,*

$$C_\phi(Z) = \min\{L(p) : \phi(p) = Z\}\,. \tag{2.56}$$

Here, p is a program that computes Z by means of the recursive function ϕ, and $L(p)$ is its length in bits. The computation is a process.

Definition 2.24 Conditional complexity *We define the complexity of an object Z, given an object Y as*

$$C_\phi(Z\,|\,Y) = \min\{L(p) : \phi(p, Y) = Z\}\,, \tag{2.57}$$

where p is the program that computes Z given Y by means of ϕ.

Definition 2.25 Prefix-code *A prefix code is a code system, in which no valid code word is a prefix ("start") of another valid code word. A prefix code is also called prefix-free. For example, a code consisting of* $\{3, 5, 35, 55\}$ *is not a prefix code, because the last two code words start with the previous ones. For prefix codes, no markers are needed between the code words.*

[a] A description is about something for someone. It is symbolic and thus requires interpretation.

Definition 2.26 Prefix machine *A so-called prefix machine is a Turing machine restricted to a set of programs, in which no program is a proper prefix of another program.*

The complexity measure C defined so far faces a number of serious problems, one being the lack of subadditivity, which would require

$$C(Z_1, Z_2) \leq C(Z_1) + C(Z_2)\,. \tag{2.58}$$

Definition 2.27 Prefix code based complexity *Most of the above deficiencies can be avoided, if the complexity definition is based on prefix machines. The complexity under this specification will be denoted by $K(Z)$ and $K(Z|Y)$, respectively. Subadditivity is fulfilled now. $K(Z)$ and $K(Z|X)$ play a major role for inductive reasoning (cf. proposition 2.4.)*

Eventually, computation is also related to the notion of games:

Definition 2.28 Games *Game theory may be considered an interactive decision theory, a mathematical discipline. The representation of games requires to specify the number of players of the game (usually two), the pertinent strategies ("procedures") based on information and actions available to each player at each decision point, and the pay-offs for each outcome. Typically, the "procedures" happen in some constrained physical state space and are "rule-based" rather than controlled by direct physical laws. Of prime interest are so-called equilibrium strategies, which are optimal in the sense that no player can profit (i.e., increase his pay-off) by deviating unilaterally.*

There are also single player games: classical examples are solitaire, tangram, or Rubik's cube. In this case, there is no explicit pay-off; the goal is to reach a certain pattern in state space.

The "quantization of games" is based on the superposition of states and the superposition of strategies [Eisert (1999)]. The full implementation requires quantum systems.

Remark 2.5 Cost of games *At first sight games appear to constitute "artificial" processes (algorithms), outside any physical realm. This is not the case, though: The games do not have to be carried out by human beings, they could alternatively be implemented by means of (computer-controlled) robots. And games may be used as tools to* *specify certain control and optimization tasks even within physics. In any case, design and control of such scenarios call for substantial physical "costs." Operationally speaking: Running games is not free of charge!*

2.3.4 *Links between Communication and Physics*

The idea of an agent-based physics is gaining popularity [Goyal (2012)]. In this view, the agent is placed center stage, thus conferring non-trivial properties on the interface between agent and physical world, cf. Section 1.5. For communication to become operative within such a picture, a kind of equivalence (mapping) has to be established.

Definition 2.29 Communication metaphor *The application of the Shannon information theory is restricted to scenarios, in which a directed interaction is involved so that the above communication principles, definition 2.14, can be shown to make sense. Typical examples refer to measurement and control. However, even scenarios may be considered, in particular from physics, for which communication would appear as a rather artificial task. Nevertheless, the "communication metaphor" as a kind of "substitute model" may still provide additional insights—as long as one is aware of its inherent limitations.*

Under any condition, one has to accept that in the Shannon approach the "symbols" a_i have no "meaning," they do not stand for something else. (If they do, this is not subject of the present investigation.) The application of the complexity theory cannot do without the interpretation of symbols—used for the description of an object Z. This (communication) task, however, requires the availability of common reference frames, which play the role of "meaning" (cf. the Peircean triad [Herrmann-Pillath (2010)]).

Definition 2.30 Limits: Unspeakable information *Shared reference frames are needed for communication but cannot be communicated themselves, unless by means of other shared reference frames. A paradigm example is the direction left/right, which has to count as unspeakable information [Chiribella (2010)]. Obviously, left/right* *might be replaced by east/west, say; but this works only if given the convention of the cardinal points. In physics, examples of unspeakable information abound: A quantum state may reliably be transferred by means of the complex amplitudes with respect to a given basis, while the basis itself would constitute unspeakable information.*

Example 2.7 Technology: Remote control *A radio remote control to open a garage door involves specific signals to be transmitted between sender and receiver. The receiver should selectively respond to the correct signal, not to others. The respective signal interpretation is possible only based on a common reference installed by the designer of the device, that is, by means of a "filter" (decoder). Finally, a servomechanism is needed to move the door. It is obvious that the physical interaction between sender and door, the radio-signal as such, could not provide the desired effect.*

Example 2.8 Biology: DNA *It is generally believed that the genetic code, see Fig. 2.2, is more than just a convenient metaphor [Yockey (2005)]. On the one hand, the code words are assigned a quantitative information measure in the sense of Shannon's communication theory. And then, given the appropriate environmental machinery, the code can provide a "description" for the built-up of certain proteins. The description would be an instant for the Kolmogorov complexity theory. How the "meaning" of the code is encoded, remains unclear, though*[a].

The following examples are taken from physics. Here the physical system with its dynamics is seen to act as a sender of "messages," the observer as the receiver. The common reference frame is assumed to be given. Contrary to explicit communication scenarios, which have been designed for that purpose, the communication interpretation may appear as a kind of substitute model only, in particular in the classical domain. Things change in the quantum domain, though.

Example 2.9 Astronomical data *Astronomy is one of the oldest sciences. Originally it has been concerned with the motion of objects in the sky visible to the naked eye. Early on the respective angular*

[a]The code may implicitly be "generated" by the observer just like he "generates" correlations via his choice of partitions, cf. Section 2.2.1.

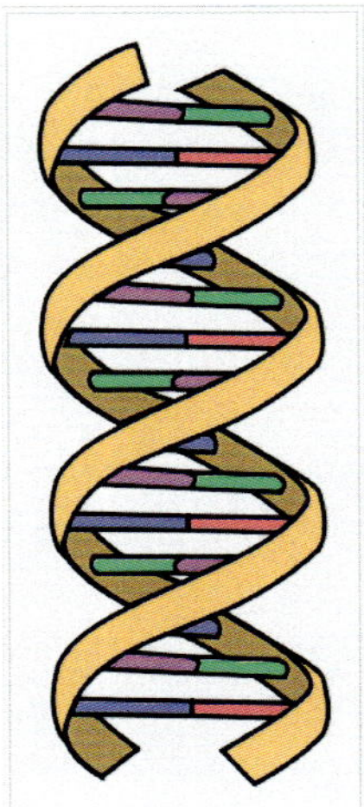

Figure 2.2 Structure of DNA double helix. Base pairs (horizontal links) connecting the two sugar-phosphate backbones represent the genetic code. Not shown is the machinery to transform the information into the intended action.

positions had been recognized to be correlated, for example, with seasonal changes on the earth. The astronomical data or pattern thus present important messages for the knowledgeable observer.

Example 2.10 Two-photon interference *The bosonic character of photons can be demonstrated by their so-called bunching behavior: Let two photons—one from each input port—be incident on a beam splitter. If arriving simultaneously, they become indistinguishable (cf. Example 3.17) and end up together randomly in either output beam [Zeilinger (2005)]. The important message here is not so much to learn which output beam has been active but rather to confirm that the two photons are always strictly correlated. This observational pattern constitutes the "sign" or fingerprint for photons behaving as Bosons.*

Example 2.11 Spin measurement *The Stern–Gerlach apparatus constitutes a fundamental quantum measurement scenario for the spin of individual atoms. There are two possible events, "spin up" and "spin down." Lacking any further knowledge, the prior probability would have to be chosen as* $P(up, down) = 0.5$, *which maximizes the* *entropy S_I. Learning the concrete outcome would then present 1 bit of information.*

Example 2.12 Blackbody radiation: information channel *For a many-particle system such as a photon-gas specified by the dispersion relation $\epsilon_j \rightarrow \hbar\omega(\vec{k})$ in thermal equilibrium is characterized by the distribution Eq. (4.60) with $\mu = 0$. A fixed direction $\vec{k}/|\vec{k}|$ is taken to specify a "channel." Based on our communication metaphor, Definition 2.29, we may consider the entropy associated with the respective distribution in terms of the one-dimensional parameter $k = |\vec{k}|$ as the information transmitted to the observer.*

It turns out that the information in bit per time is bounded from above by the associated energy flow,

$$k_2 \dot{S}_I \leq \frac{2\dot{U}}{k_B T \ln 2}, \tag{2.59}$$

where $k_2 = 1/\ln 2$, cf. Eq. (2.39).

This result is quite general and neither depends on the dispersion relation nor on the statistics (Bose or Fermi)—as shown by Pendry's inequality [Pendry (1983)]. Similar results have been obtained by a number of authors (cf. [Bekenstein (1988)]).

Proposition 2.10 Pendry's inequality *Let us quantify information based on the Shannon's communication scenario, that is, by the Shannon entropy S_I. One may then exploit the analogy between a medium carrying information and a gas of particles carrying thermodynamic entropy S (uncertainty) away from some thermal source, $S \equiv S_I k_B$. As argued in [Pendry (1983)], the optimum information flow should be obtained for thermal radiation at some temperature T_C. Of course, one has to restrict oneself to a selected current direction; in equilibrium the total current would cancel to zero.*

Under this condition, the directed information flow $k_2 \dot{S}_I$ (in bits per time) turns out to be bounded from above by the respective energy current $\dot{U}$ being dragged along by the information flow,

$$\boxed{(k_2 \dot{S}_I)^2 \leq \frac{\dot{U}\pi}{3\hbar(\ln 2)^2}.} \tag{2.60}$$

Each micro-state (also entering the definition of thermodynamic entropy) is seen here as a possible signal. Then the "communication metaphor" applies; messages are being transmitted associated with energy. This defines a noiseless quantum channel.

Proof. Consider a non-interacting many-particle system characterized by the internal energy $U = \mathrm{Tr}\{\hat{H}\,\hat{\rho}_{equ}\}$, see Definition 4.4. Let the dispersion relation be $E = \hbar\omega(k)$ and we restrict ourselves to a one-dimensional setting: The wave vector is perpendicular to the emitting surface. The density of states is $n(E) = n(k)(d\omega/dk)^{-1}\hbar^{-1}$, where the density of states in k-space, is constant, $n(k) = 1/(2\pi)$. For Fermions, cf. Definition 3.69, the average energy flow with group velocity $v_g = d\omega/dk$ away from the surface is

$$\begin{aligned}\dot{U}(T) &= \int_{-\infty}^{\infty} \frac{\hbar\omega}{\exp(\hbar\omega - \mu/k_B T) + 1} n(E) v_g d(\hbar\omega) \\ &= 2\int_{0}^{\infty} \frac{\hbar\omega}{\exp(\hbar\omega/k_B T) + 1}\frac{d\omega}{2\pi}\,. \end{aligned} \tag{2.61}$$

We realize that the group velocity has finally canceled out, the factor 2 takes care of the presence of particles and holes. For the thermal state occupation, we have applied Eq. (4.61). The chemical potential μ is the same at all points and independent of temperature, so that we can shift the energy scale such that $\mu \equiv 0$. Changing the integration variables and setting $T = T_C$ we find

$$\dot{U}(T_C) = \frac{k_B^2 T_C^2}{\pi\hbar}\int_0^{\infty}\frac{x dx}{\exp x + 1}\,. \tag{2.62}$$

Observing that (cf. Eq. (4.100))

$$dS = \frac{dQ}{T} = \frac{dT}{T}\frac{dQ}{dT}\,, \tag{2.63}$$

we get for the entropy flow

$$\begin{aligned}\dot{S} &= \int_{-\infty}^{\infty} X(E, T_C) n(E) v_g d(\hbar\omega)\,, \\ X(E, T_C) &= \int_0^{T_C}\frac{dT}{T}\frac{d}{dT}\left(\frac{|\hbar\omega - \mu|}{\exp(|\hbar\omega - \mu|/k_B T) + 1}\right)\,. \end{aligned} \tag{2.64}$$

Integration by parts leads to

$$X(E, T_C) = \frac{1}{T_C}\left(\frac{|\hbar\omega - \mu|}{\exp(|\hbar\omega - \mu|/k_B T_C) + 1}\right) - B\,, \tag{2.65}$$

where B is a positive constant. With this result and changing the integration variables one arrives at:

$$\dot{S} = \frac{2k_B^2 T_C}{\pi\hbar}\left(\int_0^\infty \frac{x dx}{\exp x + 1} - B'\right). \tag{2.66}$$

This shows that neglecting $B' > 0$,

$$\dot{S} \leq \frac{2\dot{U}}{T_C}. \tag{2.67}$$

Alternatively, T_C^2 can be eliminated by $\dot{U}$:

$$(\dot{S})^2 = \frac{4k_B^2 \dot{U}(T_C)}{\pi\hbar}(\cdots)^2 \left(\int_0^\infty \frac{x dx}{\exp x + 1}\right)^{-1}. \tag{2.68}$$

Again, neglecting B' this result simplifies to

$$(\dot{S})^2 \leq \frac{4k_B^2 \dot{U}(T_C)}{\pi\hbar}\int_0^\infty \frac{x dx}{\exp x + 1}. \tag{2.69}$$

The integral gives $\pi^2/12$. Identifying $\dot{S}$ with $\dot{S}_I k_B$ and multiplying both sides by the scale factor $k_2^2 = (1/\ln 2)^2$ we confirm the proposition. The same result obtains for Bosons; for massless Bosons the equality sign may apply. □

Physically speaking, the energy flow, connected with entropy transfer, should count as a heat current, $\dot{Q}$. From the Clausius formulation [Eqs. (2.62) and (2.67)] of the second law, Eq. (4.110), we know that $\dot{Q} \leq \dot{S} T_C$. Combined with Pendry's inequality we thus find

$$\dot{Q} \leq \frac{1}{6}\pi k_B^2 T_C^2/\hbar, \tag{2.70}$$

in agreement with the third law, cf. Section 7.5.1: The heat current goes to zero for $T_C \to 0$. Work flow, as an energy transfer without associated entropy change, would thus carry no information in the sense of Shannon (cf. LEMBAS, Section 5.2.4). Information is conserved for this process (cf. Section 2.2.2).

Remark 2.6 Information-driven interactions *The communication metaphor supports various kinds of measurement models. Under some circumstances, as exemplified by Pendry's inequality, these models can be used to derive general bounds for information transmission.*

A more explicit use of information gives rise to information-driven interactions as exemplified by remote control, Example 2.7. Typically, there are new "rules" of behavior involved. How is this possible, and how do those new rules become effective, without coming in conflict with direct physical laws?

A prominent design principle makes use of filters, which allow to select and condition specific paths of operation. Not surprisingly filters are an important formal ingredient also of Boolean logic [Primas (2007)]. Insofar as the latter underlies mathematical relations, new functions beyond the immediate physical laws can be implemented.

Chapter 3

Quantum Physics

Jonas Burgert: Sand burns leaf, 2010, based on photograph by Lepkowsky Studios [Reprinted with permission from Schreiber and Wipplinger (2011)].

Quantum Thermodynamic Processes: Energy and Information Flow at the Nanoscale
Günter Mahler

ISBN 978-981-4463-73-7 (Hardcover), 978-981-4463-74-4 (eBook)
www.panstanford.com

Summary: Quantum physics is often considered mysterious. ("I think I can safely say nobody understands quantum theory" [Feynman (1965)].) From an operational point of view, there are good reasons to prefer a more balanced attitude: Quantum mechanics does exhibit many signs of a stringent and logical scheme: What cannot be decided within a given context has to be left undecided; no implicit assumptions, no untestable constraints allowed.

Quantum physics has an abstract, formal description (valid for a detached outside view). This description specifies the concepts of system, state, and dynamics, see Section 3.3. But this abstract account has to be supplemented by connecting rules (for dealing with the concrete interaction with the observer/agent). These rules support an observer-dependent reality (participating inside view), which—unfortunately—gives rise to notorious interpretational difficulties. A prime example for the latter is the measurement problem. It renders quantum mechanics contextual and subject to fundamental uncertainties, see Section 3.4. This smells like inherent incompleteness. Nevertheless, quantum mechanics is rather strict on what can reasonably be said about physical properties given the rules of quantum mechanics and a certain context. Lack of information or information retrieval cannot be separated from its physical substrate.

Another source of controversy is the relationship between classical and quantum features. There are various routes to the so-called "classical limits." Often these relate to length-scales, energy-scales, or time-scales.

3.1 Historical Remarks

Around 1900, the atomic hypothesis for the structure of matter was widely accepted. Atoms were considered point particles, but it was unclear how various kinds of atoms could differ. While the electron had just been discovered, it remained dubious how (and if at all) electrons might be located within the atom. The origin of the different colors visible in the fluorescent decay of gas atoms was not understood. The specific heat of simple ideal gases showed severe deviations form accepted classical theory. Nevertheless, the breakthrough toward quantum theory did not occur in the realm of such atomic phenomena.

Remarkably it was thermodynamics, which eventually gave birth to quantum mechanics: The thermal radiation state resulting from an interaction between matter and optical radiation (black body radiation) had become a major challenge for theory. Max Planck in his attempt to explain his interpolation formula to fit experimental data saw himself forced to invoke the concept of quantized energy packages to be exchanged between matter and radiation field. Einstein later showed that the corresponding entropy for the radiation field looks like that of a free gas of "light particles." "Quantization" of this sort then became the recurrent theme of the rapidly developing quantum physics. Remarkably, the nature of this quantization is easily overlooked and hidden in the formalism.

It was not before 1926 when the new quantum mechanics eventually took shape in Heisenberg's matrix mechanics and Schrödinger's wave mechanics. They were soon shown to be mathematically equivalent. Nevertheless the underlying formalisms continue to support rather opposing pictures of intuition. Schrödinger had a distaste for the discontinuity in the atomic models according to Bohr and colleagues. He was convinced that the apparent atomicity of matter disguises the underlying continuum.

In fact, in the Schrödinger field (or wavefunction) approach discrete modes derived from boundary conditions [Wharton (2011)]. Such a discretization is also known for classical fields such as mechanical vibrations on a finite string (cf. a violin). But the decisive role played by the respective eigenvalue spectrum is eventually connected with the quantum measurement postulate: In this way, the discreteness of eigenvalues gets translated into the discreteness of observational values.

Till date there are no experimental findings that contradict quantum theory. And this theory continues to open new fields of research, a more recent direction being quantum information theory [Nielsen (2000)]. This remarkable success story, however, would not have been possible without a kind of working rules: Such a "connecting theory" is desperately needed to make quantum theory applicable at all. But at the same time, it creates the notorious interpretational problems.

Eventually, all these interpretational difficulties come down to the question of an objective versus observer-dependent reality. While the former clearly has dominated classical physics, the latter appears to be unavoidable in the quantum domain. Originally surfaced under the name of "wave-particle duality," observer dependence is now recognized as a much more general feature underlying all quantum phenomena [Wiseman (2011)]: Properties cannot just "exist out there." But if so, how could they materialize? This is the essence of the so-called measurement problem. To deal with this problem means to interpret those working rules.

Wikipedia lists some 20 interpretations, from the so-called Copenhagen interpretation (around 1927), the de Broglie–Bohm theory (1952), Everett's many-worlds theory (1957), consistent histories (1984), objective collapse theories (1986), up to one of the most recent ones, relational quantum mechanics (1994, see [Rovelli (1996b)] and Section 3.2.8). Some of these proposals suggest modifications of the "standard quantum theory"—Bohmian quantum mechanics is a nonlocal hidden variable theory. Remarkably, though, all these variants have the same predictive power, supporting the view expressed in Ref. [Colbeck (2011)]. No extension of quantum theory can have improved predictive power.

3.2 The Big Questions

Nonclassicality is one of the overarching issues, but also the inherent "incompleteness" of the quantum description. Most of the following questions thus relate back to information and control. The common leitmotif is that observations/descriptions cannot be isolated from the physical world; they are, in turn, connected with physical processes, which are quantum in nature.

3.2.1 *What is the Meaning of the Wave Function?*

Is the wave function "physically real"? In order to clarify the meaning of this question, Einstein et al. suggested an operational definition for the notion of "real," namely: "If, without in any way disturbing a system, we can predict with certainty . . . the value of a physical quantity, then there exists an element of physical reality corresponding to this physical quantity" [Einstein (1935)]. Along similar lines, Unruh argues that "in classical physics the relation between the mathematical theory and reality is relatively direct. The dynamical variables in the theory can be assigned values, and so can the attributes of objects in the world" [Unruh (1994)]. Such a direct correspondence is lost in quantum mechanics. The wavefunction provides only probabilistic statements about what should happen if a certain measurement was selected and performed; but most of these possible measurements are mutually exclusive. This indicates that the real space wave function of a single electron, say, cannot represent "given" local attributes of this particle.

This problem has been reexamined recently by Pusey et al. [Harrigan (2010); Pusey (2012)]—in an operational setting: Suppose that a quantum system has been prepared in a pure state, $|\phi >$. Associated with this state is a complete list $\{\lambda\}$ of physical properties (in short: "reality"). It is accepted that these are typically given in the form of probability distributions only. What is the relation between $\{\lambda\}$ and $|\phi >$?

First of all, $\psi \;\; >$ might correspond to many $\{\lambda\}$, a kind of "incompleteness" in the sense of Einstein [Harrigan (2010)]. This situation (motivating hidden variables) will be put aside here. Instead, one of the following two possibilities are being scrutinized: (i) The quantum state $|\phi >$ is uniquely determined by $\{\lambda\}$; and (ii) it is not uniquely determined by $\{\lambda\}$. In the former case, the quantum state is considered a property of the system. In the latter case, some values of $\{\lambda\}$ may be compatible with either $|\phi >$ or $|\psi >$, the quantum state is "not a definite part of reality," it is called "statistical in nature."

Harrigan et al. (2010) suggest to call the above case (i) ontic case and (ii) epistemic. This debate about the "ontic"(context-free) versus "epistemic" (descriptive) nature of the wavefunction [Atmanspacher (2005)] easily becomes confusing. Measurements are always contextual. All our theories are descriptions of nature, the concepts we introduce (like the wavefunction) are mere tools. We have no access to reality as such. "Science provides us with a modern mythology that tells us what is real.... When we look at a tree and rock, we see delicate arrangements of matter rather than spirits or the hand of God" [Hut (1997)]. "Real" is here understood as an indispensable element of our "narrative" by which we assure ourselves of our common understandings. It avoids any reference to ontological significance: whether or not the respective concepts "exist out there" is not the issue.

One easily convinces oneself that two preparation strategies 1,2 leading to the nonorthogonal states, $|\phi_1>$, $|\phi_2>$, cannot uniquely be distinguished by measuring a single object. (A distinction would be possible from measurements on a large ensemble of equally prepared systems.) The idea is now that a distinction may be possible based on an enlarged set of joint measurements on the ensemble.

For the following arguments, the basic quantum system is taken to be a two-level system (qubit). Suppose that two different preparation strategies are being used, which are (locally) indistinguishable with a finite probability q. For an independently prepared number N of qubits (resulting in a set of 2^N possible preparation methods, i.e., product states) the corresponding probability would thus be at least q^N.

For this system, there exists a set of 2^N joint mutually commuting(!) observables: As an example see the four correlation terms $\hat{K}_{jj}$ as introduced for the Bell states (N = 2), Definition 3.64. Switching from the set of local to such a set of joint (i.e., entangling) measurements to specify the dataset $\{\lambda\}$, the authors are able to show that, depending on the concrete preparation, the probabilities for certain measurement outcomes have to be zero.[a]

Orthogonal states (maximum distance) can always be uniquely distinguished with zero uncertainty, $q = 0$. It suffices to consider the case $N = 1$. But even for distances as small as one likes, it is then possible to uniquely reconstruct the total preparation strategy, if N is taken large enough, in contradiction to the assumed increase of uncertainty q^N. (In this sense, all pure states become distinguishable, a quasi-classical limit, see Section 3.3.8.)

While some of the underlying idealizations may be dropped, joint measurements as such will remain hard to implement. But this should not invalidate the conceptual significance of the proposal

---

[a] cf. destructive interference, Example 3.25.

in terms of a priori probabilities. The alternative opinion that the wavefunction would "represent nothing but our knowledge" about the system in question [Fuchs (2000)] is thus loosing ground.[a]

3.2.2 *What is the Referent of the Wavefunction?*

According to Ballentine [Ballentine (1998)], one may distinguish between two principle classes of interpretations for the state vector:

(A) A pure state provides a complete and exhaustive description of an individual system. A dynamical variable $\hat{Q}$ has a value q only, if the state happens to be an eigenstate of $\hat{Q}$.

(B) A pure state describes the statistical properties of an ensemble of similarly prepared systems.

Ballentine favors the latter position, but this comes at a price: Supporters of B also cannot escape the need to say something about individual systems. So, Ballentine argues, as pointed out by d'Espagnat [d'Espagnat (2003)], that "individual systems possess definite, although perhaps unknown, values of certain fundamental variables (such as position) which are not described by the state vector—unless it happens to be an eigenvector of that observable." Clearly this is a hidden variable assumption (cf. Section 3.2.10)—and thus subject to the severe constraints by Bell's theorem [Bell (1966)].

But it is this additional (and unjustified) assumption by which Ballentine believes to avoid the measurement problem (in terms of quantum jumps): The entangled state between system and apparatus (resulting after interaction) is simply interpreted to mean an ensemble, in which each system–apparatus pair would show a definite outcome.

Most researchers now hold that the restriction of quantum mechanics to ensembles is too narrow; one would have to put aside many of the exciting new experiments, which often are on individual quantum objects or small ensembles. Furthermore, the ensemble

---

[a] "Knowing a state" means knowing which one out of a given set (in the sense of Shannon information). We may thus know the system to be in its ground state E_0—even though the precise form of this state will not be available. In any case, knowledge can only be acquired via observation.

interpretation is intimately related to the ignorance interpretation of probabilities, which is not in accord with the objective nature of probabilities in the quantum domain [Mermin (1998)].

But this does not render the ensemble concept entirely obsolete. In fact, the statistics of measurement events is consistently interpreted by means of the ensemble interpretation of probability [Brody (1993)]. The probabilities themselves—irrespective of their origin—can only be measured based on a repetition of the respective experiment, that is, based on relative frequencies (cf. Eq. (2.24)).

3.2.3 *What is the Measurement Problem?*

"In a simplified nutshell, the quantum theory formalism distinguishes between observations made from outside a system (which yield discrete results) and interactions within a system (which are treated much more continuously). This isn't tenable because there's no objective definition of a 'system'; it can always be expanded to include any 'outside' observations. The measurement problem is how objectively different mathematical procedures could possibly be associated with an apparently subjective parsing of joint vs. separate systems" [Wharton (2011)]. In other words, when is a physical process to be regarded a measurement?[a]

This quote underlines the tension that exists between abstract and observational quantum theory, between outside and inside. Here we suggest to associate the outside view with the "detached" observer, the inside view with the "participating" observer. The measurement problem then concerns the inconsistency between the "outside view" (measurement as dynamical, unitary interaction between system and apparatus (cf. [Allahverdyan (2011a)]) and the "inside view" (measurement as projection postulate). Roger Penrose [Penrose (1989)] admits his concern that the borderline between—in his terminology—the U-process (unitary) and R-process (reduction type) still appears to represent unknown physics. The R-process is conventionally defined by a "rule": If the observable A is measured, the result will be an eigenvalue of the associated operator

[a]Bohmian quantum mechanics [Passon (2006)] tries to avoid this problem by means of additional hidden variables (particle position) in the "completed" initial state, which then predetermines the final measurement result.

$\hat{A}$ with a probability that can uniquely be calculated from the original quantum state. The system is then in the corresponding eigenstate.

The appearance of such a rule may remind oneself of the informational coupling (cf. Section 2.2.7) typical for effective descriptions. However, one should note that observational quantum dynamics has to differ from "abstract dynamics." As the observer is a part of the total system but cannot observe himself completely, the observed total dynamics can no longer be unitary. In particular with respect to the measured properties, for which a strong correlation between system and apparatus had to be prepared, unpredictable events must occur [Breuer (1995)]. The difference between U- and R-processes is consistent with this finding; one should not expect that these features could go away in a "more complete theory."

In fact, this difference does not disappear by expanding the object at the expense of the environment: Clearly, if one system interacts with a second one, the latter may serve as a premeasuring device; but there is no "jump" into one of the possible measurement outcomes. To let both these subsystems interact with a third, one, etc., does not help either: Eventually one ends up with the notorious "von Neumann chain" [d'Espagnat (2003)]. The desperately needed final jump never "occurs," unless postulated via a quantum–classical transition, which some researchers have placed at the borderline between matter and consciousness.

Eventually, even the U-process somehow needs to be confirmed by observation: A combination of two measurements, separated by some period of quasi-isolated evolution, allows to verify the unitary behavior in between. That this can indeed be achieved is a precondition for the functionality of quantum gates and quantum computation [Nielsen (2000)]. The continuous dynamics is incompatible with the idea of a basic discreteness, which seems to be at the heart of quantum mechanics, actually motivating its very name: "If one takes away all measurements, quantum theory turns out to be not quantized at all" [Wharton (2011)].

3.2.4 *What are Weak Measurements?*

The two process types U and R might be seen as two opposing limits of a wider range of possibilities: the so-called weak measurements.

Contrary to strong measurements for a given prepared (pre-selected) state weak, measurements refer to pre- and post-selection. And the coupling between the measuring device and the respective observable is so weak that the uncertainty in a single measurement would be large compared with the separation between the respective eigenvalues of the observable [Ritchie (1991)], actually so large that a single weak value may even lie far outside the range of the allowed eigenvalues. Not surprisingly this has created interpretational problems.[a]

While the spectrum can thus no longer be resolved, the weak ensemble average does yield the standard quantum expectation value of that observable. This value could have been reached alternatively via an ensemble average on normal measurements. But as the perturbation of the system goes to zero in an ideal weak measurement, one can continue to perform additional measurements on the "same state," even measurements that would otherwise be incompatible with the first one.

Indeed, the weak ensemble measurement of the position $\vec{x}$ of a particle post-selected by a subsequent normal measurement of the momentum $\vec{p} = 0$ can be shown to provide direct measurability of the wave function $\Psi(\vec{x})$ of that single particle [Lundeen (2011)]. So far, such a "direct" measurement of the wave function had been considered impossible.[b]

3.2.5 *What is Entanglement?*

The linearity of the Schrödinger equation implies the superposition principle for states. Any two (or more) solutions can be combined to give a new, valid solution. In particular, consider the original individual solution as the eigenfunction of some observable $\hat{F}$, thus corresponding to a sharp value of this observable. Then the superposition of such states (to different eigenvalues) generates a

---

[a] Most of the confusion can be avoided when considering weak measurement as a special case of the so-called contextual values [Dressel (2010)]. Here the context is defined by the concrete measurement environment.

[b] State reconstructions are routinely performed via the so-called tomographic methods. In any case, however, the "measurability" does not imply that the wavefunction with its associated probabilistic statements on all kinds of possible observables can be considered "physically real," cf. Section 3.2.1.

state in which the value of $\hat{F}$ is necessarily uncertain. Superpositions underlie coherence phenomena (like constructive or destructive interference).

However, these features depend on the basis states used: What appears as a superposition in one representation may show up as a single eigenfunction with respect to another observable $\hat{G}$, noncommuting with $\hat{F}$, and thus amount to a sharp value of this $\hat{G}$.

With respect to multipartite systems, superpositions give rise to the intriguing feature of entanglement. In this case, convenient basis states are the so-called product states; here the subsystems are completely uncorrelated. Pure states, which cannot be written as such an uncorrelated state, are called entangled: they can only be represented as superpositions of product states. For entangled states, definitely all local properties are uncertain: their local state is mixed. There is a kind of incompatibility between local and global properties. As the global observables decompose into local ones (like the total angular momentum does with respect to the local momenta), such global properties imply correlations between local properties. This means, for example, that there can be a strict correlation between the state of two spins A and B, even though their local spins are completely undetermined! There are correlations without correlata [Mermin (1998)]. Indeed, a convenient measure for entanglement (and local uncertainty) are the local von Neumann entropies, $S_{vN}(A)$, $S_{vN}(B)$. In this way, correlation shows up locally, an entirely nonclassical feature. A well-known example exhibiting such features is the famous Einstein–Podolsky–Rosen or EPR state [Einstein (1935)].

Entanglement also has far-reaching consequences in statistical physics: As already noted, the origin of local entropy does not necessarily result from subjective ignorance (as in the classical realm) but may be objective and due to entanglement. This fact is exploited in quantum thermodynamics (cf. Chapter 5).

In turn, any locally mixed state can be embedded into the larger space of a multipartite system, in which the total state appears as a pure state [Hughston (1993)].

Entangled states may violate the so-called Bell inequality [Bell (1966)]. Such a violation disproves that incompatible observables might exist independent of context.

When the total state is non-pure (total entropy $S_{vN}(A, B) > 0$), general entanglement measures are hard to come by. A convenient correlation measure is the so-called mutual information $I_{vN} = S_{vN}(A) + S_{vN}(B) - S_{vN}(A, B) \geq 0$, see Definition 3.62. Entanglement implies correlation, but not vice versa.

3.2.6 *What is Quantum Nonlocality?*

Locality derives its appeal from the human intuition that things have to touch one another in order to interact. In mechanics, nonlocality would thus have to be associated with an "action at a distance." Phenomena such as gravitational forces—which appear to be of just this type—have thus led to the idea of a field mediating across empty space. Not surprisingly, electrodynamics as a classical field theory abounds with examples of such nonlocal interactions. In all those cases, however, the mediation is not instantaneous but constrained by the speed of light.

Nonlocality is believed to show up also in observational quantum mechanics. Quantum measurements do not merely exhibit "what is," but have effects, and these effects can be nonlocal: Suppose a measurement projection is applied to subsystem A of a total system consisting of two subsystems. As a consequence of this action, the state of subsystem A will jump into an eigenstate of the respective local measurement observable. But if the original state of the total system had included some inter-subsystem correlation (cf. Section 3.2.5), the state of subsystem B would also be affected: a kind of co-jump results [Granzow (1998)]. And this remote state preparation effect is "instantaneous" in the sense that it does not depend on the real space distance between the two subsystems. In fact, the two subsystems can be so far apart that no communication at speeds below the speed of light could be made responsible for the state changes. However, as the measurement result is unpredictable, the original experimenter cannot use such a procedure to instantaneously transmit information to another distant experimenter, who has local access to subsystem B only. This is the so-called non-signalling theorem. The state after measurement is a product state, that is, a state without correlations: no further nonlocal effects will be possible.

The first experimental confirmation of this intriguing phenomenon has been provided by the study of the so-called EPR state realized in the polarization space of a photon pair [Aspect (1981)]. Since then a vast amount of experimental data has been accumulated indicating that this nonlocality is indeed a characteristic feature of (observational) quantum mechanics.

3.2.7 *How Does Time Enter Quantum Mechanics?*

In quantum mechanics, time is a classical parameter, not an observable (and thus not an operator). Time appears indirectly, that is, via the change of some observable relative to any given reference frame.[a]

Eigendynamics: In this case, the evolution is completely determined by the system (i.e., its Hamiltonian) and its (momentary) state. Necessarily, this is an "abstract view," there cannot be an external reference whatsoever. So, what evolves with respect to what? A preliminary answer is: There are basically two equivalent pictures: the Schrödinger picture and the Heisenberg picture. These pictures characterize motion in two opposite ways. In the former, the states move relative to fixed "properties" defined by the respective operators; in the latter, the properties (operators) move relative to the fixed states (cf. Section 3.3.7). The so-called interaction pictures are in between; the choice of pictures is a matter of convenience. Time-dependent expectation values are the same in either picture.

Relational dynamics: "To speak of free propagation without preparation and detection is meaningless" [Briggs (2000)]. Change is of something with respect to something else (like a pointer with respect to a dial). However in quantum mechanics, the establishment of such a relation is not for free, it constitutes a physical interaction and by that introduces relational dynamics. Only then, with change having emerged from the system's interaction with an environment, the parameter time gets a (operational) meaning as a parametrization of this very change.

---

[a]The fundamental nature of time has been challenged on general grounds [Callender (2010)].

Correlation and memory: There is no time without change—there is no change without memory. Fundamental equations of motion are deterministic in the sense that the presence allows to completely predict the future (and reconstruct the past). More formally speaking the memory time (with respect to temporal correlations) is infinite. Effective dynamics of motion typically lack this property. The (incomplete, coarse-grained) present state does no longer suffice, neither to uniquely predict the future nor to reconstruct the past. The effective memory time has become finite. This shortcoming can partly be overcome via reference to additional data as provided in measurement records. In general, the past (history) is present today in such records only—insofar as these records can still be interpreted, that is, correlated with some previous event. In short, the past has to be stored as information (cf. Section 2.2.5)!

Thus, it appears as if a physical subspace would have to be reserved to accommodate an ever increasing amount of data representing footprints of the evolution of the observed quantum world. We note, however, that those data (usually of very limited lifetime) are only a substitute for the full present state, which, if available, would allow to trace back the past without limit, cf. Section 2.2.5.)

Clock and the arrow of time: A clock is a counting device (for "identical" time differences)—not just a periodic subsystem like a pendulum. Necessarily, counting creates a record of the past associated with that very device. The recording itself will constitute a series of quantum measurements. So, right from the start time and measurement are intertwined. Axiomatic measurement theory calls for jumps, that is, events, which appear to be located on the time axis—which, in turn, derives from measurements!

If time were a quantum observable, there would have to be "absolute" measurement schemes, limited only by unavoidable technical shortcomings. (There are still attempts to construct operators with specific "functionalities": The arrow-of-time operator, for example, would show a monotonical decrease of its expectation value, independent of the initial state [Strauss (2008)].)

3.2.8 *Is Quantum Mechanics Contextual?*

No interpretation of quantum mechanics has so far succeeded to convince the majority of physicists. The main source of controversy appears to be the role played by the distinction between outside and inside view (cf. Section 3.2.3). This distinction is based on a partitioned quantum universe, it defines a context (or interface [Rössler (1998)]) and thus explicitly gives up the notion of an observer-independent state of a physical system. But what are acceptable contexts?

According to classical physics, all properties (values of observables) "coexist," whether or not they are measured (registered). According to relational quantum mechanics [Rovelli (1996a); Rovelli (1996b)], a specific property F of some system S (as associated with its eigenstate) exists at a given time with a finite probability P and *with respect to a specific reference system* O only. O can be any quantum subsystem: a single spin or a macroscopic many-particle subsystem. The "observer" is not physically special at all. Other properties may exist with respect to different references, cf. Remark 3.22. Each of these S–O relations is symmetrical in the sense that the roles of S and O can be interchanged. (The subsystems, of course, need not be of the same type.) Any interaction presents a *potential* measurement process. There is a "static" aspect as well as a "dynamic" aspect.

The *static aspect* has a flavor similar to relativistic physics in the sense that any statement (description) becomes context dependent. As a consequence, one has to accept conflicting statements: In relativity theory, two events may happen at the same time or one preceding the other: One just has to keep track of the reference frame. Rovelli tries to do for the formalism of quantum mechanics what Einstein did with the Lorentz transformations for relativity theory.

Rovelli starts from two hypothesis: (1) All systems are equivalent, nothing distinguishes a priori macroscopic systems from quantum systems. (2) Quantum mechanics provides a complete and self-consistent description of the physical world. (There is no duality between classical measuring devices and quantum systems in a strict sense.) Quantum mechanics is then taken to be a theory about the physical description of the object system S relative to the reference O.[a]

Turning to the *dynamic aspect*, the grand picture is as follows: We typically have a multipartite quantum system with various internal interactions. Its overall dynamics is unitary. The reference (one of the subsystems) is either treated as part of the total system ("inside view") or considered detached ("outside view").

But how to deal with the *inside view* in a strict sense? Obviously, O cannot have a complete description of its own intrinsic situation. One way out is by means of a "third person perspective," that is, an additional observer P, which describes the original observer O plus the system S from the outside. This does not necessarily lead to an infinite regress:

The hierarchical scenario $(O(S)P)$ allows for two options:

(i) The observer P does not interact with the system $O(S)$. P merely predicts the future state of O based on his information about its previous state and the underlying dynamics. This is abstract quantum dynamics (strict *outside view*). But how can P know? This is left open, in fact P must have interacted with $O(S)$ in the past.

(ii) The observer P is part of the total physical system ("participatory inside view").

Accessing some property q of the object $(O)S$ necessitates a physical interaction. In the process of this interaction, the state of P has to change. (Note that in quantum mechanics properties do not exist beforehand.)

But as P—despite interaction with the "rest of the world"—has been taken outside of the pertinent physical description, the remaining system is no longer complete, no longer isolated, and thus can no longer evolve unitarily. The unpredictability of the measurement results is in accord with this fact [Breuer (1995)].

---

[a]The need for a material reference is corroborated by the fact that a basis (representation) constitutes "unspeakable information" UI [Chiribella (2010)]. UI (like a direction in space) cannot be encoded in a string of bits but rather needs a material object (like a gyroscope in this case). Cf. also Definition 2.30.

3.2.9 *Are there Restrictions for the Superposition Principle?*

Even though superpositions are at the heart of quantum mechanics, there are situations for which such superpositions are not allowed (or at least not observable).

It is impossible to consider all Hermitian operators acting on the respective Hilbert space as observable quantities. The operator $\hat{Q}$ is defined to describe a so-called superselection charge, if such an operator $\hat{Q}$ is not just a multiple of the unit operator, but nevertheless commutes with all the operators associated with observable quantities [d'Espagnat (2003)]. As a consequence, the matrix elements of all observable quantities taken between eigenvectors of $\hat{Q}$ corresponding to different eigenvalues of $\hat{Q}$ ("different sectors") vanish. As the generator of time evolution is itself a physical observable, the equation of motion necessarily decouples into separate equations for each sector. The subset of such operators (commuting with all observables) are sometimes called classical observables.

Are there examples for such a fundamental superselection charge? Indeed, no physical observable can have nonvanishing matrix elements between states with integer angular momentum and half odd-integer angular momentum [Ballentine (1998)]. A related example is the symmetry property of multiparticle states with respect to permutations, cf. Section 3.3.6.

The ubiquitous mechanism of decoherence effectively restricts the local verification of coherence. Hence "effective superselection rules" emerge locally in a practically irreversible manner, albeit the correlations are actually never destroyed but merely delocalized. This is called environment-induced superselection or *"eins"-selection* [Zurek (2003)]. This concept has been applied to the question of why certain molecules occur in eigenstates of chirality rather than of energy, that is, why sectors of different chirality seem to be superselected so that chirality becomes a classical observable. This problem is related to the question of how classical behavior emerges in quantum theory [Giulini (1996)], cf. Section 3.3.8.

3.2.10 *Are there Limits to Quantum Mechanics?*

"In a complete theory there is an element corresponding to each element of reality" [Einstein (1935)]. Einstein and others stuck to the prejudice that the proper elements of reality (cf. Section 3.2.1) had to be the full set known from classical physics. They refused to accept the underlying revolution imposed on us by the new quantum concepts. As quantum mechanics is unable to include incompatible observables (corresponding to noncommuting operators) as joint theoretical elements of reality, hidden variables had been proposed to fill the gap and thus render the quantum theory complete.

However, as first argued by Bell [Bell (1966)], such hidden variables would have to be seriously constrained. In fact, local deterministic hidden variables could not violate Bell's inequality—contrary to quantum mechanics and contrary to experiments [Aspect (1981)]. The hidden variables would thus have to appear "nonclassical" by themselves, a rather unconvincing remedy.

Meanwhile, the old question of whether or not quantum theory should be considered complete has lost much of its original appeal. A more timely question has gained some attention: Are there limits of quantum theory? If so, where should we look? Remarkably, these questions are still motivated by the notorious measurement paradox—definitely not by well documented shortcomings of present quantum theory. This is in marked contrast with the crisis faced by classical physics some hundred years ago (cf. Section 3.1).

One option is to look for superpositions in the macroscopic domain. R. Penrose is hoping to see fundamental limits as the mass of the quantum object increases [Marshall (2003)]—up to now without clear results. In a similar vein, A. Leggett [Leggett (2002)] suggests to analyze the quantum behavior at various hierarchical levels defined by the number of constituting elementary particles. He notes that superpositions have meanwhile been confirmed with systems composed of 10^3 particles in the case of molecular diffraction, some 10^{10} in the case of Josephson devices, and 10^{14} electrons for counterruning currents [Hime (2006)]. While this still falls short of the macroscopic limit in a strict sense (10^{24} particles, say), the prospects to find unequivocal fundamental limits in this way appear rather dim.

There are still attempts to overcome the alleged classic/quantum dichotomy by postulating continuous spontaneous localization of the wavefunction [Adler (2009)]. However, potential experiments, which might prove or disprove such modifications of the Schrödinger equation, are still many orders of magnitude below the required sensitivity.

A more likely scenario could be that the explicit quantum behavior of macroscopic systems simply becomes practically invisible—without indicating any formal limit of its applicability (cf. [Kofler (2010); Gerlich (2011)]). Another pertinent example is the emergence of thermodynamics from quantum mechanics (cf. Section 4.2.2).

In the following, the quantum formalism will be developed in two parts: outlining the abstract and the contextual (observational) approaches, respectively.

3.3 The Formalism I

In classical mechanics, there is no need to separate observables and states: A complete set of observables uniquely specifies the mechanical state. In the Hamiltonian description of a single particle, the basic observables are position and momentum, $\{\vec{r}, \vec{p}\}$. All other observables F are then functions of these, $F = F(\vec{r}, \vec{p})$.

The same would be true in quantum mechanics, if we restrict ourselves to the joint eigenstates of a complete set of commuting operators, $\hat{F}^{(m)}$, $m = 1, 2, \ldots$, cf. Eq. (3.33). Because there is a one-to-one correspondence between (sharp) eigenvalue and eigenstate, the latter becomes redundant.

However, a restriction to such states would render quantum mechanics almost useless: On the one hand, we are also interested in the observables described by noncommuting (incompatible) operators; and on the other hand, we would also like to deal with the dynamics, which cannot be restricted to occur within the subset of states with sharp eigenvalues of one and the same set of operators only. If discrete, there would be no way to have a continuous dynamics at all.

In quantum mechanics, states and observables have thus to be treated separately: Observables become operators acting on the state. Any exposition of quantum mechanics has thus to be based on operator algebra. This tool is then applied to quantum systems, quantum states, and quantum dynamics.

Classical mechanics and quantum mechanics relate to each other in a number of ways: The so-called "canonical quantization" starts from some preexisting classical theory, and, vice versa, selected quantum features (of a given model) should "disappear" in some appropriate "classical limit," cf. Section 3.3.8.

3.3.1 *Operators in a Complex Vector Space*

We consider a complex vector space or Hilbert space $\mathcal{H}$ of dimension d, where d may become infinite. Its elements $\vec{\phi}$ and their duals $\vec{\phi}^*$ (complex conjugate vectors) are conveniently written in the so-called Dirac notation.

Definition 3.1 Dirac notation

$$\vec{\phi} \mapsto |\phi>, \tag{3.1}$$

$$\vec{\phi}^* \mapsto <\phi|. \tag{3.2}$$

For the Hilbert space, a so-called inner product (or scalar product) is defined.

Definition 3.2 Inner product

$$\vec{\phi}^* \cdot \vec{\chi} = <\phi|\chi>= c, \tag{3.3}$$

$$(\vec{\phi}^* \cdot \vec{\chi})^* = <\chi|\phi>= c^*.$$

The conjugate of the first vector together with the second one is associated with a scalar, the complex number c—therefore also called scalar product. c^* denotes complex conjugation.

The norm (or length squared) of the vector is $<\phi|\phi>$. Usually we require

$$<\phi|\phi>= 1. \tag{3.4}$$

If the scalar product of two vectors of nonzero norm is zero, the vectors are denoted as orthogonal. A set of orthogonal vectors is thus specified by

$$< e^{(j)}|e^{(k)} >= \delta_{jk} \quad j, k = 1 \ldots d\,. \tag{3.5}$$

With respect to this basis the vector components of $\vec{\phi}$ are

$$(\vec{\phi})_j = (\vec{e}^{(j)})^* \cdot \vec{\phi} =< e^{(j)}|\phi >= \phi_j\,. \tag{3.6}$$

Definition 3.3 Operator *An operator is a prescription in the sense that its application on a vector $|\phi >$ of Hilbert space leads, in general, to a different vector $|\chi >$ within the same space, $\hat{C}|\phi >= |\chi >$. Here we restrict ourselves to the so-called linear operators, $\hat{A}|\phi + \chi >= \hat{A}|\phi > + \hat{A}|\chi >$.*

Definition 3.4 Expectation value *Given $|\chi >= \hat{C}|\phi >$ the inner product,*

$$< \phi|\chi >=< \phi|\hat{C}|\phi >\equiv < C >, \tag{3.7}$$

will be called expectation value of $\hat{C}$ with respect to $|\phi >$.

Definition 3.5 Outer product *Let $|e^{(1)} >$ and $|e^{(2)} >$ be two basis vectors of some Hilbert space $\mathcal{H}$. Then*

$$|e^{(1)} > \otimes |e^{(2)} > \tag{3.8}$$

is an element of $\mathcal{H} \otimes \mathcal{H}$ with dimension d^2. In general, the vectors entering the direct product may be taken from different vector spaces.

Definition 3.6 Outer product as linear operator *Let $|\phi >$ and $|\chi >$ be two vectors in $\mathcal{H}$. Then*

$$\vec{\phi} \otimes \vec{\chi}^* = |\phi >< \chi| \equiv \hat{C}\,, \tag{3.9}$$

where $\hat{C}$ denotes an operator. Indeed, the scalar product of $\hat{C}$ with some state vector $|\psi >$ can be given a new interpretation:

$$\hat{C}|\psi >= |\phi >< \chi|\psi >, \tag{3.10}$$

that is, the action of $\hat{C}$ is to give the input vector $|\psi >$ a new direction ($|\phi >$) and a new length, characterized by the scalar product; this is all what an operator acting on a vector can do.

The outer product associates two vectors with a tensor:

$$(\vec{\phi} \otimes \vec{\chi}^*)_{jk} = (\vec{e}^{(j)})^*(\vec{\phi} \otimes \vec{\chi}^*)\vec{e}^{(k)} =< e^{(j)}|\phi >< \chi|e^{(k)} >= \phi_j \chi_k^*\,. \tag{3.11}$$

The outer product is therefore also called tensor product.

Definition 3.7 Basis operators *Applying the outer product to the orthonormalized set,* $|e^{(i)}>, i = 1, 2, \ldots d$, *we obtain the transition operators,*

$$\hat{P}^{(ik)} = |e^{(i)}><e^{(k)}|, \tag{3.12}$$

and the projection operators,

$$\hat{P}^{(ii)} = |e^{(i)}><e^{(i)}|. \tag{3.13}$$

Finally, the unit operator,

$$\hat{1} = \sum_{k=1}^{d} \hat{P}^{(kk)}, \tag{3.14}$$

can alternatively be seen as the completeness relation of the basis $|e^{(j)}>, j = 1, 2, \cdots d$. The completeness (and orthogonality) of this basis allows for a representation of any vector:

Definition 3.8 Vector representation

$$\boxed{|\phi>=\hat{1}|\phi>=\sum_{k=1}^{d}|e^{(k)}>\phi_k \quad , with\ \phi_k=<e^{(k)}|\phi>.} \tag{3.15}$$

Here, ϕ_k *are the respective vector components, that is, d complex numbers. The state* $|\phi>$ *appears as a superposition in terms of (or a decomposition into) the complete set of d orthogonal basis vectors,* $|e^{(k)}>$. *The expectation value of the projector* $\hat{P}^{(jj)}$ *with respect to this* $|\phi>$ *is given by*

$$<\phi|\hat{P}^{(jj)}|\phi>=|<e^{(j)}|\phi>|^2=|\phi_j|^2. \tag{3.16}$$

The above unit operator can also be conveniently used to rewrite the scalar product of two vectors in terms of their vector components

$$<\phi|\chi>=<\phi|\hat{1}\chi>=\sum_{j=1}^{d}<\phi|e^{(j)}><e^{(j)}|\chi>=\sum_{j=1}^{d}\phi_j^*\chi_j. \tag{3.17}$$

This defines an operational procedure; note that each component consists of a real and imaginary part.

Definition 3.9 Operator representation

$$\hat{F} = \hat{1}\hat{F}\hat{1} = \sum_{k,l}^{d} \hat{P}^{(kl)} F_{kl} \quad , with\ F_{kl} =< e^{(k)}|\hat{F}|e^{(l)} > . \tag{3.18}$$

The F_{kl} are the respective matrix elements (d^2 complex numbers). In analogy to Eq. (3.15), this operator representation can be seen as a decomposition of $\hat{F}$ into the set of d^2 basis operators, the transition and projection operators $\hat{P}^{(kl)}$. They constitute a complete, orthogonal set in the sense that

$$\mathtt{Tr}\{\hat{P}^{(kl)}\hat{P}^{(mn)}\} = \delta_{kn}\delta_{lm}, \tag{3.19}$$

where the trace relation means

Definition 3.10 Trace

$$\mathtt{Tr}\{\ldots\} = \sum_{j=1}^{d} < e^{(j)}|\ldots|e^{(j)} > . \tag{3.20}$$

The dots stand for any operator defined in this d-dimensional space. The trace does not depend on the basis, $|e^{(j)}>$, chosen. Alternatively, the matrix elements in Eq. (3.18) can be rewritten as

$$F_{kl} = \mathtt{Tr}\{\hat{P}^{(lk)}\hat{F}\}. \tag{3.21}$$

Definition 3.11 Combination of operators *Addition and multiplication of operators are defined as follows:*

$$(\hat{F} + \hat{G})|\phi> = \hat{F}|\phi> + \hat{G}|\phi>, \tag{3.22}$$

$$\hat{G}\hat{F}|\phi> = \hat{G}|\hat{F}\phi> = \hat{G}|\chi> = |\psi> . \tag{3.23}$$

Thus, multiplication specifies a sequential order for the application of operators.[a]

Definition 3.12 Commutator *Typically, the changed order of two operators has a different effect on the state; this difference is quantified by the commutator:*

$$[\hat{F}, \hat{G}] = \hat{F}\hat{G} - \hat{G}\hat{F} = \hat{K}. \tag{3.24}$$

[a] Sequential order presupposes a basic notion of time, even though parameter time does not show up explicitly here.

The commutator is, again, an operator. Given the result $\hat{K}$ only, its origin as a commutator is no longer explicit.

The trace of two operators is commutative, $\mathtt{Tr}\{\hat{A}\hat{B}\} = \mathtt{Tr}\{\hat{B}\hat{A}\} = \sum_{ij} A_{ij}B_{ji}$. This trace relation is closely related to the scalar product of vectors:

Definition 3.13 Vectorization *Based on the matrix representation of the operator $\hat{A}$, we introduce*

$$A_{ij} \to \vec{A} \equiv \{A_{11}, A_{21}, \ldots A_{d1}, A_{12}, \ldots A_{d2}, \ldots A_{1d}, \ldots A_{dd}\}. \tag{3.25}$$

The trace between two operators can thus be rewritten as $\mathtt{Tr}\{\hat{A}^\dagger\hat{B}\} = (\vec{A})^*\vec{B}$, that is as a scalar product.

The inverse operator is the "undo" operation: If $\hat{F}|\phi>=|\chi>$, then $\hat{F}^{-1}|\chi>=|\phi>$. This means that

$$\hat{F}\hat{F}^{-1} = \hat{F}^{-1}\hat{F} = \hat{1}. \tag{3.26}$$

Definition 3.14 Adjoint operator *The operation $\hat{F}^\dagger$ is adjoint to $\hat{F}$, if for any pair of vectors*

$$<\chi|\hat{F}\phi>=<\hat{F}^\dagger\chi|\phi>. \tag{3.27}$$

Example 3.1 *Let $\hat{F} = \hat{P}^{(kl)}$. Then $(\hat{P}^{(kl)})^\dagger = \hat{P}^{(lk)}$. Indeed, applying Eq. (3.27),*

$$\begin{aligned}(<\hat{F}^\dagger\chi|\phi>)^* &= <\phi|\hat{F}^\dagger\chi>=<\phi|l><k|\chi>,\\ <\hat{F}^\dagger\chi|\phi> &= <\chi|k><l|\phi>=<\chi|\hat{F}\phi>.\end{aligned} \tag{3.28}$$

Definition 3.15 Self-adjoint operator *A self-adjoint operator is characterized by $\hat{F} = \hat{F}^\dagger$ and thus*

$$<\chi|\hat{F}|\phi>=<\phi|\hat{F}|\chi>^* \tag{3.29}$$

This means that $F_{jk} = F^*_{kj}$. Typical operators of physical interest are self-adjoint or Hermitian[a]*.*

For a special subset of vectors, the so-called eigenvectors, the action of the operator $\hat{F}$ simply reproduces the respective vector:

---

[a]For bounded operators self-adjointness and Hermiticity are equivalent. Examples for unbounded operators are differential operators.

Definition 3.16 Eigenvalue equation

$$\boxed{\hat{F}|f^{(j)}>= F_j|f^{(j)}> .} \tag{3.30}$$

F_j is the eigenvalue associated with the eigenstate $|f^{(j)}>$, the set of eigenvalues $\{F_j\}$ the spectrum of $\hat{F}$. Self-adjoint operators have real eigenvalues. For an eigenstate, the expectation value is identical with its eigenvalue, $< f^{(j)}|\hat{F}|f^{(j)} >= F_j$. Different eigenvectors may have the same eigenvalue ("degeneracy").

The eigenvectors constitute an orthogonal set,

$$< f^{(j)}|f^{(k)} >= \delta_{jk}, \tag{3.31}$$

which can be used as a basis to represent other vectors in that Hilbert space.

Definition 3.17 Eigen representation *According to Eq. (3.18), the operator $\hat{F}$ can be represented in its own eigenbasis as*

$$\hat{F} = \sum_j F_j|f^{(j)} >< f^{(j)}| . \tag{3.32}$$

Definition 3.18 Complete set of commuting operators *A set of operators $\{\hat{F}^{(m)}\}$, $m = 1, 2\ldots$ is complete (for a given Hilbert space), if it holds that any other operator $\hat{G}$ is either a function of those $\hat{F}^{(m)}$, that is, $\hat{G} = \hat{G}(\hat{F}^{(1)}, \hat{F}^{(2)}, \ldots)$, or $\hat{G}$ does not commute with all the $\hat{F}^{(m)}$. Commuting operators have a joint set of eigenvectors[a]. There is a one-to-one correspondence between the eigenvalues of a complete set of commuting operators and the eigenstate, here indexed by the set of eigenvalues,*

$$\{F_j^{(1)}, F_k^{(2)}, \ldots\} \leftrightarrow |j, k, \cdots > . \tag{3.33}$$

In particular, let $\{\hat{F}^{(m)}\}$ be a complete set of commuting operators. Then the eigenvalue equation for the operator function $\hat{G} = \hat{G}(\hat{F}^{(1)}, \hat{F}^{(2)}, \ldots)$ reads

$$\hat{G}(\hat{F}^{(1)}, \hat{F}^{(2)}, \ldots)|j, k, \cdots >= G(F_j^{(1)}, F_k^{(2)}, \ldots)|j, k, \cdots > . \tag{3.34}$$

The eigenvalue $G(F_j^{(1)}, F_k^{(2)}, \ldots)$ reproduces the original classical relation between the respective classical observables.

---

[a]By definition, the operators of such a complete set are simultaneously measurable, cf. Section 3.4.1; this is also an operational requirement. It involves a rudimentary notion of time, cf. Definition 3.11

Definition 3.19 Unitary operator *A unitary operator is defined by*

$$\hat{U}^{\dagger}\hat{U} = \hat{U}\hat{U}^{\dagger} = \hat{1}\,. \tag{3.35}$$

Example 3.2 Unitary operators *Basic examples for $\hat{U}$ acting on some vector are:*

- *orthogonal matrices in real vector space;*
- *in particular, rotations in a two-dimensional real vector space;*
- *multiplications by* $\exp(i\theta)$ *within complex vector space, where θ defines a (real) phase;*
- *Fourier-transformations;*
- *permutation matrices on a finite vector space (which thus shuffle around the respective orthonormal basis).*

Definition 3.20 Unitary transformation *Let $\hat{U}$ be a unitary operator. Then*

$$\boxed{\begin{aligned} \hat{A}' &= \hat{U}\hat{A}\hat{U}^{\dagger}\,, \\ |\phi'> &= \hat{U}\,|\phi>\,. \end{aligned}} \tag{3.36}$$

These transformations give access to remarkable invariance properties: the scalar product of two vectors and the trace of an operator is invariant. In particular

$$<\phi'|\hat{A}'|\chi'> = <\phi|\hat{A}|\chi>\,. \tag{3.37}$$

Another example for a unitary transformation is the change of representation (basis) between $\{|e^{(j)}>\}$ and $\{|f^{(m)}>\}$:

Definition 3.21 Change of basis

$$\begin{aligned} <f^{(m)}|\phi> = <f^{(m)}|\hat{1}\phi> &= \sum_j <f^{(m)}|e^{(j)}><e^{(j)}|\phi> \\ &= \sum_j (\hat{S})_{mj} <e^{(j)}|\phi>\,. \end{aligned} \tag{3.38}$$

Note that this operator $\hat{S}$ does not exist in an abstract way: It defines a specific transformation matrix between the two representations. With $(\hat{S}^{\dagger})_{kn} = <e^{(k)}|f^{(n)}> = S^*_{nk}$ *this matrix is unitary,*

$$\sum_k (\hat{S})_{mk}(\hat{S}^{\dagger})_{kn} = \delta_{mn}\,. \tag{3.39}$$

Correspondingly, the matrix elements of operator $\hat{G}$ in the $|f^{(m)}>$-basis are related to those in $|e^{(j)}>$-basis by

$$< f^{(m)}|\hat{G}|f^{(n)} >= \sum_{jk}(\hat{S})_{mj} < e^{(j)}|\hat{G}|e^{(k)} > (\hat{S}^{\dagger})_{kn} . \quad (3.40)$$

The unitarity of such a basis change guarantees that the invariance properties apply, that is, the underlying "physics does not change." Special unitary transformations of the state *or* the operators specify the time-evolution according to the so-called Schrödinger or Heisenberg picture, respectively (cf. Section 3.3.7).

3.3.2 *Quantum Systems*

Hilbert spaces are vector spaces. But what is their physical origin? There is not just one to encompass all quantum phenomena; rather there are as many different spaces as there are different quantum models.

The system model, that is, the Hamiltonian, directly allows to define a representation for the states. Indeed, a convenient basis is the eigenbasis of $\hat{H}$. The nature of the model, for example, the indistinguishability of subsystems, cf. Section 3.3.6, further constraints the respective Hilbert space.

A convenient way to introduce quantum models is by reference to their classical counterparts. Indeed, the canonical quantization procedure presupposes classical physics with its respective wealth of well-tested models.

Definition 3.22 Lagrange function *A one-particle Lagrange system (particle position $\vec{r}$, velocity $\dot{\vec{r}}$) is specified by*

$$L(\vec{r}, \dot{\vec{r}}) = \frac{1}{2}m\dot{\vec{r}}^2 - V(\vec{r}) , \quad (3.41)$$

$$\vec{p} = \frac{\partial L}{\partial \dot{\vec{r}}} = m\dot{\vec{r}} . \quad (3.42)$$

$\vec{p}$ is the conjugate momentum, $V(\vec{r})$ the potential energy term.

Such a model is easily generalized to interacting multiparticle systems. From the Lagrange function L, we arrive at the Hamilton function H by means of a Legendre transformation.

Definition 3.23 Legendre transformation *Let f(x) be a function with $z(x) = df/dx$ and $d^2f/dx^2 = dz/dx \neq 0$. Then the function $z(x)$ can be inverted, $x = x(z)$ and*

$$\mathcal{L}f(x) = f(x) - x\frac{df}{dx} = f(x(z)) - x(z)\cdot z = h(z) \qquad (3.43)$$

is called the Legendre transformation of $f(x)$. The variable x has been replaced by the variable $z = df/dx$.

It immediately follows that $\mathcal{L}h(z) = f(x)$. Legendre transformations can be easily generalized to functions with several independent variables.

Definition 3.24 Hamilton function *The Hamilton function is now introduced as the (negative) Legendre transformation of the Lagrange function, by which we replace $x = \dot{\vec{r}}$ by $z = \vec{p}$ according to Eq. (3.42),*

$$\boxed{\begin{aligned} -\mathcal{L}L(\vec{r}, \dot{\vec{r}}) &= \tfrac{\partial L}{\partial \dot{\vec{r}}}\dot{\vec{r}} - L(\vec{r}, \dot{\vec{r}}(\vec{p})) \equiv H(\vec{p}, \vec{r}), \\ H(\vec{p}, \vec{r}) &= \tfrac{\vec{p}^2}{2m} + V(\vec{r}). \end{aligned}} \qquad (3.44)$$

Explicit time dependence of L and thus H may occur, but will be postponed to Section 3.3.7. The Hamilton function serves as a system definition; it has the meaning of a total energy, being composed of kinetic, $\vec{p}^2/(2m)$, and potential energy, $V(\vec{r})$.

Definition 3.25 Γ-space *The space of six coordinates, $\{\vec{p}, \vec{r}\}$, is called Γ-space. For N particles, the respective space is the direct sum, that is, it is 6N-dimensional.*

Any observable F (including the Hamiltonian) is a unique function on this space, $F = F(\vec{p}, \vec{r})$.

Definition 3.26 Poisson bracket *Let F and G be two observables. Then the Poisson bracket is defined as*

$$\{F, G\} \equiv \sum_\alpha \frac{\partial F}{\partial r_\alpha}\frac{\partial G}{\partial p_\alpha} - \frac{\partial F}{\partial p_\alpha}\frac{\partial G}{\partial r_\alpha} = -\{G, F\}. \qquad (3.45)$$

One easily convinces oneself that ($\alpha, \beta = 1, 2, \ldots 3N$)

$$\{r_\alpha, r_\beta\} = \{p_\alpha, p_\beta\} = 0 \quad \{r_\alpha, p_\beta\} = \delta_{\alpha,\beta}. \qquad (3.46)$$

Proposition 3.1 Classical observable dynamics *Given a mechanical model defined by the Hamilton function H, then the equation of* *motion for some observable F reads as*

$$\frac{dF}{dt} = \{F, H\} + \frac{\partial F}{\partial t} \,. \tag{3.47}$$

The Hamilton formulation of classical mechanics is now essential for quantization:

Definition 3.27 Canonical quantization *A classical Hamilton model is systematically quantized using*

$$H(\vec{p}, \vec{r}) \mapsto \hat{H}(\hat{\vec{p}}, \hat{\vec{r}}) \,, \tag{3.48}$$

subject to the fundamental commutator

$$\boxed{[\hat{p}_\alpha, \hat{r}_\beta] = \frac{\hbar}{i}\hat{1}\delta_{\alpha\beta} \,,} \tag{3.49}$$

and all other basic commutators being zero, $[\hat{p}_\alpha, \hat{p}_\beta] = [\hat{r}_\alpha, \hat{r}_\beta] = 0$. *Contrary to the corresponding Poisson brackets given above, the fundamental commutator has the dimension of an action, specified by* $\hbar$*, the famous Planck constant.*

This quantization procedure only works, after the model coordinates $\vec{r}$ and their conjugate momenta $\vec{p}$ have been identified. All the existing relations between classical observables can then be reinterpreted as operator relations.

Definition 3.28 Quantum observable *Let G be a classical observable with* $G = G(\vec{r}, \vec{p})$*. This function translates into an operator function,* $\hat{G} = \hat{G}(\hat{r}_\alpha, \hat{p}_\beta)$*. All these operators are Hermitian and thus have real eigenvalues.*[a] *The commutators involving* $\hat{G}$ *can be reduced to those basic ones, Eq. (3.49).*

Definition 3.29 Eigenvectors and representations *Let the eigenvalue equation for some observable* $\hat{G}$ *be given by*

$$\boxed{\hat{G}|g^{(n)} >= G_n|g^{(n)} > \,.} \tag{3.50}$$

If this operator is complete with respect to some Hilbert space $\mathcal{H}$ *(cf. Definition 3.18), its eigenvectors* $|g^{(n)} >$ *can be used to represent any other state or operator; this constitutes the so-called G-representation.*

An example for $\hat{G}$ is the angular momentum.

[a] It is argued that all measurement results have to be real in the classical limit.

Definition 3.30 Angular momentum *According to classical physics, the angular momentum is defined by* $\vec{L}(\vec{r}, \vec{p}) = \vec{r} \times \vec{p}$. *When interpreted as an operator relation the commutators of the angular momentum components can easily be derived from the basic commutator between real space and linear momentum coordinates. One finds*

$$[\hat{L}_i, \hat{L}_j] = i\hbar \sum_{k=1}^{3} \epsilon_{ijk} \hat{L}_k \,, \tag{3.51}$$

where

$$\epsilon_{ijk} = \begin{cases} +1 \text{ for } (ijk) = (123) \text{ and cyclic permutations,} \\ -1 \text{ for } (ijk) = (132) \text{ and cyclic permutations,} \\ \;\;0 \;\; \text{for } (ijk) \text{ otherwise.} \end{cases} \tag{3.52}$$

is the totally antisymmetric tensor. From $\hat{L}^2 = \sum_j \hat{L}_j^2$ *and*

$$[\hat{L}^2, \hat{L}_j] = 0 \tag{3.53}$$

it follows that a joint set of eigenvectors must exist for $\hat{L}^2$ *and one of the* $\hat{L}_j$. *These are, selecting* $j = 3$,

$$\hat{L}^2 |l, m> = \hbar^2 l(l+1) |l, m> \,, \tag{3.54}$$

$$\hat{L}_3 |l, m> = \hbar m |l, m> \,, \tag{3.55}$$

l, m *have to be determined yet.*

Definition 3.31 Angular momentum ladder operators *The respective eigenstates can be constructed based on the eigen-equations, Eqs. (3.54) and (3.55), and the non-Hermitian ladder operators*

$$\hat{L}_\pm = \hat{L}_1 \pm i\hat{L}_2 \,, \tag{3.56}$$

$$(\hat{L}_+)^\dagger = \hat{L}_- \,, \tag{3.57}$$

$$\hat{L}^2 = \hat{L}_+\hat{L}_- + \hat{L}_3^2 - \hat{L}_3 = \hat{L}_-\hat{L}_+ + \hat{L}_3^2 + \hat{L}_3 \,, \tag{3.58}$$

$$[\hat{L}_3, \hat{L}_\pm] = \pm\hat{L}_\pm \,, \tag{3.59}$$

$$[\hat{L}^2, \hat{L}_\pm] = 0 \,. \tag{3.60}$$

Just using the commutator relations one convinces oneself that for $|\phi_\pm> \equiv \hat{L}_\pm |l, m>$

$$\hat{L}_3 |\phi_\pm> = (m \pm 1) |\phi_\pm> \,, \tag{3.61}$$

that is, the eigenvalue for $|\phi_\pm>$ *flips by* ± 1 *compared with* $|l, m>$. *Requiring the norm of this function to be positive implies*

$$<\phi_\pm|\phi_\pm> = <l, m|\hat{L}_\mp \hat{L}_\pm|l, m> = \hbar^2(l(l+1) - m^2 \pm m \geq 0\,. \quad (3.62)$$

Here we have used $<l, m|l, m> = 1$ *and Eq. (3.58). The range of possible values for m is thus bounded from above and from below:* $m_{max/min} = \pm l$. *Allowed values for m are:* $-l, -l+1, \ldots l-1, l$. *For given l, there are* $2l+1$ *values of m; this number must be integer, so that l itself has to be either integer or half integer. The allowed values of l run from zero to infinity; they may be integer or half integer.*

Definition 3.32 Angular momentum representation *The set of operators* $\{\hat{L}^2, \hat{L}_3\}$ *are complete on the unit sphere of a three-dimensional real space. The completeness relation reads (l has to be integer here),*

$$\sum_{l=0}^{\infty} \sum_{m=-l}^{+l} |l, m><l, m| = \sum_l \hat{1}(l) = \hat{1}\,. \qquad (3.63)$$

It may be seen to decompose into a direct sum of $(2l+1)$*-dimensional subspaces* $\mathcal{H}_l$ *for each given l.*

Definition 3.33 Spin *Subspaces* $\mathcal{H}_l$ *do occur separately as a description of a spin with* $l \equiv s$ *fixed. Contrary to the orbital angular momentum, such a spin quantum number s may also be half-integer. The simplest case of a spin results for* $s = 1/2$, *applicable for an electron or proton, say.*

However, such a spin algebra does not only apply to spins in a strict sense but to any effective two-level quantum system. The latter may be called "operational spin," as the experimental conditions must be such that, indeed, only two out of the (usually) infinitely many states can participate. In that sense, the spin quantum mechanics can be implemented by (otherwise) physically different systems. This is the simplest case of one quantum system being able to "simulate" another one ("quantum simulator"). From the point of view of applications, in particular within quantum information theory, such a spin-1/2 system is also called "qubit."

Example 3.3 Qubit: Spin algebra for s = 1/2 *Introducing the dimensionless operators*

$$\hat{\sigma}_j = \frac{2}{\hbar}\hat{L}_j\,, \qquad (3.64)$$

the commutator relation, Eq. (3.51), can be rewritten as

$$[\hat{\sigma}_i, \hat{\sigma}_j] = 2i \sum_{k=1}^{3} \epsilon_{ijk} \hat{\sigma}_k , \tag{3.65}$$

and the eigenvalue equations are

$$\hat{\sigma}^2 |m> = 3|m> , \tag{3.66}$$

$$\hat{\sigma}_3 |m> = \sigma_m |m> , \tag{3.67}$$

with $m = \pm 1$ *and* $\hat{\sigma}^2 = \hat{\sigma}_1^2 + \hat{\sigma}_2^2 + \hat{\sigma}_3^2$. *With* $s = l = 1/2$ *being fixed (and suppressed in the notation of the eigenvectors), the first eigen-equation becomes dispensable.*

For this two-dimensional Hilbert space, we use the notation $|e^{(1)} >= |m = -1 >$ *and* $|e^{(2)} >= |m = +1 >$.

There are four basis operators, which follow modulo phase factors, from the commutation relations. Based on the notation

$$\hat{P}_{ij} = |e^{(i)} > < e^{(j)}| \tag{3.68}$$

a convenient choice are the Pauli operators

$$\boxed{\begin{aligned} \hat{\sigma}_0 &= \hat{1} = \hat{P}_{11} + \hat{P}_{22} , \\ \hat{\sigma}_1 &= \hat{P}_{12} + \hat{P}_{21} , \\ \hat{\sigma}_2 &= i(\hat{P}_{12} - \hat{P}_{21}) , \\ \hat{\sigma}_3 &= \hat{P}_{22} - \hat{P}_{11} , \end{aligned}} \tag{3.69}$$

with $\hat{\sigma}_j^2 = \hat{1}$. *These basis operators are orthogonal,*

$$\mathtt{Tr}\{\hat{\sigma}_i \hat{\sigma}_j\} = 2\delta_{ij} \quad i, j = 0, 1, 2, 3 . \tag{3.70}$$

Consequently, any operator in this space can be expanded in the form

$$\hat{F} = \frac{1}{2} \sum_{j=0}^{3} F_j \hat{\sigma}_j \quad F_j = \mathtt{Tr}\{\hat{F} \hat{\sigma}_j\} . \tag{3.71}$$

This is reminiscent of the operator representation, Definition 3.9; we merely use the set of operators $\hat{\sigma}_j$ *instead of the transition operators* $\hat{P}_{jk}$. *If the Hamiltonian is taken to commute with* $\hat{\sigma}_3$, *its representation simplifies to*

$$\hat{H} = \frac{1}{2}(H_0 \hat{1} + H_3 \hat{\sigma}_3) . \tag{3.72}$$

The eigenstates are $|m >$ *and associated with the energy eigenvalues*

$$E_m = 1/2(H_0 + H_3 m) \quad m = \pm 1 . \tag{3.73}$$

Definition 3.34 Position operator *The eigenvalue equation is*

$$\hat{\vec{r}}|\vec{r}>=\vec{r}|\vec{r}> . \tag{3.74}$$

The spectrum is continuous.

Definition 3.35 Real space representation *The representation based on the eigenstates $|\vec{r}>$ requires special attention. One first notes that the basic commutator relation between position and momentum can be fulfilled by*

$$<\vec{r}|\hat{\vec{r}}|\phi> = \vec{r}<\vec{r}|\phi>\equiv\vec{r}\phi(\vec{r}) , \tag{3.75}$$

$$<\vec{r}|\hat{\vec{p}}|\phi> = \frac{\hbar}{i}\frac{\partial}{\partial\vec{r}}\phi(\vec{r}) . \tag{3.76}$$

The corresponding operators become numbers and differential operators, respectively. The latter are "unbounded." As a consequence, the single particle Hamilton operator reads (cf. Eq. (3.44))

$$\boxed{<\vec{r}|\hat{H}|\phi>=\left(-\frac{\hbar^2}{2m}\frac{\partial^2}{\partial\vec{r}^2}+V(\vec{r})\right)\phi(\vec{r}) .} \tag{3.77}$$

The term in parenthesis can be identified as the Hamilton operator, $\hat{H}(\hat{\vec{r}},\hat{\vec{p}})$, applied to the wavefunction $\phi(\vec{r})$. The energy eigenvalue equation,

$$<\vec{r}|\hat{H}|e^{(n)}>=E_n<\vec{r}|e^{(n)}>=E_n e_n(\vec{r}) , \tag{3.78}$$

thus becomes a differential equation for the eigenfunctions $e_n(\vec{r})$.

In so far as the underlying models are taken over from classical physics, they are necessarily defined in $(\vec{r}, \vec{p})$-space (Γ-space), involving concrete potential terms and concrete interparticle interactions. The real space representation establishes a direct link between quantum mechanics and classical mechanics. In this representation, the spatial coordinates show up as eigenvalues, that is as (real) numbers. This is reminiscent of classical mechanics, the quantum state appears to be redundant. But this is not the case: The wave function is needed to bridge the gap with the subset of noncommuting operators.

In the following, we will mainly be concerned with single particle, $N = 1$, and bipartite systems, $N = 2$. Each such model is characterized by a set of parameters, which defines the respective model space.

Remark 3.1 Single-particle models with analytical solutions *There are some 20 well-known Hamilton models,* $\hat{H}$, *which allow for simple, analytic solutions of the respective eigenvalue equation. Among these are*

(1) *free particle,*
(2) *hydrogen atom,*
(3) *particle in a box,*
(4) *linear rigid rotor,*
(5) *harmonic oscillator,*
(6) *particle in a one-dimensional lattice.*

These models depend on one or two parameters defining a one- or two-dimensional model space.

Example 3.4 Hydrogen atom as an effective one-particle model *A particle bound via Coulomb forces to a charged center is a paradigm three-dimensional model. Because of the spherical symmetry a convenient set of complete operators is* $\{\hat{H}, \hat{L}^2, \hat{L}_3\}$. *For the discrete bound states, the respective joint eigenvectors are* $|n, l, m\ >$, *the associated eigenvalues being the "good quantum numbers" (constants of motion) for the energy,* $E_n \sim -1/n^2, n = 1, 2, \ldots \infty$ *and the angular momentum with* $l \leq n-1$. *Pertinent transition series are indicated in Fig. 3.1. The spin of the electron and proton requires additional attention, though.*

Also this model can be used as an effective theory to describe, for example, localized impurities or weakly bound excitons in semiconductors, say.

Example 3.5 Particle in a box *We consider a particle of mass* m_0 *in a box (with infinite walls at* $x = 0, L$*), that is,*

$$V(x) = \begin{cases} 0 \ \textit{for} & 0 < x < L \\ \infty \ \textit{otherwise}. \end{cases} \tag{3.79}$$

Inside the box, the potential $V(x)$ *is zero, so that in real space representation, the energy eigenvalue equation reads:*

$$\left(-\frac{\hbar^2}{2m_0}\frac{d^2}{dx^2} + V(x)\right)\phi(x) = E\phi(x). \tag{3.80}$$

Defining $\kappa^2 = 2m_0 E/\hbar^2$, *this differential equation can be rewritten as*

$$\phi(x)'' = -\kappa^2\phi(x) \quad \textit{for} \quad 0 < x < L. \tag{3.81}$$

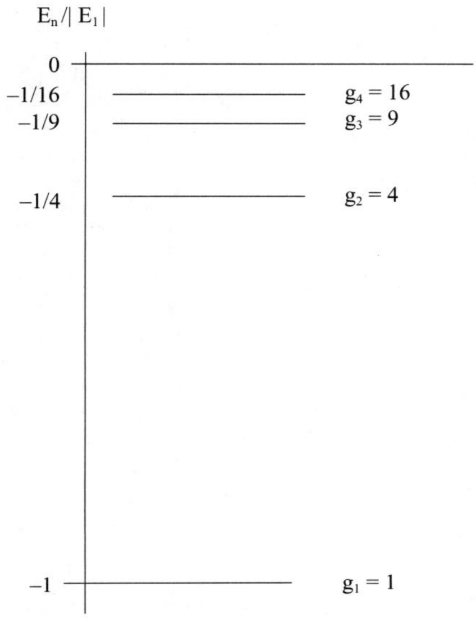

Figure 3.1 Hydrogen spectrum: Eigenvalues and degeneracies.

Observing the boundary conditions $\phi(0) = \phi(L) = 0$, *the solutions (normalized eigenvectors) are restricted to* $\kappa = k\pi/L$, *so that*

$$\phi_k(x) = \sqrt{\frac{2}{L}} \sin(k\pi x/L), \tag{3.82}$$

and the discrete energy spectrum is given by (see Fig. 3.3)

$$\boxed{E_k = \frac{\hbar^2\pi^2k^2}{2m_0L^2} \quad k = 1, 2, \ldots \infty.} \tag{3.83}$$

Example 3.6 Harmonic oscillator *Being taken over from classical mechanics, the quantum model starts from*

$$\hat{H}(\hat{p}, \hat{x}) = \frac{1}{2m_0}\hat{p}^2 + \frac{1}{2}m_0\omega^2\hat{x}^2, \tag{3.84}$$

with $[\hat{p}, \hat{x}] = (\hbar/i)\hat{1}$, ω *being the angular frequency, and* m_0 *the mass of the particle. The energy eigenvalue equation in real space representation would thus amount to a second-order differential equation. Instead, it is convenient to solve the eigenvalue equation after transforming to the non-Hermitian creation/destruction operators*

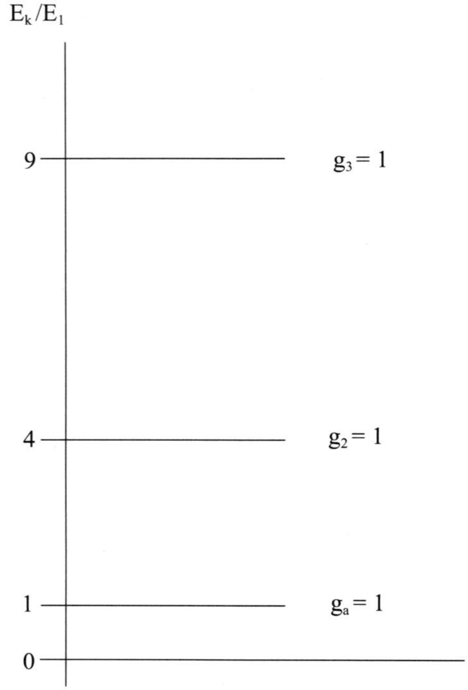

Figure 3.2 Particle in box: Eigenvalues and degeneracies.

$\hat{b}^\dagger$, $\hat{b}$,

$$\hat{x} = \sqrt{\frac{\hbar}{2m_0\omega}}\left(\hat{b} + \hat{b}^\dagger\right) , \tag{3.85}$$

$$i\,\hat{p} = \sqrt{\frac{\hbar m_0\omega}{2}}\left(\hat{b} - \hat{b}^\dagger\right) . \tag{3.86}$$

The commutator for the new operators is easily shown to be

$$[\hat{b}, \hat{b}^\dagger] = \hat{1} , \tag{3.87}$$

so that

$$\boxed{\hat{H}\,(\hat{b}^\dagger, \hat{b}) = \hbar\omega\left(\hat{b}^\dagger\hat{b} + \frac{1}{2}\hat{1}\right) .} \tag{3.88}$$

Obviously, $\hat{N} = \hat{b}^\dagger\hat{b}$ *commutes with* $\hat{H}$*. We are thus looking for their joint eigenstates,*

$$\hat{N}|n >= n|n > . \tag{3.89}$$

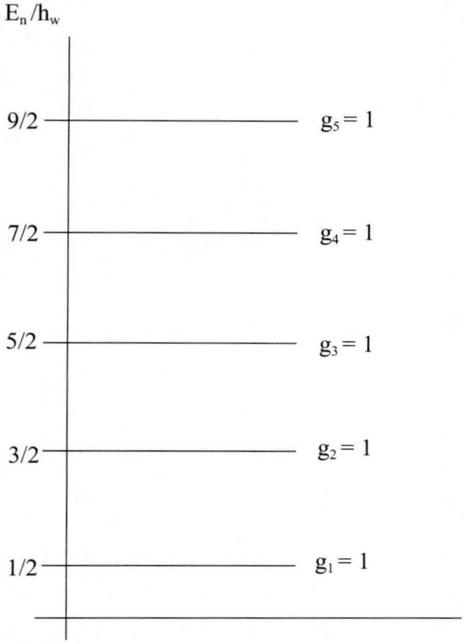

Figure 3.3 Harmonic oscillator: Eigenvalues and degeneracies.

Based on the commutator relation

$$[\hat{b}^s, \hat{N}] = s\hat{b}^s \tag{3.90}$$

with s being a positive integer, one convinces oneself that

$$\hat{N}|\hat{b}^s n >= (n - s)|\hat{b}^s n > , \tag{3.91}$$

that is, $\hat{b}$, indeed, functions as a destruction operator, each application lowering the original eigenvalue n by 1,

$$\hat{b}|n > = \gamma|n - 1 > , \tag{3.92}$$

$$< \hat{b}n|\hat{b}n > = \gamma^2 =< n|\hat{N}|n >= n\,. \tag{3.93}$$

We thus identify $\gamma = \sqrt{n}$. Correspondingly, $\hat{b}^\dagger$ works as a creation operator, $\hat{b}^\dagger|n >= \gamma'|n + 1 >$ with $\gamma' = \sqrt{n+1}$. But because the norm must always be positive,

$$< \hat{b}^{s+1}n|\hat{b}^{s+1}n >=< \hat{b}^s n|\hat{N}\hat{b}^s n >= (n-s) < \hat{b}^s n|\hat{b}^s n >\geq 0, \tag{3.94}$$

the lowering of eigenvalues has to come to an end. This happens if n is an integer: Because then the right-hand side becomes the null vector for $s = n$, and the lowering procedure comes to a final halt. The energy eigen-spectrum thus reads:

$$\boxed{E_n = \hbar\omega\left(n + \frac{1}{2}\right) \qquad n = 0, 1, \ldots \infty\,.} \tag{3.95}$$

The harmonic oscillator model, just like its classical counterpart, has many effective applications, from phonons in insulators to photons in vacuum.

Definition 3.36 Coherent states *Coherent states ("Glauber states") of the quantum oscillator are defined as a special superposition of energy eigenstates, $|n>$,*

$$|\alpha> \equiv \exp(-|\alpha|^2/2) \sum_n \frac{\alpha^n}{\sqrt{n!}} |n> \,. \tag{3.96}$$

α is a complex parameter, $\alpha = |\alpha| \exp(-i\phi)$. The relative phase between the number states $|n>$ and $|n+k>$ is thus given by $\exp(-ik\phi)$. Relative phases are, in principle, measurable.

The probability to find a specific energy n in this state is Poisson distributed (cf. Example 2.3), $P(n) = \exp(-|\alpha|^2)|\alpha|^{2n}/n!$, the energy expectation value is

$$< E > = < \alpha|\hat{H}|\alpha > = \hbar\omega(|\alpha|^2 + 1/2)\,. \tag{3.97}$$

The phase ϕ does not enter here nor in other expectation values. The same result would thus be obtained from the phase-mixed state

$$\hat{\rho} = \int_0^{2\pi} \frac{d\phi}{2\pi} ||\alpha| \exp(-i\phi) >< |\alpha| \exp(-i\phi)| = \sum_n P(n)|n >< n|\,. \tag{3.98}$$

Remark 3.2 Non-classical oscillator states *Coherent states can be used as a (nonorthogonal and over complete) basis:*

$$\hat{\rho} = \int d^2\alpha\, P(\alpha)|\alpha > < \alpha|\,. \tag{3.99}$$

If $P(\alpha)$ has the form of a classical probability density, the corresponding quantum state is said to have a classical analog [Titulaer (1965)], otherwise it is nonclassical. Unfortunately, this criterion does not work as a general test of nonclassicality.

Example 3.7 Extension to higher dimensions *The harmonic oscillator as well as the particle in a box can be easily generalized to higher dimensions. This is so because these Hamilton models are additive in terms of the Cartesian coordinates,* (p_j, r_j), $j = 1, 2, \ldots D$ *in the sense that*

$$\hat{H} = \sum_{j=1}^{D} \hat{H}^{(j)}(\hat{p}_j, \hat{r}_j)\,. \tag{3.100}$$

In this case, it immediately follows that the total eigenfunction can simply be written as the product state, $\phi_{n_1,n_2,\ldots}(\vec{r}) = \phi^{(1)}_{n_1}(r_1)\phi^{(2)}_{n_2}(r_2)\ldots$, *and the corresponding eigenvalues as the sum,* $E_{n_1,n_2,\ldots} = E^{(1)}_{n_1} + E^{(2)}_{n_2} + \ldots$. *General states are superpositions of such product states.*

If the model parameters, ω_j *or* L_j, *respectively, do not depend on* j, *the eigenvalues are proportional to* $n = n_1 + n_2 + \ldots$ *(oscillator) or* $n^2 = n_1^2 + n_2^2 + \ldots$ *(particle in box). Various eigenvectors may thus be associated with the same eigenvalue, we have degeneracy. As a result of symmetry, the Hamiltonian alone is no longer a complete operator.*

Definition 3.37 Energy representation *For a discrete, nondegenerate energy spectrum, the Hamiltonian itself serves as a complete operator, and its eigenvectors can thus conveniently be used as an energy representation. In this case,*

$$\hat{H}\,|e^{(n)}> = E_n|e^{(n)}>\,, \tag{3.101}$$

$$|\phi> = \sum_{n=1}^{d} |e^{(n)}> \phi_n \quad \phi_n =< e^{(n)}|\phi>\,. \tag{3.102}$$

Subjective ignorance may enter not only with respect to states, cf. Definition 3.52, but also with respect to the underlying Hamiltonian system. In fact, when comparing model calculations with "reality," we can hardly ever be sure about the model parameters specifying those Hamiltonian parts we decided to consider explicitly, let alone about the possible influence of those parts we have to or want to neglect. Nevertheless, the model studies are intended to help us in understanding what is "really" going on. That this expectation is not entirely futile, must be due to the fact that those models are comparatively robust, that is, physical phenomena do not disappear once parameters have been changed just a little bit from their "correct" value. Different models may thus be equivalent; they form classes.

Definition 3.38 Random matrices *The eigen representation of an Hamiltonian $\hat{H}$ leads to a very special (diagonal) matrix. Let us start here, instead, from a given basis, $|e^{(l)}>$, which should in no way be preferred with respect to $\hat{H}$. According to Eq. (3.18)*

$$\hat{H} = \hat{1}\hat{H}\hat{1} = \sum_{k,l}^{d} \hat{P}^{(kl)} H_{kl} \quad , \textit{with } H_{kl} = < e^{(k)}|\hat{H}|e^{(l)} > . \tag{3.103}$$

The matrix entries are thus treated as random, specified by some distribution function $w(H_{ij})$. If these distributions do not depend on the basis, that is, are required to be invariant under any unitary transformation, it follows that [Haake (1991)]

$$w(H_{11}, H_{12}, \ldots H_{21}, H_{22}, \ldots) \sim \exp\left(-\frac{1}{4}\mathrm{Tr}\{\hat{H}^2\}\right). \tag{3.104}$$

This distribution defines the Gaussian unitary ensemble (GUE). Hermiticity requires $H_{ij} = H_{ji}^*$ so that for Hilbert space dimension $d = 2$, as an example,

$$\begin{aligned} \mathrm{Tr}\{\hat{H}^2\} &= H_{11}^2 + H_{22}^2 + 2H_{12}H_{12}^* \\ &= H_{11}^2 + H_{22}^2 + 2\Re^2(H_{12}) + 2\Im^2(H_{12}) . \end{aligned} \tag{3.105}$$

We can thus apply (for any dimension d) independent zero mean Gaussian distributions

$$\boxed{w(x) = \frac{1}{\sigma\sqrt{(2\pi)}} \exp\left(-\frac{x^2}{2\sigma^2}\right)} \tag{3.106}$$

with variances $\sigma_d = \sqrt{2}\Delta E$ for the diagonal matrix elements and $\sigma_{od} = \Delta E$ for the real, $\Re(H_{ij})$, and imaginary parts, $\Im(H_{ij})$, of the off-diagonal elements; ΔE defines some appropriate energy scale.

Such ensembles of Hamiltonians (in terms of random matrices) will also be essential for the concept of thermal properties: We do not want to study special models, which happen to exhibit thermal behavior; rather do we hope to show that this behavior is "typical" for a whole class of similar models.

3.3.3 *Composite Systems*

As already mentioned, systems bring their own Hilbert space: Unlike the three-dimensional real space (as seen in a nonrelativistic setting) Hilbert space is not "out there" to be filled with objects. When we start, for example, with one spin, Hilbert space is two dimensional, when we add a second and third spin, the dimension increases to 2^2, 2^3, etc. These are so-called tensor-spaces, cf. Section 3.3.5.

Definition 3.39 Composite systems *Multicomponent systems abound. There are now global operators (acting on the total system) and local operators (acting on individual subsystems only).*

(i) Local operators (while defined in the total Hilbert space) act on a single subsystem only,

$$\hat{F}(\nu) \equiv \hat{1}(1) \otimes \cdots \hat{F}(\nu) \otimes \cdots \hat{1}(N)\,. \tag{3.107}$$

(The left-hand side is a short-hand notation.) Local operators involving different subsystems $\mu, \nu = 1, 2, \ldots N$ *always commute,*

$$[\hat{F}(\nu), \hat{G}(\mu)] = 0 \quad \text{for } \nu \neq \mu\,. \tag{3.108}$$

(ii) Cluster operators are composed of local operators, but affect more than one subsystem; a typical pair operator would be

$$\hat{F}(\mu, \nu) \equiv \hat{1}(1) \otimes \cdots \hat{F}(\mu) \otimes \cdots \hat{F}(\nu) \otimes \cdots \hat{1}(N)\,. \tag{3.109}$$

Triple operators, etc., would be defined correspondingly.

(iii) Global operators affect all subsystems. Typically, they are sums of local or pair operators,

$$\hat{F} = \sum_{\nu=A,B,\ldots} \hat{F}(\nu)\,, \tag{3.110}$$

$$\hat{F} = \sum_{\mu,\nu=A,B,\ldots} \hat{F}(\mu, \nu)\,. \tag{3.111}$$

Examples for the former are the total momentum or the total angular momentum. Examples for the latter are "two-body" interactions.

For systems composed of indistinguishable particles all pertinent operators have to be invariant under permutations of the particle index. This implies, in particular, that all operators are global operators; there cannot be individually addressable local subsystems or specific clusters. Indistinguishability has operational consequences, cf. Definition 3.69.

Example 3.8 Born–Oppenheimer approximation (BOA) *Typical many-body systems are composed of electrons (e) and nuclei (n). Both these subgroups consist of indistinguishable particles. The BOA tries to exploit the fact that the electrons are light and thus fast, the nuclei heavy and thus slow particles; so it is assumed that the correct Hamiltonian can be replaced by*

$$\hat{H}^{BOA}(\vec{R}_n) = \hat{T}_e + \hat{V}_{ee} + \hat{V}_{en}\,, \tag{3.112}$$

with fixed nuclear positions, $\vec{R}_n$. This electronic Hamiltonian (subsystem e) defines an effective Schrödinger equation for the electrons parametrized by a given nuclear configuration (subsystem n). The total wave function is taken to be a tensor product of the form

$$< \vec{r}_e; \vec{R}_n|\Psi >= \Psi(\vec{r}_e; \vec{R}_n) = \phi(\vec{r}_e; \vec{R}_n) \times \chi(\vec{R}_n)\,. \tag{3.113}$$

This is an example of a factorization approximation (FA), see proposition 5.11. There is no entanglement between electronic and nuclear substates. For more details see Section 5.2.1.

Remark 3.3 Bipartite Hamilton models *Simple models for bipartite systems are as follows:*

(1) *interacting two-spin systems*
(2) *spin-oscillator models*
(3) *coupled oscillators*
(4) *hydrogen molecule*

These depend on a small number of parameters defining the respective model spaces.

Example 3.9 Two-spin Heisenberg model

$$\boxed{\begin{aligned}\hat{H}(AB) = {} & \tfrac{\hbar\Gamma B_3}{2}(\hat{\sigma}_3(A) + \hat{\sigma}_3(B)) \\ & + \textstyle\sum_{j=1}^{3} J_j \hat{\sigma}_j(A) \otimes \hat{\sigma}_j(B)\,.\end{aligned}}$$

Here, Γ is the gyromagnetic ratio, B_3 the external magnetic field in direction 3, J_j are the respective coupling constants. Simplifying to the *so-called XXZ model, $J_1 = J_2 = J$, and using the ladder operators, cf. Definition 3.31,*

$$\hat{\sigma}^{\pm} = \hat{\sigma}_1 \pm \hat{\sigma}_2 \,, \tag{3.114}$$

$$\hat{\sigma}^{+} = 2\hat{P}_{21} \quad \hat{\sigma}^{-} = 2\hat{P}_{12} \,, \tag{3.115}$$

the first two interaction terms can be rewritten as

$$\begin{aligned} J\,\hat{F}(A,B) &= J\,\hat{\sigma}_1(A) \otimes \hat{\sigma}_1(B) + J\,\hat{\sigma}_2(A) \otimes \hat{\sigma}_2(B) \\ &= \frac{J}{2}(\hat{\sigma}_+(A) \otimes \hat{\sigma}_-(B) + c.c.) \,. \end{aligned} \tag{3.116}$$

In the product basis, $|11>$, $|12>$, $|21>$, $|22>$ the total Hamiltonian reads

$$H(AB) = \begin{pmatrix} J_3 - \hbar\Gamma B_3 & 0 & 0 & 0 \\ 0 & -J_3 & 2J & 0 \\ 0 & 2J & -J_3 & 0 \\ 0 & 0 & 0 & J_3 + \hbar\Gamma B_3 \end{pmatrix} . \tag{3.117}$$

Eigenstates and eigenvalues are easily obtained:

$$\begin{array}{ll} |\psi_1> = |11> & E_1 = J_3 - \hbar\Gamma B_3 \,, \\ |\psi_2> = \frac{1}{\sqrt{2}}(|21> - |12>) & E_2 = -J_3 - 2J \,, \\ |\psi_3> = \frac{1}{\sqrt{2}}(|21> + |12>) & E_3 = -J_3 + 2J \,, \\ |\psi_4> = |22> & E_4 = J_3 + \hbar\Gamma B_3 \,. \end{array} \tag{3.118}$$

$|\psi_2>$, $|\psi_3>$ are maximally entangled; they are known as Bell states, cf. Definition 3.64. For $J_3 = 0$, one obtains the so-called XX model, for $J_3 = J$ the isotropic Heisenberg model.

The spectrum depends on three parameters, which may be subject to external control. For independent spins, $J = J_3 = 0$, the parameter dependence reduces to "spectral deformation" as a function of B_3, see Definition 4.6. Alternatively, this holds for $B_3 = 0$ with $cJ_3 = J$ as control, c being a fixed real constant.

Example 3.10 N-spin network *A noninteracting network of N (distinguishable) spins of equal level splitting, ΔE, has the finite, equidistant spectrum*

$$E_s = \Delta E\, s \quad s = 0, 1, \ldots N \,, \tag{3.119}$$

with degeneracy

$$g_s = \binom{N}{s} = \frac{N!}{(N-s)!s!} \tag{3.120}$$

The eigenstates are simple tensor product states.

Example 3.11 N one-dimensional oscillators *Another simple example for a multipartite system would be a noninteracting network of N one-dimensional oscillators. Again, simple tensor product states form the eigenbasis of this Hamiltonian. If degeneracies occur, superpositions can be introduced, which will lead to entanglement. Limitations as a result of indistinguishability, cf. Definition 3.71, might apply, though.*

As noted in Example 3.7, such a basis also occurs as the eigenstates of a single D-dimensional oscillator. Here the individual factor states are indexed by the Cartesian coordinates $1, 2, \ldots D$. However, the choice of these coordinates is entirely ambiguous; furthermore, no spatial separation is possible. Superposition states should thus not be discussed in terms of entanglement in this case—despite formal equivalence.

The combination of a single spin (A) interacting with a single oscillator (B) gives rise to surprisingly versatile and useful models. The following two differ by the type of interaction:

Example 3.12 Jaynes–Cummings model *Based on the so-called rotating wave approximation (RWA), this model is defined as*

$$\boxed{\begin{aligned} \hat{H} &= \hat{H}_0 + \hat{V}(AB), \\ \hat{H}_0 &= \frac{1}{2}\hbar\omega(A)\hat{\sigma}_3(A) + \hbar\omega(B)\hat{a}^\dagger(B)\hat{a}(B), \\ \hat{V}(AB) &= \frac{1}{2}\hbar g(\hat{\sigma}^+(A) \otimes \hat{a}(B) + \hat{\sigma}^-(A)\hat{a}^\dagger(B)). \end{aligned}} \tag{3.121}$$

RWA means to neglect terms such as $\hat{\sigma}^-(A) \otimes \hat{a}(B)$. The (real) parameter g describes the coupling strength. This model, Eq. (3.121), is easily solved in terms of the so-called dressed states. We first note that the eigenfunctions of $\hat{H}_0$ are product states $|j(A); n(B) >$ with $j(A) = 1, 2$. Introducing the detuning $\delta = \omega(A) - \omega(B)$ the respective nonzero matrix elements are

$$< 2(A); n(B)|\hat{H}_0|2(A); n(B) > = \frac{\hbar}{2}\delta + \left(n + \frac{1}{2}\right)\hbar\omega(B), \tag{3.122}$$

$$< 1(A); n+1(B)|\hat{H}_0|1(A); n+1(B) > = -\frac{\hbar}{2}\delta + \left(n + \frac{1}{2}\right)\hbar\omega(B), \tag{3.123}$$

$$< 2(A); n(B)|\hat{V}|1(A); n+1(B) > = \frac{\hbar}{2} g\sqrt{(n+1)} \equiv \frac{\hbar}{2} g^{(n+1)} . \tag{3.124}$$

In the last equation, we have used Eq. (3.92). We now restrict ourselves to the two-dimensional subspaces $j(A) = 1, 2$ for fixed parameter $n+1$. Then the matrix representation of $\hat{H}$ is

$$H^{(n+1)} = \left(n+\frac{1}{2}\right)\hbar\omega(B)\begin{pmatrix} 1 & 0 \\ 0 & 1 \end{pmatrix} + \frac{\hbar}{2}\begin{pmatrix} -\delta & g^{(n+1)} \\ g^{(n+1)} & \delta \end{pmatrix} . \tag{3.125}$$

Diagonalization leads to the eigenvalues

$$E_{\pm}^{(n+1)} = \left(n+\frac{1}{2}\right)\hbar\omega(B) \pm \frac{\hbar}{2}\Omega_R^{(n+1)} , \tag{3.126}$$

$$\Omega_R^{(n+1)} = \sqrt{(g^{(n+1)})^2 + \delta^2} . \tag{3.127}$$

These spin states are said to be "dressed" by the oscillator; $\Omega_R^{(n+1)}$ is the Rabi frequency. The corresponding eigenstates are entangled; for $\delta = 0$ and any parameter n + 1 one finds

$$|\pm >^{(n+1)} = \frac{1}{\sqrt{2}}\left(|1(A); n+1(B) > \pm |2(A); n(B) >\right) . \tag{3.128}$$

Reduced to the spin subspace these states are totally mixed, the respective entropy is maximal. A typical application is a two-level system interacting with a single electromagnetic mode.

Example 3.13 Spin-Boson model

$$\begin{aligned} \hat{H} &= \frac{1}{2}\hbar\omega(A)\hat{\sigma}_3(A) + \hbar\omega(B)\hat{a}^{\dagger}(B)\hat{a}(B) + \hat{V}_1(AB) , \\ \hat{V}_j(AB) &= \frac{1}{2}\gamma\hat{\sigma}_j(A)(\hat{a}(B) + \hat{a}^{\dagger}(B)) . \end{aligned} \tag{3.129}$$

This model (for j = 1) is also known as a Jahn–Teller System.

Identifying

$$\frac{\gamma}{2} \equiv L\frac{1}{\sqrt{2m\omega(B)}} , \tag{3.130}$$

(γ or L characterizes the coupling strength) and with the help of

$$\hat{x}(B) = \sqrt{\frac{\hbar}{2m(B)\omega(B)}}(\hat{a} + \hat{a}^{\dagger}) \equiv \frac{\hbar^{1/2}}{m^{1/2}(B)}\hat{q}(B) , \tag{3.131}$$

(cf. 3.85), we are able to rewrite the original model definition, Eq. (3.129), in the form [Hines (2004); Levine (2004)]

$$< q|\hat{H}|q >= \frac{\hbar\omega(A)}{2}\hat{\sigma}_3 + L\sqrt{\frac{\hbar}{m(B)}}q\hat{\sigma}_1 - \frac{1}{2}\left(\frac{d^2}{dq^2} - \hbar^2\omega^2(B)q^2\right). \tag{3.132}$$

We have thus expressed the oscillator subsystem B in its real space representation. For $\omega(A) = 0$, each of the resulting eigenvalues, [Hines (2004)]

$$E_n = \hbar\omega(B)n - \frac{L^2}{m(B)\hbar\omega^2(B)}, \tag{3.133}$$

corresponds to two (degenerate) product states,

$$< q|\psi_n^L > = \chi_n^L(q)|1(A)>, \tag{3.134}$$

$$< q|\psi_n^R > = \chi_n^R(q)|2(A)> . \tag{3.135}$$

The general ground state would be a superposition thereof; such superposition states are entangled.

The $\omega(A)\hat{\sigma}_3$-term breaks the above degeneracy and forces the ground state to exhibit a fixed superposition between $|1(A)>$ and $|2(A)>$, resulting in an entangled state for all $\omega(A) > 0$. For very large L, one approaches equal superposition and thus maximum entanglement.

It is interesting to note that in the massive limit of the oscillator B, $m \to \infty$ with $m(B)\omega^2(B)$ kept constant, the onset of entanglement becomes discontinuous with respect to the parameter $\alpha = L^2/(m(B)\omega^2(B)\omega(A))$, that is, with respect to the coupling strength. The critical value is $\alpha_c = 1$. Above this bifurcation point $< q >= \pm q_0 \neq 0$. and $< \sigma_1 > \neq 0$. This shows, for example, how "resonating" electronic ground states (interpreted as two alternatives $|1(A)>, |2(A)>$) can develop entanglement with the nuclear degrees of freedom (beyond the Born–Oppenheimer approximation). Benzene or ammonia are well-known molecules to exhibit such resonances [McKemmish (2009)]. The coupling strength with the pertinent vibrational mode defines whether or not a well-defined molecular structure is stabilized (symmetry breaking).

3.3.4 *Quantum States*

The most general concept of a state is the density operator, $\hat{\rho}$. This state description includes pure states—the states considered so far and represented as vectors, $|\phi>$.

Definition 3.40 Pure state as projector *A pure state is a state that cannot be written as a convex combination (mixture) of other states.*

$$\hat{\rho} = |\phi >< \phi| \,. \tag{3.136}$$

The expectation value, Eq. (3.7), is recovered using

$$< F >= \mathtt{Tr}\{\hat{\rho}\hat{F}\} =< \phi|\hat{F}|\phi > \,. \tag{3.137}$$

For the last step $|\phi\ >$ has been taken as one member of the orthogonal basis set $|e^{(j)} >$ used to carry out the trace.

Definition 3.41 Density operator $\hat{\rho}$ in eigen representation

$$\hat{\rho}|j> = \rho_{jj}|j> \quad j = 1, 2 \cdots d \,, \tag{3.138}$$

$$\hat{\rho} = \sum_{j=1}^{d} \rho_{jj}|j >< j| \,. \tag{3.139}$$

(cf. Definition 3.17.) For the case of a pure state, all eigenvalues ρ_{jj} are zero, except one, which is 1. This case will now be generalized.

Under the condition that

$$\mathtt{Tr}\{\hat{\rho}\} = \sum_k \rho_{kk} = 1 \,, \tag{3.140}$$

and ρ_{kk} to be real and non-negative, one concludes that $0 \leq \rho_{kk} \leq 1$. The eigen representation, Eq. (3.139), thus appears as a weighted sum of projection operators, $\hat{P}^{(jj)} = |j >< j| \equiv \hat{\rho}^{(j)}$, of the generic form

$$\hat{\rho} = p\hat{\rho}^{(1)} + (1 - p)\hat{\rho}^{(2)} \tag{3.141}$$

(mixed state). Here p may describe subjective ignorance. The density operator is therefore also called statistical operator.

Correspondingly, the expectation value, Definition 3.4, generalizes to

$$< F >= \sum_{k=1}^{d} < k|\hat{F}|k > \rho_{kk} \,. \tag{3.142}$$

$<F>$ can thus be interpreted as a statistical average over the individual pure state expectation values, indexed by k.

A (unitary) transformation to a general basis,

$$|j >= \sum_n |e^{(n)} >< e^{(n)}|j > \,, \tag{3.143}$$

leads to

$$\hat{\rho} = \sum_{n,m} \rho_{nm} |e^{(n)} >< e^{(m)}| \,. \tag{3.144}$$

With $\rho_{nm} = \sum_j \rho_{jj} < e^{(n)}|j >< j|e^{(m)} >$, it follows that $\rho_{nm} = \rho^*_{mn}$: The density matrix is an Hermitian matrix, that is, $\hat{\rho} = \hat{\rho}^\dagger$, cf. Definition 3.15.

Definition 3.42 Trace terms *Let us introduce*

$$C(d, q) \equiv \mathrm{Tr}\{\hat{\rho}^q\}, \tag{3.145}$$

where d is the dimension of the respective Hilbert space, q = 1, 2, . . . d. One easily shows, based on the eigen representation of $\hat{\rho}$, that

$$C(d, q+1) \le C(d, q) \le 1 \,. \tag{3.146}$$

All these trace terms are invariant under unitary transformation; together they are equivalent to the d eigenvalues of the density matrix.

Definition 3.43 Purity

$$\boxed{P = C(d, 2) = \mathrm{Tr}\{\hat{\rho}^2\} \le 1 \,.} \tag{3.147}$$

$P = 1$ holds for pure states only. The lower bound is $P_{min} = 1/d$; it corresponds to states with maximum von Neumann entropy, Eq. (3.159).

Remark 3.4 *If we keep only the first term in the expansion, Eq. (3.162), we get approximately*

$$\hat{\rho} \ln \hat{\rho} \approx \hat{\rho}^2 - \hat{\rho} \,. \tag{3.148}$$

The so-called linearized entropy is thus defined by

$$S^{lin.}_{vN} \equiv -\mathrm{Tr}\{\hat{\rho}^2 - \hat{\rho}\} = 1 - P \le S_{vN} \,. \tag{3.149}$$

Definition 3.44 Distance between states *There are various possibilities to introduce distance measures (cf., e.g., [Dajka (2011)]).*

A convenient form is the so-called Schmidt–Hilbert distance

$$D^2(\hat{\rho}, \hat{\rho}') \equiv \mathrm{Tr}\{(\hat{\rho} - \hat{\rho}')^2\} \ge 0 \,. \tag{3.150}$$

It holds that $D^2 \le 2$; the upper limit is obtained for orthogonal pure states. D is bounded from above by the so-called trace distance D_{Tr},

$$D_{Tr} \equiv \frac{1}{2}\mathrm{Tr}\left\{\sqrt{(\hat{\rho} - \hat{\rho}')^2}\right\}, \tag{3.151}$$

$$D \le 2 D_{Tr} \,. \tag{3.152}$$

An alternative distance is the so-called quantum fidelity, F. For two pure states F is defined as the overlap in terms of the scalar product,

$$F(\psi, \phi) = | < \psi|\phi > |^2 . \tag{3.153}$$

For mixed states, the fidelity is generalized to

$$F(\hat{\rho}, \hat{\rho}') = \left(\mathrm{Tr}\sqrt{\sqrt{\hat{\rho}}\hat{\rho}'\sqrt{\hat{\rho}}}\right)^2 . \tag{3.154}$$

All these distance measures are invariant under unitary transformation; in particular, they do not change under unitary time evolution, cf. Eq. (3.227).

Rather than discussing the full details of a state, it is sometimes sufficient to provide some characteristic measure—as a means of classification. Such a convenient measure is the entropy.

It is remarkable that quantum mechanics and information theory and thermodynamics share this fundamental concept.

Definition 3.45 Quantum Renyi entropies *Generalizing the Definition (2.13) from information theory one is led to [Baez (2011)]*

$$S_\alpha[\hat{\rho}] = \frac{\ln(\mathrm{Tr}\hat{\rho}^\alpha)}{1-\alpha} . \tag{3.155}$$

Here $0 < \alpha$ is the respective order.

Definition 3.46 von Neumann entropy

$$\boxed{S_{vN}(\hat{\rho}) = -\mathrm{Tr}\{\hat{\rho}\ln\hat{\rho}\} \geq 0 .} \tag{3.156}$$

This functional assigns a real number S_{vN} to each quantum state $\hat{\rho}$. It may be identified with the quantum Renyi entropy for $\alpha \to 1$.

The trace is conveniently calculated in the eigen representation of $\hat{\rho}$. In this case, $\ln\hat{\rho} = \sum_j \ln\rho_{jj}|j><j|$, and thus

$$S_{vN} = -\sum_k^d \rho_{kk}\ln\rho_{kk} . \tag{3.157}$$

Note that this definition of entropy refers to the set of orthogonal state alternatives, which could operationally be distinguished in the eigenbasis $\{|j>\}$ of $\hat{\rho}$: The eigenvalues $0 \leq \rho_{jj} \leq 1$ denote the probabilities for finding states $|j>$ in $\hat{\rho}$,

$$\rho_{kk} = P_k . \tag{3.158}$$

The entropy appears as a kind of uncertainty measure. The value of S_{vN} runs between zero for pure states (peaked distribution) and the maximum value

$$S_{vN}^{max} = \ln d \tag{3.159}$$

for the broadest distribution possible, $\rho_{kk} = 1/d$, where d is the dimension of Hilbert space, that is, the number of independent basis state vectors. The (possible) subjective ignorance about the eigenbasis itself is irrelevant. This means, in particular, that the d complex coefficients, which would be needed to specify a pure state in a general basis, are not of any concern, cf. Definition 3.51. Note that those complex coefficients would be "meaningless" unless the reference basis was "known" beforehand, cf. Definition 2.30 on unspeakable information.

The entropy is a function of the state $\hat{\rho}$. This functional dependence is constrained by:

Proposition 3.2 Concavity of entropy *A function $f(x)$ is concave, if $f(cx_1 + (1-c)x_2) \geq cf(x_1) + (1-c)f(x_2)$ for any $0 \leq c \leq 1$ being a real number. Let $\hat{\rho}_1$ and $\hat{\rho}_2$ be two density operators (within the same Hilbert space). Then*

$$S_{vN}(c\hat{\rho}_1 + (1-c)\hat{\rho}_2) \geq cS_{vN}(\hat{\rho}_1) + (1-c)S_{vN}(\hat{\rho}_2)\,. \tag{3.160}$$

The equal sign holds for $\hat{\rho}_1 = \hat{\rho}_2$ only.

For a proof see, for example, [Diu (1989)].

Proposition 3.3 *Let $\hat{U}$ be a unitary operator. Then*

$$\ln(\hat{U}\hat{\rho}\hat{U}^\dagger) = \hat{U}\ln\hat{\rho}\hat{U}^\dagger\,. \tag{3.161}$$

Proof. Consider the operator expansion

$$\ln\hat{\rho} = (\hat{\rho} - \hat{1}) - \frac{1}{2}(\hat{\rho} - \hat{1})^2 + \frac{1}{3}(\hat{\rho} - \hat{1})^3 \pm \ldots\,. \tag{3.162}$$

For $\hat{\rho}' = \hat{U}\hat{\rho}\hat{U}^\dagger$ and $\hat{1} = \hat{U}\hat{U}^\dagger$, the general term is

$$\begin{aligned}(\hat{\rho}' - \hat{1})^n &= (\hat{U}\hat{\rho}\hat{U} - \hat{U}\hat{U}^\dagger)^n \\ &= \hat{U}\left(\hat{\rho} - \hat{1})\hat{U}^\dagger\hat{U}(\hat{\rho} - \hat{1})\hat{U}^\dagger\ldots\hat{U}(\hat{\rho} - \hat{1})\hat{U}^\dagger\hat{U}(\hat{\rho} - \hat{1}\right)\hat{U}^\dagger \\ &= \hat{U}(\hat{\rho} - \hat{1})^n\hat{U}^\dagger\,,\end{aligned} \tag{3.163}$$

from which the proposition follows term by term. □

Proposition 3.4 Unitary invariance of von Neumann entropy

$$\boxed{S_{vN}(\hat{U}\,\hat{\rho}\hat{U}^{\dagger}) = S_{vN}(\hat{\rho})\,.} \tag{3.164}$$

Proof. Based on Eq. (3.161), we obtain

$$\begin{aligned} S_{vN}(\hat{U}\,\hat{\rho}\hat{U}^{\dagger}) &= -\mathtt{Tr}\{\hat{U}\,\hat{\rho}\hat{U}^{\dagger}\ln(\hat{U}\,\hat{\rho}\hat{U}^{\dagger})\} \\ &= -\mathtt{Tr}\{\hat{U}\,\hat{\rho}\hat{U}^{\dagger}\hat{U}\ln\hat{\rho}\hat{U}^{\dagger}\} = -\mathtt{Tr}\{\hat{\rho}\ln\hat{\rho}\}\,. \end{aligned} \tag{3.165}$$

Here we have used $\mathtt{Tr}\{\hat{F}\hat{G}\} = \mathtt{Tr}\{\hat{G}\hat{F}\}$ and $\hat{U}^{\dagger}\hat{U} = \hat{1}$. □

Definition 3.47 Quantum relative entropy *Let $\hat{\rho}_1$ and $\hat{\rho}_0$ both be density operators. Then the relative entropy of $\hat{\rho}_1$ to $\hat{\rho}_0$ is given by [Nielsen (2000)]*

$$\boxed{S_{vN}(\hat{\rho}_1\|\hat{\rho}_0) = -\mathtt{Tr}(\hat{\rho}_1\ln\hat{\rho}_0) + \mathtt{Tr}(\hat{\rho}_1\ln\hat{\rho}_1)\,.} \tag{3.166}$$

This is the quantum version of the classical definition (cf. Def. (2.20)).

According to Klein's inequality [Nielsen (2000)], the quantum relative entropy is non-negative. It can therefore be used as an alternative distance measure.

Proposition 3.5 Klein's inequality

$$\boxed{S_{vN}(\hat{\rho}_1\|\hat{\rho}_0) \geq 0\,.} \tag{3.167}$$

Proof.

$$S_{vN}(\hat{\rho}_1\|\hat{\rho}_0) = \mathtt{Tr}(\hat{\rho}_1\ln(\hat{\rho}_1/\hat{\rho}_0)\,. \tag{3.168}$$

Using the expansion $\ln x \geq 1/x$ for $x = \hat{\rho}_1/\hat{\rho}_0$, we obtain

$$\hat{\rho}_1\ln(\hat{\rho}_1/\hat{\rho}_0) \geq \hat{\rho}_1 - \hat{\rho}_0 \tag{3.169}$$

and thus

$$S_{vN}(\hat{\rho}_1\|\hat{\rho}_0) \geq \mathtt{Tr}\{\hat{\rho}_1 - \hat{\rho}_0\} = 0\,. \tag{3.170}$$

□

Definition 3.48 Real space representation of states: Wave function *In the case of a one-particle model in three dimensions, possible complete sets of commuting operators are the three position operators $\hat{\vec{r}} = \{\hat{r}_1, \hat{r}_2, \hat{r}_3\}$ or the three momentum operators $\hat{\vec{p}} = \{\hat{p}_1, \hat{p}_2, \hat{p}_3\}$. The respective joint eigenvalue equations are,*

$$\hat{\vec{r}}|\vec{r}> = \vec{r}|\vec{r}>\,, \tag{3.171}$$

$$\hat{\vec{p}}|\vec{p}> = \vec{p}|\vec{p}>\,. \tag{3.172}$$

The corresponding eigenvectors define the real space and the momentum representation, respectively, cf. Eq. (3.15):

$$|\phi> = \int d^3r|\vec{r} > \phi(\vec{r}) \qquad \phi(\vec{r}) =< \vec{r}|\phi > , \tag{3.173}$$

$$|\phi> = \int d^3 p|\vec{p} > \phi(\vec{p}) \qquad \phi(\vec{p}) =< \vec{p}|\phi > . \tag{3.174}$$

These are continuous representations: The vector components now come as functions, the so-called wavefunctions; normalization means

$$< \phi|\phi >= 1 = \int d^3r|\phi(\vec{r})|^2 . \tag{3.175}$$

Obviously, the wave function has the dimension $(volume)^{-1/2}$.

Definition 3.49 Quantum coherence *The original concept of coherence has been borrowed from (classical) wave phenomena, where superpositions give rise to interference phenomena. Indeed, the real space representation of state* $|\phi >$*, cf. Eq. (3.173),*

$$|\phi >= \int d^3r|\vec{r} > \phi(\vec{r}) , \tag{3.176}$$

can be interpreted as a superposition of states $|\vec{r}\;\;>$*. In this representation, the corresponding density matrix would contain off-diagonal matrix elements,*

$$\rho(\vec{r}, \vec{r'}) \sim \phi(\vec{r})\phi^*(\vec{r'}) \neq 0 , \tag{3.177}$$

which are, in general, complex, that is, involve a relative phase θ*,*

$$\rho(\vec{r}, \vec{r'}) = |\rho(\vec{r}, \vec{r'})| \exp{(i\theta(\vec{r}, \vec{r'})} . \tag{3.178}$$

Now, taking such off-diagonal matrix elements as the characteristic feature of coherence, this concept can immediately be generalized to any quantum state and in any representation and any Hilbert space dimension: For a state $\hat{\rho}$ *to be (at least partly) coherent with respect to a given discrete basis,* $|e^{(n)}\;\;>$ *we simply have to require some of its off-diagonal matrix elements,* $\rho_{mn}\;\;=< e^{(m)}|\hat{\rho}|e^{(n)}\;\;>; m \neq n$*, to be non-zero. Obviously, this property depends on the representation chosen. Density matrices can always be diagonalized; in the diagonal basis there is no coherence at all.*

Definition 3.50 Typicality *What are "typical properties" of pure states? Let us start from subjective ignorance about the actual* *state, which is conveniently represented by an ensemble. Typical—in agreement with its colloquial meaning—are then properties, which occur for "almost all" ensemble members. In order to quantify such a behavior, one needs to define a pure state distribution function (cf. Section 5.1.1, [Gemmer (2009)]).*

There are various ways to construct such a function. A convenient way to proceed is to introduce a specific parametrization for the pure states in question, with respect to a given basis set (like the energy representation, if the Hamiltonian is a complete operator).

Definition 3.51 Parametrization of pure states *Any pure state $|\phi>$ in a given d-dimensional Hilbert space can be written in some basis $|j>$ as*

$$\boxed{|\phi>=\sum_{j=1}^{d}(\eta_j+i\xi_j)|j>\ .} \tag{3.179}$$

Here, η_j, ξ_j are $2d$ real parameters. The normalization condition is

$$G(\eta_1,\xi_1,\eta_2,\xi_2,\ldots\eta_d,\xi_d)=\sum_{j=1}^{d}(\eta_j^2+\xi_j^2)=1\,. \tag{3.180}$$

Definition 3.52 Pure state distribution w *The corresponding pure state distribution does not relate to a concrete physical scenario like a thermal equilibrium state. There should be no bias, except symmetry (see below).*

The distribution function parametrized as $w(\eta_1,\xi_1,\ldots\eta_d,\xi_d,)$ may now be considered a "prior" in the sense of Bayesian statistics (cf. Proposition 2.4). A simple way to arrive at this distribution is to apply the maximum entropy principle subject to the single macroconstraint (norm)

$$\overline{G}=\int d^d\eta d^d\xi\, w(\eta_1,\xi_1,\ldots\eta_d,\xi_d)\sum_{j=1}^{d}(\eta_j^2+\xi_j^2)=\text{const.} \tag{3.181}$$

This procedure immediately leads to the Gaussian (cf. Proposition 4.1 and Eq. (2.37))

$$\boxed{w(\eta_1,\xi_1,\ldots\eta_d,\xi_d)=\left(\frac{1}{\sqrt{\pi\gamma}}\right)^{2d}\exp\left(-G(\eta_1,\xi_1,\ldots\eta_d,\xi_d)/\gamma\right).} \tag{3.182}$$

We note that this distribution fulfills the normalization condition on average only. This deficiency can be cured in the asymptotic limit for large d, if we set $\gamma = 1/d$. In this limit, the fluctuations of G become negligible.

More rigorously, the above result can be obtained by requiring the distribution function $w(\eta_1, \xi_2, \ldots \eta_d, \xi_d)$ to be invariant under unitary transformations [Gemmer (2009)].

This distribution function underlies the so-called Hilbert space average method (HAM), cf. Proposition 5.3.

3.3.5 *Product Hilbert Space*

Definition 3.53 Γ-space and classical pure state *A classical particle is described by $d = 6$ coordinates, three position and three momentum coordinates. For a composite system (N particles), the respective space is a direct sum, the so-called Γ-space. Its dimension is*

$$D = \sum_{\mu=1}^{N} d(\mu) = 6N\,. \tag{3.183}$$

It scales linearly with N. A pure classical state is a point in this Γ-space; all these pure states are distinguishable (given unlimited measurement accuracy). Mixed states can be introduced via probability measures.

Definition 3.54 Tensor product *In quantum mechanics, each subsystem μ brings its own Hilbert space of dimension $d(\mu)$. For a composite system, the total Hilbert space is thus a direct product (or tensor-) space of dimension*

$$D = \prod_{\mu=1}^{N} d(\mu)\,. \tag{3.184}$$

For $d(\mu) = d$, one finds $D = d^N$, that is, the total dimension scales exponentially with N.

Definition 3.55 Product representation *A convenient set of basis vectors for a composite system are product states, cf. Definition 3.5.*

For $N = 2$ and if $\hat{F}(A)$ and $\hat{F}(B)$ are complete commuting operators within their subspace, the completeness relation reads

$$\hat{1} = \hat{1}(A) \otimes \hat{1}(B) = \sum_{i=1}^{d(A)} \sum_{j=1}^{d(B)} | f^{(i)}(A), f^{(j)}(B) >< f^{(i)}(A), f^{(j)}(B)| . \tag{3.185}$$

Any pure state for such a bipartite system can thus be represented in terms of those product states, $| f^{(i)}(A) > \otimes | f^{(j)}(B) >= | f^{(i)}(A), f^{(j)}(B) >$, as

$$|\psi(A, B) > = \sum_{i=1}^{d(A)} \sum_{j=1}^{d(B)} | f^{(i)}(A), f^{(j)}(B) > \psi_{ij}$$

$$\psi_{ij} = < f^{(i)}(A), f^{(j)}(B) | \psi(A, B) > . \tag{3.186}$$

This is easily generalized to $N = 3, 4, \ldots$.

The opposite to composition is partitioning:

Definition 3.56 "Virtual" partitioning *Consider a total Hilbert space of dimension d. If d is a prime number, the system is necessarily elementary. Otherwise it has a nontrivial prime factorization [Zanardi (2001)],*

$$d = \prod_{j=1}^{r} q_j^{n_j} . \tag{3.187}$$

The number of such elementary subsystems, each of prime factor dimension, would be given by

$$N_d = \sum_{j=1}^{r} n_j . \tag{3.188}$$

Definition 3.57 Operationally accessible partitioning *However, for a concrete system $\hat{H}$ living in such a Hilbert space of dimension d only a fraction of those elementary subsystems will be "real," that is, correspond to operationally well-defined and accessible subunits. Its number, $N \leq N_d$, and type is not an absolute property of the system but contextual. It defines a kind of reference frame. Each such unit is characterized by a (classical) index, $\nu = A, B, C, \ldots$. The index serves as a kind of address.*

For example, a Hilbert space with $N_d = 3$ may be described as a single system, $N = 1$, (ABC), as a bipartite system, $N = 2$, (A)(BC) *or (AB)(C) or as* $N = N_d$, *(A)(B)(C). This is more than a formal aspect. While all these decompositions are formally equivalent, that is, span the same total space, they give rise to different phenomena in terms of inter-subsystem correlations (entanglement). Different decompositions also lead to different notations for states:* $|k(ABC)>$ *or* $|k(A); m(BC)>$ *or* $|k(A); m(B); n(C)>$.

Remark 3.5 What are typical indices? *As a classical reference, no superposition of indices is allowed. An index may specify different (distinguishable) particles or different modes of a quantum field (like a photon field with given boundary conditions). In case of a frozen-in structure, the real space location (even of in principle indistinguishable atoms, say) may also serve as such an index. In any case, we have to come up with a procedure by which we can select or filter out the respective addressee.*

For $N = 2$, the operator $\hat{G}(A, B)$ can be represented as

$$\hat{G}(A, B) = \sum_{ij;kl} G_{ij;kl} |i(A), j(B) >< k(A), l(B)| , \quad (3.189)$$

$$G_{ij;kl} = < i(A), j(B)|\hat{G}|k(A), l(B) > .$$

The expectation value for a local operator $\hat{F}(A, B) = \hat{F}(A) \otimes \hat{1}(B)$ is given by

$$\overline{F}(A) = \mathrm{Tr}\{\hat{\rho}(\hat{F}(A) \otimes \hat{1}(B))\} = \sum_{ij;kl} \rho_{ij;kl}(A, B) F_{ki}(A) \delta_{lj}(B)$$

$$\equiv \sum_{ik} \rho_{ik}(1) F_{ki}(A) . \quad (3.190)$$

It can thus conveniently be expressed by the

Definition 3.58 Reduced density operator

$$\boxed{\begin{aligned} \hat{\rho}(A) &= \mathrm{Tr}_B\{\hat{\rho}\} , \\ \rho_{ik}(A) &= \textstyle\sum_j \rho_{ij;kj}(A, B) . \end{aligned}} \quad (3.191)$$

Here Tr_B *means partial trace, that is, trace over subsystem B only. The formulas for observables in subsystem B are correspondingly. The definition is easily generalized for multipartite systems.*

Definition 3.59 Correlation function in tensor space *Let us consider the pair operator* $\hat{F}(\mu, \nu) = \hat{F}(\mu) \otimes \hat{F}(\nu)$ *as defined in Eq. (3.109). The respective correlation function is then expressed as*

$$C(F(\mu), F(\nu)) = \mathrm{Tr}\{\hat{\rho}\hat{F}(\mu) \otimes \hat{F}(\nu)\} . \quad (3.192)$$

For product states, this correlation function factorizes into the product of local expectation values.

Definition 3.60 Joint entropy *The joint entropy of a bipartite system can be written as*

$$S_{vN}(AB) = -\mathrm{Tr}\{\hat{\rho}(AB)\ln(\hat{\rho}(AB))\}. \qquad (3.193)$$

Unfortunately, its eigenbasis is not necessarily the set of product states. We could have a mixed state of Bell states, for example.

Definition 3.61 Reduced von Neumann entropies *The local entropies ($\nu = A, B$)*

$$S_{vN}(\nu) = -\mathrm{Tr}\{\hat{\rho}(\nu)\ln\hat{\rho}(\nu)\} \qquad (3.194)$$

fulfill the subadditivity condition

$$S_{vN}(A) + S_{vN}(B) \geq S_{vN}(A, B) \qquad (3.195)$$

and a further remarkable inequality, the theorem according to Araki and Lieb,

$$|S_{vN}(A) - S_{vN}(B)| \leq S_{vN}(A, B). \qquad (3.196)$$

Here, $S_{vN}(A, B)$ is the total entropy. For a proof of these inequalities, see [Nielsen (2000)].

The Araki–Lieb theorem implies that the reduced entropies must be the same, if the total entropy is zero (i.e., the total state is pure).

Example 3.14 Typical reduced entropies in bipartite quantum systems *Consider a bipartite system in a pure state $|\psi>$ parametrized by $\{\eta_1, \xi_1, \ldots \eta_d, \xi_d\}$. The reduced entropy of subsystem A can then be expressed as a unique function of those parameters. When averaged over the pure state distribution Eq. (3.182), one finds for $1 < d(A) \ll d(B)$ [Page (1993)]*

$$\boxed{\overline{S}_{vN}(A) \approx \ln d(A) - \frac{d(A)}{2d(B)}}\,. \qquad (3.197)$$

This means that a subsystem (A) embedded in a much larger subsystem (B) is most likely totally mixed, that is, its entropy S_{vN} is maximal. As the reduced entropy is a measure of entanglement, entangled states (under the conditions as specified) must count as "typical," they are nothing special. So, if we pick at random an *unknown pure state of a bipartite system, we should bet that the smaller part is maximal entangled.*

Definition 3.62 Quantum mutual information

$$\boxed{I_{vN}(\hat{\rho}(A,\,B);\hat{\rho}(A),\,\hat{\rho}(B)) = S_{vN}(A) + S_{vN}(B) - S_{vN}(A,\,B) \geq 0\,.} \tag{3.198}$$

For the classical counterpart, cf. Def. (2.19), the quantum mutual information is said to describe the total amount of correlations contained in a bipartite state.

Indeed, it holds that $I_{vN}(\hat{\rho}(A,\,B);\hat{\rho}(A),\,\hat{\rho}(B)) = 0$, if $\hat{\rho} = \hat{\rho}(A) \otimes \hat{\rho}(B)$, that is, if $\hat{\rho}(A,\,B)$ is uncorrelated: In this case, the local entropies simply add up to the joint entropy.

Definition 3.63 Entanglement measures *For a total pure state, the local states of a bipartite system (A, B) are called entangled, if*

$$S_{vN}(A) = S_{vN}(B) > 0\,. \tag{3.199}$$

The local (reduced) entropies serve as an entanglement measure, with maximum entropy indicating maximal entanglement. Entangled states are specific superposition states in product space.

For nonpure total states, the definition of entanglement requires special attention. For a mixed state of 2 qubits, the so-called concurrence $C(\hat{\rho})$ or the PPT criterion [Peres (1996)] can be used. (PPT stands for "positive under partial transposition.") For un-entangled states the concurrence is zero, maximum entanglement obtains for $C = 1$. For general composite systems, simple entanglement measures are not available, though.

Example 3.15 EPR state *For a spin pair, $\{\hat{\sigma}_j(A), \hat{\sigma}_k(B)\}$ the superposition state*

$$\boxed{|\psi^{(EPR)}(A,\,B)> = \frac{1}{\sqrt{2}}\left(|e^{(2)},\,e^{(1)} > -|e^{(1)},\,e^{(2)}>\right)} \tag{3.200}$$

implies maximal (anti-)correlation, cf. Definition (3.59),

$$C(\hat{\sigma}_j(A),\,\hat{\sigma}_k(B)) = -\delta_{jk}\,. \tag{3.201}$$

The reduced density operators are for $\nu = A,\,B$

$$\hat{\rho}(\nu) = \frac{1}{2}\hat{1}(\nu)\,. \tag{3.202}$$

These are mixed states with the maximum reduced entropies $S_{vN}(\nu) = \ln 2 = 1 bit$.

In this state, all local spin observables are maximally uncertain—despite the fact that the total state is pure (zero total entropy). Nevertheless, there is a strict anticorrelation between those local observables! This bewildering feature has been termed "correlation without correlata" [Mermin (1998)]. Furthermore, the local uncertainty is an objective fact here, not due to subjective ignorance.

In quantum mechanics—contrary to classical physics—a nonzero entropy may thus be of completely different origin: due to objective uncertainty or due to subjective ignorance. The former possibility is highlighted by the following model states:

Definition 3.64 Two-spin systems: Bell states *The four pair operators* $\hat{K}_{jj}(A, B) = \hat{\sigma}_j(A) \otimes \hat{\sigma}_j(B); j = 0, 1, 2, 3$ *are global observables in the sense of Eq. (3.111). They form a complete set of commuting operators. Note that* $\hat{K}_{00}(A, B) = \hat{1}(A, B)$. *The respective eigen-equations are*

$$\hat{K}_{jj}(A, B)|\Phi^{(k)}(A, B) >= K_j^{(k)}|\Phi^{(k)}(A, B) > . \tag{3.203}$$

The set of four joint eigenvectors are orthogonal and all maximally entangled:

$$|\Phi^{(1)} > = \frac{1}{\sqrt{2}}(|e^{(2)}, e^{(2)} > +|e^{(1)}, e^{(1)} >) \quad K_j^{(1)} = \{1, 1, -1, 1\}, \tag{3.204}$$

$$|\Phi^{(2)} > = \frac{1}{\sqrt{2}}(|e^{(2)}, e^{(2)} > -|e^{(1)}, e^{(1)} >) \quad K_j^{(2)} = \{1, -1, 1, 1\}, \tag{3.205}$$

$$|\Phi^{(3)} > = \frac{1}{\sqrt{2}}(|e^{(2)}, e^{(1)} > +|e^{(1)}, e^{(2)} >) \quad K_j^{(3)} = \{1, 1, 1, -1\}, \tag{3.206}$$

$$|\Phi^{(4)} > = \frac{1}{\sqrt{2}}(|e^{(2)}, e^{(1)} > -|e^{(1)}, e^{(2)} >) \quad K_j^{(4)} = \{1, -1, -1, -1\}. \tag{3.207}$$

Here we have used the notation $|e^{(1)} >= |m = 1 >; |e^{(2)} >= |m = -1 >$. *The last state,* $\Phi^{(4)}$, *is the EPR state, Eq. (3.200).*

It is noteworthy that product states can now alternatively be interpreted as special superpositions of Bell states.

Definition 3.65 Change of partition *Change of representations are common methods in the description of quantum states. There are nontrivial generalizations in terms of partitions. As an example, we consider a four-spin system, $\mu = A, B, C, D$. Its pure state $|\Psi >$ can be represented in the form of three alternative Bell state decompositions,*

$$|\Psi > = \sum_{ij} d_{ij} |\Phi^{(i)}(AB) > \otimes |\Phi^{(j)}(CD) > , \qquad (3.208)$$

$$|\Psi > = \sum_{ij} f_{ij} |\Phi^{(i)}(AD) > \otimes |\Phi^{(j)}(BC) > , \qquad (3.209)$$

$$|\Psi > = \sum_{ij} g_{ij} |\Phi^{(i)}(AC) > \otimes |\Phi^{(j)}(BD) > . \qquad (3.210)$$

Specializing on

$$|\Psi >_0 = |\Phi^{(4)}(AB) > \otimes |\Phi^{(4)}(CD) > \qquad (3.211)$$

one identifies, for example,

$$d_{44} = 1 \quad \text{and zero otherwise,} \qquad (3.212)$$

$$f_{11} = f_{44} = -0,5 \quad f_{22} = f_{33} = 0,5 \quad \text{and zero otherwise.} \qquad (3.213)$$

The single Bell state pair in (AB)(CD) appears as a superposition of Bell pairs in (AD)(BC). Such ambiguities can be exploited for "entanglement swapping," cf. Definition 3.89.

A different change of partition can be introduced by the so-called beam splitter transformation. Originally, it relates to the different modes within the input and output channel of a linear optical device, that is, modes defined in different spatial regions. Insofar they are mutually exclusive. Mode indices are good subsystem indices, cf. Remark 3.5; the corresponding sets of subsystems are thus also mutually exclusive in an operational sense.

Definition 3.66 Beam splitter transformation *A beam splitter is a passive optical device that transforms a pair of input modes, $\mu = A, B$ specified by the destruction operators $\hat{a}(A)$, $\hat{a}(B)$ into a pair of output*

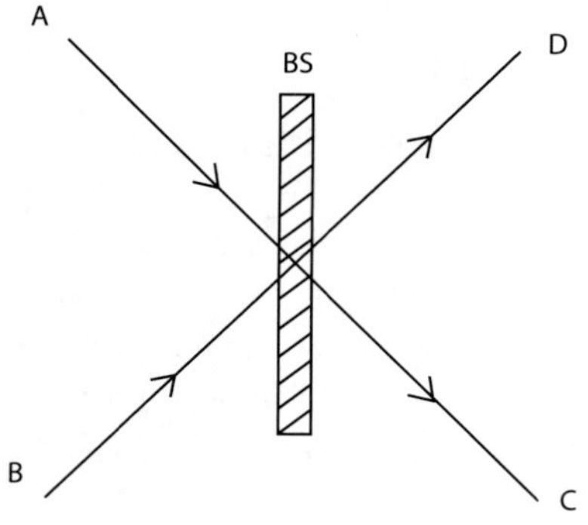

Figure 3.4 Beam splitter (BS) scenario: A, B, input modes, C, D, output modes.

modes $\nu = C, D$ with the destruction operators $\hat{b}(C)$, $\hat{b}(D)$:

$$[\hat{a}(\mu), \hat{a}^\dagger(\mu')] = \hat{1}\delta_{\mu\mu'}, \tag{3.214}$$

$$[\hat{b}(\nu), \hat{b}^\dagger(\nu')] = \hat{1}\delta_{\nu\nu'}, \tag{3.215}$$

$$\hat{b}(\nu) = X_{\nu A}\hat{a}(A) + X_{\nu B}\hat{a}(B), \tag{3.216}$$

$$\hat{b}^\dagger(\nu) = X^*_{\nu 1}\hat{a}^\dagger(A) + X^*_{\nu B}\hat{a}^\dagger(B). \tag{3.217}$$

The transformation matrix $X_{\mu\nu}$ is constrained; it can be written as

$$\mathcal{X} = \begin{pmatrix} t & r \\ r^* & -t^* \end{pmatrix}. \tag{3.218}$$

with $|t|^2 + |r|^2 \equiv T + R = 1$, implying loss-less behavior. With the proper definition of mode indices, see Fig. 3.4, T is the transmission, R the reflection coefficient; these are characteristic material properties of the device. The inverse transformation is

$$\mathcal{X}^{-1} = \begin{pmatrix} t^* & r \\ r^* & -t \end{pmatrix}. \tag{3.219}$$

Example 3.16 Entangler *A single photon state in input channel A, say, $|\Psi >= |1(A), 0(B) >= \hat{a}^\dagger(A)|0>$ would thus be transformed into the entangled state $|\Psi >= t|1(C)0(D) > +r^*|0(C)1(D) >$ in the space of the output modes C,D. The device acts as an "entangler."*

Mathematically the beam splitter transformation is merely a special (parameter-dependent) operator transformation. The transformation might thus be applied in physical scenarios other than the original setting in terms of a beam splitter, cf. Section 8.3.2.

Definition 3.67 Quantum conditional entropy

$$\boxed{S_{vN}(A|B) \equiv S_{vN}(A, B) - S_{vN}(B)\,.} \tag{3.220}$$

(For the classical counterpart cf. Eq. (2.52)). As exemplified by the Bell states, this conditional entropy can become negative.

In the quantum domain, a more direct definition has been suggested based on the so-called conditional quantum states; the latter result from local measurements on bipartite systems, see Definition 3.92.

Proposition 3.6 Embedding theorem (purification) *Any mixed state $\rho(A)$ can be considered a reduced state of an appropriately chosen pure composite state [Nielsen (2000); Hughston (1993)].*

Proof. Let

$$\hat{\rho}(A) = \sum_i \rho_{ii}(A)|i(A)><i(A)|\,. \tag{3.221}$$

We then introduce a supplementary system (B) with the same state space as (A) and a basis $|k(B)\;>$. Taking the pure state for the combined system as

$$|\psi(A, B)>= \sum_j \sqrt{\rho_{jj}(A)}|j(A), j(B)>\,, \tag{3.222}$$

then its local trace immediately confirms the proposition:

$$\mathtt{Tr}_B\{|\psi(A, B)><\psi(A, B)|\} = \sum_{ij} \sqrt{\rho_{ii}(A)\rho_{jj}(A)}|i(A)><j(A)|$$

$$\mathtt{Tr}_B\{|i(B)><j(B)|\} = \sum_i \rho_{ii}(A)|i(A)><i(A)|\,. \tag{3.223}$$

□

The embedding theorem motivates the following remark:

Remark 3.6 State of an isolated system *From knowing some density operator it is undecidable whether its nonpurity is due to subjective ignorance ("extrinsic") or resulting from an embedding into a larger quantum system ("intrinsic").*

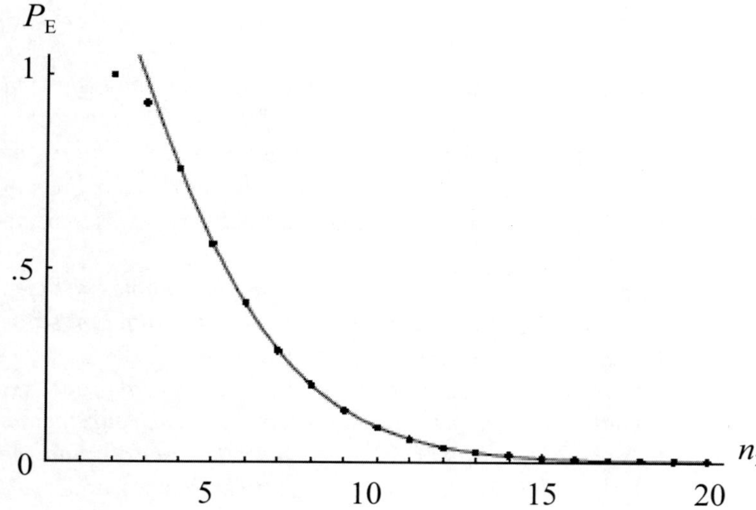

Figure 3.5 Probability P_E for composite subsystem (AB) with $d_A = d_B = 2$ to be entangled when embedded into subsystem C with $d_C = n_2$. P_E is calculated based on the PPT criterion, cf. Definition 3.93. Solid line: exponential fit.

For an isolated system, it makes no sense to talk about extrinsic aspects; as there is nothing left to embed the system in, also the intrinsic origin of nonpurity cannot apply. It thus appears reasonable to tentatively assume that an isolated system should be in a pure state.

Remark 3.7 Competition for entanglement *Bipartite spaces are certainly not the only ones of interest. So what should we expect, if a given bipartite system (AB) is embedded into another system (C)? We know that entanglement measures are becoming more involved, in particular, there are measures for the entanglement between (A) and (B) and (AB) and (C). Which embedding wins the competition?*

One should expect that (AB) becomes mixed if (C) acts as a large environment. But in this case, loss of coherence also means loss of internal entanglement between (A) and (B). This can be demonstrated to hold at least for $d_A = d_B = 2 \ll d_C$. (As the subsystem (AB) is no longer pure, its entanglement with (C) is needed to generate the local entropy.) Pertinent results [Otte (2001)] are shown in Fig. 3.5.

3.3.6 *Indistinguishability*

Remark 3.8 How to count product states? *The system is defined by its Hamiltonian, the Hamiltonian allows for a set of eigenstates, and any linear superposition thereof describes a possible state. For composite systems, product states built from the local eigenstates of the respective subsystems are a convenient basis; there seems to be no fundamental problem.*

But there is, if the system consists of indistinguishable particles. Because then one has to clarify what should count as a different state. And whether there could be further restrictions.

Particles can be distinguished by their intrinsic physical properties only (mass, charge, spin, etc.). Particles with the same such parameters are thus to be considered identical—if for no other reason than the operational impossibility to tell apart one particle from the other.

Example 3.17 Classical indistinguishability *Consider an ensemble of an equal number of white and black balls (i.e., each ball has one of the $g = 2$ different properties). The probability p to draw one black(b) or one white(w) ball is 1/2.*

We now proceed to a pair of independent drawings; it is assumed that after each drawing the ball has been put back. For $N = 2$ balls, there are $g^N = 4$ states of equal probability $p = 1/4$: bb, bw, wb, ww.

In order to distinguish bw and wb, we have to conceptually mark the balls: the first ball is black (b) and the second ball is white (w) or vice versa. We may postulate that these two states are indistinguishable, so we have just a single state of unequal color. But what is the probability to get this state, is it $p = 1/2$, because we now have two realizations out of four? In the classical domain (and in the quantum case for distinguishable particles), we would conclude just this. In the quantum domain of indistinguishable particles, this no longer holds. What are the correct probabilities, assuming equipartition among all the accessible states?

Definition 3.68 Corrected Boltzmann statistics *A first answer could be based on combinatorics: We correct the above number of states for permutations. If there are N balls, there are $N!$ permutations (if the particles are not in the same state). This procedure leads* *to the so-called corrected Boltzmann statistics: The number of indistinguishable states is*

$$\boxed{n_{CB} = g^N/N!,} \tag{3.224}$$

an approximation valid if the probability to find two or more particles in the same state is negligible. (This is certainly not a good approximation for the above example, where the probability to find two particles in the same state is $p = 1/2$*. The situation rapidly improves if we increase g for fixed N.)*

The corrected state counting has consequences in thermodynamics: If two gases of different types are mixed, the result is an increase of entropy, known as the mixing entropy. The so-called Gibbs paradox results, if two gases of the same type are "mixed," apparently leading to the same mixing entropy—even though there is clearly no mixing at all. This paradox can be removed based on the corrected Boltzmann counting. The necessity for this correction underlines the importance of quantum mechanical concepts (here: indistinguishability) already in the macroscopic domain.

But the above counting correction is not the complete story yet—and this is where full quantum mechanics enters the scene.

Definition 3.69 Hilbert space for indistinguishable particles
In quantum mechanics, indistinguishability has a much deeper operational meaning. Not only that we post festum *correct the number of states, we have to make sure that there is not a single observable (including the Hamiltonian to define the system under consideration), which would allow to single out (address) specific particles: Everything has to be invariant under particle permutations! Mathematically, this means that any observable has to commute with all permutation operators. But this very requirement defines a superselection rule (cf. Section 3.2.9).*

A technical problem with the set of permutation operators has far reaching consequences, though: As the permutation operators do not all commute among themselves, there is an, in principle, infinite number of symmetry classes (sectors). Fortunately, only the two most basic of them turn out to be physically relevant, the Fermi–Dirac (FD) and the Bose–Einstein (BE) statistics. The space for Fermions (half integer spin), $\mathcal{H}$*(antisymmetrical), contains only states that* *reproduce with a minus sign under each particle permutation. The space for Bosons (integer spin), $\mathcal{H}(symmetrical)$ contains only states that reproduce under particle permutation.*

The respective number of distinguishable states differ from the corrected Boltzmann value, n_{CB}: One finds

$$\boxed{\begin{aligned} n_{FD} &= g!/(N!(g-N)!)\,, \\ n_{BE} &= (g+N-1)!/(N!(g-1)!)\,, \end{aligned}} \tag{3.225}$$

respectively. In the so-called classical limit (large energy, small particle density), these differences disappear.

Definition 3.70 Decomposition of product Hilbert space *The full set of partitionings for distinguishable subsystems are no longer allowed. Indistinguishability implies serious constraints for the acceptable product states. In fact, the total Hilbert space considered so far splits into three sectors,*

$$\mathcal{H} = \mathcal{H}(symmetrical) + \mathcal{H}(antisymmetrical) + \mathcal{H}(rest)\,, \tag{3.226}$$

where $\mathcal{H}(rest)$ is physically unrealizable, the first two sectors are mutually exclusive (for Bosons, Fermions, respectively).

Example 3.18 Spin pair *For a composite system of $N = 2$ two-level systems ($g = 2$) with states $m = \pm 1$, the subspace $\mathcal{H}$(symmetrical) is n_{BE}-dimensional; it contains the three states*

$$|1(A), 1(B)>;$$
$$|-1(A), -1(B)>;$$
$$1/\sqrt{2}(|1(A), -1(B)> + |-1(A), 1(B)>).$$

The sub-space $\mathcal{H}$(antisymmetrical) is n_{FD}-dimensional; it contains the single state

$$1/\sqrt{2}(|1(A), -1(B)> - |-1(A), 1(B)>).$$

This is the EPR state, Eq. (3.200). $\mathcal{H}(rest)$ is empty in this case.

Definition 3.71 Operational distinguishability *While specific subsystems* per se *cannot be singled out under the condition of indistinguishability, individual states or modes may, nevertheless, be* *addressable via appropriate experimental design. This is the basis for operational distinguishability. Here are two examples:*

(i) *We may prepare a single photon pair (Bosons) to be emitted from a localized source with zero total momentum. By design one photon is going to the right, and the other going to the left. Right/left becomes the "name" of the respective photon. Correlations can thus be tested via local measurements (detector left/right). This possibility has been exploited in the first experimental investigation of the Bell's inequality [Aspect (1981)].*

(ii) *In a spin lattice, the spins (while indistinguishable to begin with) become operationally addressable by their position in real space. Excluding spin motion between the sites, there is one and only one spin at a given set of discrete positions,* $\vec{r}_j$. *The positions serve as a name for the spins. A similar example is the localization of adatoms, say on a material surface. The atoms become distinguishable by their position as long as any energy, which happens to be available in the given experimental situation, does not suffice to induce position changes.*

Nominally indistinguishable particles can thus be made distinguishable if certain properties can be treated as "frozen-in" parameters (i.e., classical).

3.3.7 *Quantum Dynamics*

For isolated systems the respective Hamiltonian specifies the complete repertoire of possible motions—just like in classical Hamiltonian dynamics. The operator $\hat{H}$ is thus said to be the "generator" of the so-called eigendynamics with respect to parameter time t.[a]

There are two basic and equivalent pictures, the Schrödinger picture and the Heisenberg picture, cf. Section 3.2.7. In the former, the states move relative to fixed "properties" defined by the

[a]The parameter time has to be related to an external clock time. Whether or not such a time may express a fundamental aspect of nature is under debate [Smolin (2001)]. Cf. also Section 5.2.1.

respective operators; in the latter, the properties (operators) move relative to the fixed states.[a]

Definition 3.72 Schrödinger equation *In the Schrödinger picture, the temporal evolution is carried by the state*

$$\boxed{i\hbar\frac{d}{dt}|\psi(t)>= \hat{H}(t)|\psi(t)> \ .} \tag{3.227}$$

Here we have allowed for an explicit time dependence of $\hat{H}$. Such a time dependence renders the system nonautonomous, energy is not conserved. It is convenient to introduce the time evolution operator $\hat{S}$ by

$$|\psi(t)>= \hat{S}(t, t_0)|\psi(t_0)> \ . \tag{3.228}$$

When substituted into the Schrödinger equation, one finds

$$i\hbar\frac{d}{dt}\hat{S}(t, t_0) = \hat{H}(t)\hat{S}(t, t_0) \quad \text{or} \tag{3.229}$$

$$-i\hbar\frac{d}{dt}\hat{S}^\dagger(t, t_0) = (\hat{H}(t)\hat{S}(t, t_0))^\dagger = \hat{S}^\dagger(t, t_0)\hat{H}^\dagger(t) = \hat{S}^\dagger(t, t_0)\hat{H}(t) \, . \tag{3.230}$$

In the last step, we have used the fact that $\hat{H}(t)$ is a self-adjoint operator cf. Definition 3.14.

Proposition 3.7 Unitarity *The time evolution operator generates a unitary transformation, that is,* $\hat{S}(t, t_0)^\dagger\hat{S}(t, t_0) = \hat{1}$, *cf. Definition 3.19.*

Proof. From the constancy of the norm, it follows that

$$\begin{aligned} 0 &= <\psi(t)|\psi(t)> - <\psi(t_0)|\psi(t_0)> \\ &= <\hat{S}(t, t_0)\psi(t_0)|\hat{S}(t, t_0)\psi(t_0)> - <\psi(t_0)|\psi(t_0)> \\ &= <\psi(t_0)|(\hat{S}^\dagger(t, t_0)\hat{S}(t, t_0) - \hat{1})\psi(t_0)> \ . \end{aligned} \tag{3.231}$$

As this must hold for any state and any time, the proposition follows. □

[a]It is not possible to determine the motion of states relative to observables or vice versa; this relative motion is operationally not accessible, cf. Section 3.4.1. We are restricted to measure observables and by that update states.

Remark 3.9 Deterministic evolution *The Schrödinger equation is deterministic. In particular, we can return from state* $|\psi(t) >$ *back to the initial state by means of the inverse transformation* $\hat{S}(t, t_0)^{-1} = \hat{S}(t, t_0)^{\dagger}$.

Proposition 3.8 Time evolution operator for conservative systems *For conservative systems,* $(\partial/\partial t)\hat{H} = 0$*, one easily solves Eq. (3.229) by*

$$\hat{S}(t, t_0) = \exp\left(-i\hat{H}(t - t_0)/\hbar\right). \tag{3.232}$$

Remark 3.10 Linearity *The Schrödinger equation is linear in the state vector. There have been speculations that this should be approximate only. However, there are no experiments whatsoever calling for nonlinear corrections. And it turns out that their inclusion would necessarily come in conflict with other desirable properties or even fundamental constraints [Gisin (1990); Abrams (1998)].*

Such nonlinear extensions should be confused neither with effective Schrödinger equations accounting for self-interaction (cf. the Gross–Pitaevskii equation [Leggett (2001)]) nor with the so-called nonlinear terms in the representation of the respective Hamiltonian. A pertinent example for the latter are the anharmonic terms within phonon models.

Proposition 3.9 Eigenstates as stationary states *All eigenstates of an autonomous Hamilton system are stationary (modulo phase factor).*

Proof. Consider the eigenvalue equation

$$\hat{H}\,|e^{(n)} > E_n|e^{(n)} > . \tag{3.233}$$

Applying the time evolution operator with $t_0 = 0$ to such eigenstates immediately leads to

$$\hat{S}(t)|e^{(n)} >= \exp\left(-iE_n t/\hbar\right)|e^{(n)} >\equiv |e^{(n)}(t) > . \tag{3.234}$$

The phase factor is not measurable. Its absolute value has no meaning as the absolute value of energy has no meaning: the zero point can be shifted at will. However, relative phases, which appear in superpositions of such stationary states for different energies, do have physical significance. □

Proposition 3.10 Nonstationary states *Based on Eq. (3.228), the general time-dependent state in energy representation reads for an autonomous system*

$$|\psi(t)>=\sum_n c_n \exp(-iE_n t/\hbar)|e^{(n)}> . \tag{3.235}$$

Here, $c_n =< e^{(n)}|\psi(0) >$ *are complex amplitudes, which act as initial conditions. When dealing with dynamics, coherence becomes unavoidable: It cannot be transformed away by choosing a specific representation.*

Example 3.19 Coherent oscillations of a two-level system *Consider the two lowest energy levels of a double well potential. The Hamiltonian can be written in the local basis (left* $=|1>$*/right* $=|2>$*) as*

$$\hat{H} = \sum_{j,k=1}^{2} H_{jk}|e^{(j)} > < e^{(k)}| . \tag{3.236}$$

Assuming $H_{11} = H_{22} = E_0$ *and* $H_{12} = H_{21} = K$ *we find the eigenvalues and eigenvectors*

$$|+> = \frac{1}{\sqrt{2}}(|1> + |2>) \quad E^+ = E_0 + |K| , \tag{3.237}$$

$$|-> = \frac{1}{\sqrt{2}}(|1> - |2>) \quad E^- = E_0 - |K| . \tag{3.238}$$

The general time-dependent solution then is

$$|\psi(t)> = \exp(-iE_0 t/\hbar)(|+> \exp(-i|K|t/\hbar) < +|\psi(0)> + \\ |-> \exp(i|K|t/\hbar) < -|\psi(0)>) . \tag{3.239}$$

Choosing the initial state

$$|\psi(0)>= \cos\theta|1> + \sin\theta|2> \tag{3.240}$$

one readily shows that

$$| < \psi(0)|\psi(t)> |^2 = 1 + (\sin^2(2\theta) - 1)\sin^2(|K|t/\hbar) . \tag{3.241}$$

For $\theta = 0$ *we have* $|\psi(0)>= |1>$ *and* $| < 1|\psi(t)> |^2 = \cos^2(|K|t/\hbar)$*. For* $\theta = \pi/4$*, we have* $|\psi(0)>= |+>$ *and* $| < +|\psi(t)> |^2 = 1$*, which, is stationary.*

Remark 3.11 Entanglement oscillations *The above model also directly applies to two identical spins, if the two participating states* *(out of the four-dimensional Hilbert space) are taken to be* $|1> \rightarrow |1(A)2(B)>$ *and* $|2> \rightarrow |2(A)1(B)>$. *For* $\theta = 0$ *and* $K \neq 0$, *an energy transfer oscillation results connected with entanglement oscillations.*

Local coherence and entanglement are relative (representation dependent) concepts. For time-independent states, one can thus always get rid of those features. Obviously, this is no longer possible under time evolution—if described with respect to a fixed reference frame.

Let us come back to nonautonomous systems:

Example 3.20 Driven system *A driven two-level system can be described by the time-dependent Hamiltonian*

$$\hat{H}(t) = \hat{H}_0 + \hat{V}(t), \tag{3.242}$$

$$\hat{H}_0 = E_1 \hat{P}_{11} + E_2 \hat{P}_{22}, \tag{3.243}$$

$$\hat{V}(t) = \frac{g\hbar}{2}(\hat{P}_{12} \exp(-i\omega t) + \hat{P}_{21} \exp(i\omega t)). \tag{3.244}$$

Here, g is the coupling constant; the interaction is written in the so-called rotating wave approximation (RWA), in which nonresonant terms are neglected. By means of the time-dependent unitary transformation [Mahler (1998)],

$$\hat{U}(t) = \exp(i\omega t \hat{P}_{11}), \tag{3.245}$$

$$\hat{H}' = \hat{U}^{\dagger} \hat{H}(t) \hat{U}, \tag{3.246}$$

$$|\psi'> = \hat{U}|\psi>, \tag{3.247}$$

one easily transforms into the effective time-dependent Schrödinger equation

$$i\hbar \frac{d}{dt}|\psi'> = \hat{H}_{eff}|\psi'>, \tag{3.248}$$

$$\hat{H}_{eff} = (E_1 + \hbar\omega)\hat{P}_{11} + E_2 \hat{P}_{22} + \frac{\hbar g}{2}(\hat{P}_{12} + \hat{P}_{21}). \tag{3.249}$$

Note that $\hat{H}_{eff}$ *is now time independent. We are in a "rotating frame."*

Proposition 3.11 Discretization of dynamics *Consider the time-dependent Hamiltonian with*

$$\hat{H}(t) = \hat{H}_j \quad \text{for} \quad t_j \leq t \leq t_j + \Delta t = t_{j+1}, \tag{3.250}$$

and $j = 1, 2, \ldots n$ and $\Delta t = t/n$. For this piecewise constant Hamiltonian, it follows that

$$\hat{H}(t)t = \Delta t \sum_{j=1}^{n} \hat{H}_j , \tag{3.251}$$

and the time evolution operator becomes

$$\hat{S}(t) = \exp\left(-i\Delta t \sum_{j=1}^{n} \hat{H}_j/\hbar\right) . \tag{3.252}$$

Based on the Suzuki–Trotter decomposition [Hatano (2005)],

$$\boxed{\exp(\delta(\hat{A} + \hat{B})) = \lim_{\delta \to 0}[\exp(\delta\hat{A})\exp(\delta\hat{B}) + O(\delta^2)] ,} \tag{3.253}$$

we thus find for $n \to \infty$ (and t finite)

$$\hat{S}(t) \approx \prod_j \hat{S}^{(j)}(\Delta t) , \tag{3.254}$$

$$\hat{S}^{(j)}(\Delta t) = \exp(-i\Delta t\hat{H}_j/\hbar) . \tag{3.255}$$

For a time-independent Hamiltonian, such a decomposition is exact for any n.

Rather than decomposing a given evolution into a series of transformations one may look instead at individual transformations with specific properties.

Definition 3.73 Quantum gate *We restrict ourselves to the special piecewise constant Hamiltonian with*

$$\hat{H}(t) = \begin{cases} \hat{H} \ \text{for } 0 \leq t \leq \Delta t \\ 0 \ \text{for } t \text{ otherwise} \end{cases} . \tag{3.256}$$

The respective time evolution operator is a "one-shot" (unitary) transformation. In technical parlance, this constitutes a quantum gate. Combinations of such quantum gates can be used to transform states in a prescribed manner, as is required, for example, in quantum information processing. Time does no longer show up explicitly, except as a sequential order.

Example 3.21 Pauli gate *A Hamilton operator $\hat{H} \sim \hat{\sigma}_j$, $j = 0, 1, 2, 3$ acting for some finite time period, Δt, generates the local gate:*

$$\hat{S}^{(j)}(\alpha_j) = \exp(i\alpha_j\hat{\sigma}_j) = \hat{1}\cos\alpha_j + i\hat{\sigma}_j \sin\alpha_j . \tag{3.257}$$

This identity follows by comparing the respective expansions for the exponential and the sine and cosine functions, combined with the properties $\hat{\sigma}_j^{2n} = \hat{1}$, $n = 1, 2, \ldots$. Here, the phase α_j is proportional to the time period chosen. The gate operates locally on a single spin; it can be represented as a 2×2-matrix.

A convenient way to characterize the action of a gate is by means of input-output tables. Such a table would read for the Pauli gate:

$$\hat{S}^{(1)}(\alpha_j)|\psi_{in}> = |\psi_{out}>, \tag{3.258}$$

$$|e^{(1)}> \to \cos\alpha_1|e^{(1)}> + i\sin\alpha_1|e^{(2)}>, \tag{3.259}$$

$$|e^{(2)}> \to i\sin\alpha_1|e^{(1)}> + \cos\alpha_1|e^{(2)}>. \tag{3.260}$$

We use such tables now for 2-qubit gates:

Example 3.22 SWAP gate *For a spin pair A, B, a SWAP gate performs a cyclical permutation of states,*

$$|e^{(2)}, e^{(2)}> \to |e^{(2)}, e^{(2)}>, \tag{3.261}$$

$$|e^{(2)}, e^{(1)}> \to |e^{(1)}, e^{(2)}>, \tag{3.262}$$

$$|e^{(1)}, e^{(2)}> \to |e^{(2)}, e^{(1)}>, \tag{3.263}$$

$$|e^{(1)}, e^{(1)}> \to |e^{(1)}, e^{(1)}>. \tag{3.264}$$

Here we have used the notation $|e^{(1)} >= |m = 1>$; $|e^{(2)} >= |m = -1>$ for both subsystems. The gate may be implemented by an appropriately chosen two-spin Hamiltonian. It can easily be represented as a 4×4 matrix (in the product space representation).

Example 3.23 CNOT-gate *This is another 2-qubit gate. The function of controlled NOT is to flip qubit B, if qubit A (the "control") is in state 2, and do nothing otherwise,*

$$|e^{(2)}, e^{(2)}> \to |e^{(2)}, e^{(1)}>, \tag{3.265}$$

$$|e^{(2)}, e^{(1)}> \to |e^{(2)}, e^{(2)}>, \tag{3.266}$$

$$|e^{(1)}, e^{(2)}> \to |e^{(1)}, e^{(2)}>, \tag{3.267}$$

$$|e^{(1)}, e^{(1)}> \to |e^{(1)}, e^{(1)}>. \tag{3.268}$$

Again, this gate can be represented as a 4×4 matrix.

Proposition 3.12 No cloning *A standard task in information processing is to copy a bit of information, that is, an unknown state $|e^{(X)}>$, $X = 1, 2$. This operation can conveniently be implemented* *using the CNOT gate with a fixed input $|e^{(1)}>$ for qubit B; the state to be copied enters control qubit A. The copy is then found as the output of qubit B, cf. Eq. (3.266) and Eq. (3.268). However, this does no longer work, if the state to be copied is a general superposition state: [Nielsen (2000)] This is so, because with*

$$a|e^{(1)}> +b|e^{(2)}> \otimes |e^{(1)}> \rightarrow a|e^{(1)}> \otimes |e^{(1)}> +b|e^{(2)}> \otimes |e^{(2)}> \tag{3.269}$$

the output state turns out to be an entangled state, not the desired product of two identical superposition states in A and B. General (unknown) quantum states cannot be "cloned."

A set of mixed states can be cloned only if they are mutually orthogonal [Barnum (1996)]. We can measure the eigenstate of a beam of identically prepared particles or photons. This theorem prevents us from knowing the eigenstate also of a single quantum object, because copying is not allowed: The latter problem cannot be mapped onto the former.

Remark 3.12 Quantum circuits *A quantum circuit is a decomposition of a unitary matrix into a sequence of single and pair-wise quantum logic gates. Such a decomposition underlies any quantum computing or quantum simulation process [Nielsen (2000)]. A simple example is discussed in Section 6.5. In general, extensive error corrections are mandatory; they often imply a prohibitively large overhead.*

The Schrödinger equation can be generalized to deal with density operators as the most general states.

Proposition 3.13 Liouville–von Neumann equation

$$\boxed{i\hbar \frac{d}{dt}\hat{\rho}(t) = [\hat{H}(t), \hat{\rho}(t)] .} \tag{3.270}$$

Proof. In the following, we suppress the parameter $t_0 \equiv 0$.

$$\hat{\rho}(t) = \hat{S}(t)\hat{\rho}\hat{S}^{\dagger}(t) , \tag{3.271}$$

$$\begin{aligned} \hat{\rho}(t) &= \sum_{n,m} |e^{(n)}(t)> \rho_{nm} < e^{(m)}(t)| , \\ &= \sum_{n,m} \hat{S}(t)|e^{(n)}> \rho_{nm} < e^{(m)}|\hat{S}^{\dagger}(t) . \end{aligned} \tag{3.272}$$

From this expression, the time derivative is

$$i\hbar\frac{d}{dt}\hat{\rho} = \sum_{n,m}(i\hbar\frac{d}{dt}\hat{S}(t))|e^{(n)} > \rho_{nm} < e^{(m)}|\hat{S}^{\dagger}(t)$$
$$+\hat{S}(t)|e^{(n)} > \rho_{nm} < e^{(m)}|(i\hbar\frac{d}{dt}\hat{S}^{\dagger}(t)) . \qquad (3.273)$$

According to Eqs. (3.229, 3.230), the first expression in parenthesis is $\hat{H}(t)\hat{S}(t)$, and the last expression in parenthesis is $-\hat{S}^{\dagger}(t)\hat{H}(t)$. We thus have

$$i\hbar\frac{d}{dt}\hat{\rho} = \sum_{n,m}\hat{H}(t)|e^{(n)}(t) > \rho_{nm} < e^{(m)}(t)|$$
$$-\sum_{n,m}|e^{(n)}(t) > \rho_{nm} < e^{(m)}(t)|\hat{H}(t) . \qquad (3.274)$$

The right-hand side of the last equation can be written as the commutator, as stated in the proposition. □

Proposition 3.14 Stationary mixed states *For autonomous Hamiltonians (i.e., no explicit time dependence), the state $\hat{\rho}$ is stationary, if it commutes with $\hat{H}$. This means that both operators must have the same eigenstates. A case in point are not only the pure energy eigenstates, cf. Proposition 3.9, but also any mixed state, which is diagonal in that basis.*

Definition 3.74 Heisenberg operators *In the Heisenberg picture, the evolution is carried by the operators (observables), similar to the situation in classical mechanics. The two pictures have to be equivalent in the sense that they produce the same time-dependent expectation values:*

$$< F > (t) = \mathrm{Tr}\{\hat{\rho}(t)\hat{F}\} = \mathrm{Tr}\{\hat{S}(t)\hat{\rho}(t_0)\hat{S}^{\dagger}(t)\hat{F}\}$$
$$= \mathrm{Tr}\{\hat{\rho}(t_0)\hat{S}^{\dagger}(t)\hat{F}\hat{S}(t)\} = \mathrm{Tr}\{\hat{\rho}(t_0)\hat{F}_H(t)\} , \qquad (3.275)$$

where we have defined the Heisenberg operator

$$\hat{F}_H(t) = \hat{S}^{\dagger}(t)\hat{F}\hat{S}(t) . \qquad (3.276)$$

Note the difference to the respective transformation of the density operator, Eq. (3.271), in the Schrödinger picture! Here the state is time independent, that is, $\hat{\rho}_H = \hat{\rho}(t_0)$.

Proposition 3.15 Heisenberg equation of motion *The equation of motion for Heisenberg operators reads:*

$$i\hbar\frac{d}{dt}\hat{F}_H(t) = -[\hat{H}(t), \hat{F}_H(t)] + \left(i\hbar\frac{\partial \hat{F}_H}{\partial t}\right)_{expl} . \tag{3.277}$$

This equation might be compared with Definition Eq. (3.45), the classical observable dynamics.

Proof.

$$i\hbar\frac{d}{dt}(\hat{S}^\dagger\hat{F}\hat{S}) = \left(i\hbar\frac{d\hat{S}^\dagger}{dt}\right)\hat{F}\hat{S} + \hat{S}^\dagger\hat{F}\left(i\hbar\frac{d\hat{S}}{dt}\right) + \hat{S}^\dagger\left(i\hbar\frac{\partial\hat{F}}{\partial t}\right)\hat{S}$$

$$= -\hat{H}\hat{S}^\dagger\hat{F}\hat{S} + \hat{S}^\dagger\hat{F}\hat{S}\hat{H} + \left(i\hbar\frac{\partial \hat{F}_H}{\partial t}\right)_{expl} . \tag{3.278}$$

In the last line, we have applied Eqs. (3.229, 3.230). The first two terms then combine with the commutator, as suggested in the proposition. The last term only contributes if the observable $\hat{F}$ is explicitly time dependent, that is, already in the Schrödinger picture. Such an explicit time dependence would render the system nonautonomous (perturbation from the outside). □

Schrödinger and Heisenberg pictures are compared in Fig. 3.6 based on the vectorization of state and observable, cf. Definition 3.13.

Definition 3.75 Ehrenfest theorem *From the Heisenberg equation of motion, Eq. (3.277), it follows that*

$$\frac{d}{dt} < F > = \frac{i}{\hbar} < [\hat{H}, \hat{F}] > + < \frac{\partial\hat{F}}{\partial t} >_{expl} . \tag{3.279}$$

For the single-particle model, $\hat{H} = \hat{p}^2/(2m) + \hat{V}(\hat{x})$, one thus obtains [Ballentine (1998)]

$$m\frac{d}{dt} < x > = < p > , \tag{3.280}$$

$$\frac{d}{dt} < p > = - < dV/dx > . \tag{3.281}$$

A certain form of classical limit is obtained if $< dV/dx > \approx dV(< x >)d < x >$. Then the equations of the expectation values have the form of classical canonical equations.

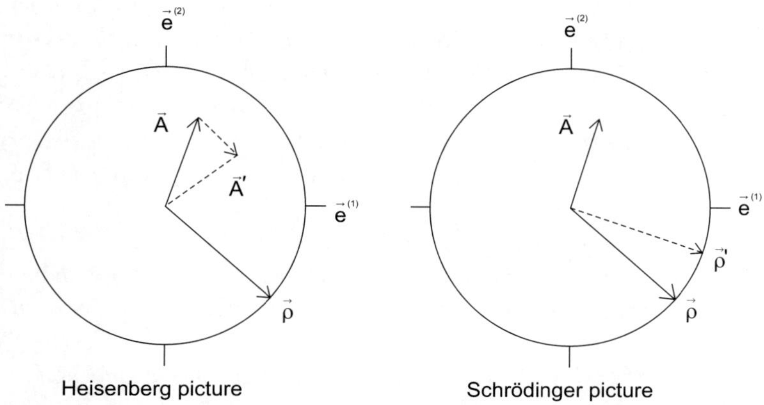

Figure 3.6 Heisenberg versus Schrödinger dynamics. The observable $\hat{A}$ is assumed to be explicitly time independent. To support intuition, state and observable are vectorized.

Unfortunately, the above condition of classicality is not sufficient: While it is always fulfilled, for example, for the harmonic oscillator, this model can nevertheless show nonclassical behavior; its discrete, stationary eigenstates are pertinent examples.

Definition 3.76 Two-time correlation function *Let us consider two observables, $\hat{F}_i(t)$, $\hat{F}_j(t')$, in the Heisenberg picture; the respective state, $\hat{\rho}(0)$, is then time independent. The expectation value*

$$G_{ij}(t, t') = \mathrm{Tr}\{\hat{F}_i(t)\hat{F}_j(t')\hat{\rho}(0)\} \tag{3.282}$$

parametrized by the two times t, t' characterizes the temporal correlation of the two observables. For $i = j$, this function describes the respective autocorrelation.

If for $\tau = t' - t \gg \tau_c$, such a correlation approaches zero, τ_c is called correlation time. For times $t' > t + \tau_c$, the present value of $\hat{F}_i(t)$ looses any predictive power about its future value. The correlation time allows to quantify temporal coherence, which tells us, how monochromatic a wave is.

Proposition 3.16 Conservation laws *Whenever there is change, it is also interesting to identify constants of motion. These can be of different types:*

(i) *Unconditional constants of motion: Examples are the von Neumann entropy* S_{vN} *(see Proposition 3.4), the distance between states, D, see Definition 3.44, and the trace expressions,* $C(d, q)$, $q = 1, 2, \ldots d$, *Eq. (3.145), including the purity P. All these terms are conserved, because they are invariant under any unitary transformation, for which the time evolution is but a special class.*[a]

(ii) *Conditional constants of motion: From the Heisenberg equation, it immediately follows that all those operators* $\hat{F}$ *do not change, which are not explicitly time dependent, and which commute with* $\hat{H}$.

(iii) *From the Liouville–von Neumann equation, one concludes that a state* $\hat{\rho}$ *is stationary, if it commutes with* $\hat{H}$.

3.3.8 *Classical Limit*

It is generally believed that quantum mechanics should contain classical mechanics in some appropriate limit ("correspondence principle"). Nevertheless, there is no universal classical limit of quantum mechanics [Primas (2007)].

Various criteria have been suggested to identify such classical fingerprints.

(1) Large length scales: System size, L, is large compared with the thermal de Broglie wavelength, λ_{dB}, see Definition 4.11. Related to length scales is the conception of coarse graining, cf. Remark 5.7.
(2) Large action: Characteristic variables of the system combined to produce the dimension "action" are in size large compared with the Planck constant $\hbar$, cf. Definition 3.27.
(3) Limit of large particle numbers: It has been shown in Ref. [Lieb (1973)] that under some mild conditions the one-particle distribution function of a system of N spin 1/2-particles (electrons) approaches the classical Thomas–Fermi result.

[a]According to the Noether theorem, any conservation law is connected with a (continuous) symmetry. For the examples above, the underlying symmetry is related to time inversion, cf. Remark 3.9.

(4) Limit of a small uncertainty product: Confer the coherent states for the harmonic oscillator according to Definition 3.85. Measured values for momentum and position are both as sharp as possible.
(5) Limit of large quantum numbers: High energy states, $|e^{(n)}>$, of an harmonic oscillator have a probability density in real space, $p_n(x)$, similar to a classical oscillator.
(6) Ehrenfest limit, cf. Definition 3.75: Expectation values should satisfy the classical equation of motion.
(7) An oscillator state is said to be classical, if its Glauber–Sudarshan representation, Eq. (3.99), features a classical probability density.
(8) $\hbar \to 0$, a rather formal "trick"; in fact, $\hbar$ is not dimensionless and a constant of nature.
(9) Suppression of entanglement: Any bi-partite state reduces to a unique tensor product, if and only if one of its subsystems is classical (see [Janotta (2011); Barnum (2007)]). A classical system cannot be entangled.
(10) Small density limit for many-particle system: Quantum statistics approaching "classical statistics" (Boltzmann corrected, see Definition 3.68).
(11) Distinguishability of quantum states: Pure quantum states that cannot uniquely be discriminated are eigenstates of noncommuting observables, cf. Section 3.2.1. They are nonorthogonal and expressed as superpositions. Any classical state as a point in Γ space is pure, cf. Definition 3.53; a system may thus be termed classical, if all its pure states are uniquely distinguishable.[a]

The conditions cited are neither necessary nor sufficient, they lack generality.

The "classical observables" associated with superselection rules (cf. Section 3.2.9), though not related with a classical limit in a strict sense, also provide a strong case for classical fingerprints.

[a] Strategies have been proposed to make even nonorthogonal states distinguishable. This is based on joint measurements on ensembles in a kind of macroscopic limit [Pusey (2012)].

A dynamical way to distinguish between classical and quantum regime can be based on the phase space formulation of quantum mechanics. We close with a brief outline of the main ideas.

Definition 3.77 Density operator in real space representation *We restrict ourselves to a single particle problem in one dimension. The eigenfunctions of* $\hat{x}$ *are* $|x>$ *and thus*

$$\rho(x_1, x_2) =< x_1|\hat{\rho}|x_2> . \tag{3.283}$$

For the following application, we introduce the new variables

$$r = \frac{x_1 + x_2}{2} \qquad s = x_1 - x_2 \tag{3.284}$$

and the new density matrix

$$\sigma(r, s) = \rho(r + s/2, r - s/2) . \tag{3.285}$$

Definition 3.78 Fourier transformation *The function* $\tilde{f}$ *is the Fourier transform of* f *and vice versa, if*

$$\tilde{f}(\xi) = \int_{-\infty}^{\infty} f(x) \exp(-2\pi i x\xi) dx , \tag{3.286}$$

$$f(x) = \int_{-\infty}^{\infty} \tilde{f}(\xi) \exp(2\pi i x\xi) d\xi . \tag{3.287}$$

Definition 3.79 Wigner distribution *The Wigner distribution is introduced as the Fourier transform of* $\sigma(r, s)$ *with respect to* s*, while keeping* r*: Identifying* $x = s$ *and*

$$\xi = p/(2\pi\hbar) \tag{3.288}$$

one gets

$$w(r, p) = \int ds\, \sigma(r, s) \exp(-ips/\hbar) , \tag{3.289}$$

$$\sigma(r, s) = \int \frac{dp}{2\pi\hbar} w(r, p) \exp(ips/\hbar) , \tag{3.290}$$

s *and* ξ *both have dimension of a length,* ps *has the dimension of an action. As* $\hat{\rho}$ *is self-adjoint,* $\rho(x_1, x_2) = \rho^*(x_2, x_1)$*, it follows that* $\sigma(r, s) = \sigma^*(r, -s)$ *and* $w(r, p)$ *is real. However, the Wigner distribution can become negative; it is merely a "quasi-probability distribution."*

In the same way, one can define for the operator $\hat{G}$ the Wigner transform $G_w(r, p)$ of the real space representation $G(x_1, x_2)$. This allows to rewrite the expectation value

$$< G >= \mathrm{Tr}\{\hat{G}\hat{\rho}\} = \int\int \frac{drdp}{2\pi\hbar} G_w(r, p) w(r, p) . \tag{3.291}$$

Proposition 3.17 Evolution equation for the Wigner function *Starting from the Liouville–von Neumann equation for $\hat{\rho}$ in real space representation one finds*

$$\frac{\partial w(r, p; t)}{\partial t} = \{H(r, p), w(r, p; t)\}_{MB} . \tag{3.292}$$

The right-hand side is the so-called Moyal bracket (MB). This bracket can be expanded to yield the Poisson bracket (cf. Eq. (3.45)) with quantum corrections [Zurek (1996)]

$$\{H(r, p), w(r, p; t)\}_{MB} = \{H(r, p), w(r, p; t)\} + \sum_n C_n \partial_r^{(2n+1)} V \partial_p^{(2n+1)} w(r, p) . \tag{3.293}$$

C_n are fixed coefficients. As long as these quantum corrections can be neglected, classical dynamics is obtained from the underlying basic quantum picture.

For an initial state starting as a Gaussian packet, which is smooth over scales much larger than $\hbar$, quantum corrections can safely be neglected.[a] However, this cannot last for ever: After some time $t_\hbar$, the state will look different from its classical counterpart. It will exhibit long-range superpositions in real space. Such a "correspondence-breakdown" will occur at

$$t_\hbar \sim \begin{cases} \left(\dfrac{A_0}{\hbar}\right)^\alpha & \text{if system integrable,} \\ \ln\left(\dfrac{A_0}{\hbar}\right) & \text{if system non-integrable.} \end{cases} \tag{3.294}$$

Here, A_0 denotes the respective action. This breakdown time can be surprisingly short in the case of nonintegrable (i.e., classically chaotic) systems—even in the macrodomain. For the moon Hyperion tumbling around Saturn, the time $t_\hbar$ has been estimated to be around 20 years [Zurek (1996)].

[a] If we started from a nonclassical state, the following arguments would not apply.

It has thus been argued that a proper classical limit should not be formulated in terms of isolated quantum objects. If an appropriate environment is included, effective "position measurements" are postulated to result in spontaneous localization and thus destroy those nonclassical features. Decoherence would be an essential ingredient of the classical limit [Zurek (1996)].

Remark 3.13 Hyperion as a thermal system? *A tension remains, though, decoherence also underlies thermal behavior as advocated by quantum thermodynamics. But as a thermalized subsystem Hyperion would not be localized, it would eventually relax (at a relaxation time of 20 years) into a stationary high entropy state with uncertain position. However, quantum thermodynamics constitutes a detached "outside view," there is no physical contact with the quantum system being described.*

Both pictures could be reconciled via observation: This leads to observational quantum thermodynamics, see Section 5.5. Indeed, the effective dynamics of Hyperion might then resemble a kind of "Zitterbewegung."

3.4 The Formalism II

So far, quantum mechanics has been discussed without reference to any means of observation. In that sense, the theory has been "abstract." Nevertheless, a "theoretician" had always been on duty: It has been up to him to select representations and to contemplate issues of distinguishability (of particles, subsystems). But he remained "physically detached": no interactions allowed.

Observational tools—and with it a fundamental partition between observed and observing subsystem[a]—have thus to be added. Such an interface underlies the so-called "agent-based" view of physics, particularly relevant in the quantum domain [Goyal (2012)], cf. Section 2.3.4. As a consequence, quantum mechanics attains a fundamental statistical flavor: There must be uncertainties about various observables even if there is no

[a] It is generally understood that the observing subsystem should be "localized," that is, carry an index that refers to a position in real space.

uncertainty about the state. Quantum measurement theories typically presuppose the coexistence of quantum objects with classical measurement devices, but this has been challenged.

Can one avoid the dualistic view expressed by the two alternative modes of description, the "abstract" quantum mechanics and the "observational" quantum mechanics? Not really, so it seems, but, at least, this tension can be softened, in quite opposite directions by

(i) accepting the asymmetry imposed by the quantum/classical dichotomy (as postulated already in the Copenhagen interpretation) and
(ii) removing this very asymmetry

A modern variant of the first option leads to the information-theoretical approach to quantum mechanics. In this operational approach, [Barnum (2007)] states are understood to specify all probabilities for measurement outcomes. The measurement process as such is not considered, neither is the origin of the necessarily classical measurement device. Nevertheless, this option has gained some popularity; it even offers to formalize general nonclassical models, which may behave "more quantum" than quantum models in a strict sense [Janotta (2011)].

The second option underlies the relational quantum mechanics, cf. Section 3.2.8.

3.4.1 *Observational Quantum Mechanics*

Experience is telling us that any quantum embedding will, in general, have a significant impact on the workings of the quantum system under consideration: It is no longer closed, back actions from the observer on the observed cannot be neglected. In turn, information transfer from the observed to the observer has to be described as a physical process based on explicit strategies ("logic").

The analysis of observation significantly contributes to the understanding and to the interpretation of quantum mechanics. Many concepts thus start playing double roles: they support abstract as well as information theoretical narratives.

Remark 3.14 Free choice *In all what follows we tacitly assume that the various measurement settings C can freely be chosen, there is no "conspiracy." The choice of the setting is free in that it is uncorrelated with "anything in the past" [Colbeck (2011)]. Nonlocal hidden variables (as assumed in the Bohmian approach to quantum mechanics) have to be excluded; there are superluminal effects on the level of individual particles.*

Definition 3.80 Born rule *We consider a quantum system in state $\hat{\rho}$. Let $\hat{F}$ denote an observable with the eigenvalue equation*

$$\hat{F}\,|f^{(j)}>= F_j\,|f^{(j)}> \qquad j = 1, 2, \ldots \tag{3.295}$$

and a nondegenerate spectrum ($\hat{F}$ is a complete operator). Then the measurement is complete and described by the projector $\hat{P}^{(kk)} = |f^{(k)}>< f^{(k)}|$.

The Born rule can then be stated as follows: Upon measurement of $\hat{F}$, one finds the eigenvalue F_k with probability

$$\boxed{P(k|\hat{\rho}) = \mathrm{Tr}\{\hat{P}^{(kk)}\hat{\rho}\hat{P}^{(kk)}\} = \mathrm{Tr}\{\hat{P}^{(kk)}\hat{\rho}\} = \rho_{kk}\,.} \tag{3.296}$$

The probability P for finding eigenvalue F_k is conditioned by the state $\hat{\rho}$, cf. Definition 2.2. This condition is indispensable; and it must not be supplemented by further conditions in terms of values for other observables. The latter would be either redundant or inconsistent.

Remark 3.15 Origin of probability *As already remarked in Section 2.3.1, the fundamental origin of the quantum probabilities $P(k|\hat{\rho})$ is remarkable, indeed. No other situation in physics is known, in which this happens. In the classical domain, probabilities are subjective. Appropriate quantum devices have thus been suggested to function as true random number generators [Symul (2011)].*

However, the choice of which observable $\hat{F}$ to measure (or which subsystem to address) does not follow from the state $\hat{\rho}$: These specifications may be taken as a "reference frame" O (cf. Section 3.2.8) needed to uniquely define the random experiment, see Remark 2.1. The need for O is stressed by relational quantum mechanics, see Remark 3.22. Quantum probabilities are inherently contextual.

Definition 3.81 Projection postulate (PVM) *After measurement the system is updated ("collapse of the quantum state") according to*

$$\hat{\rho}' = \frac{\hat{P}^{(kk)}\hat{\rho}\hat{P}^{(kk)}}{P(k|\hat{\rho})} = |f^{(k)}>< f^{(k)}|\,, \tag{3.297}$$

that is, leads (for a complete measurement) to a new pure state[a]*. This updating (projection-valued measure, PVM) is not part of the "event" in the sense of Kolmogorov probability theory, but rather a physical route toward subsequent probability predictions. It guarantees that an immediate repetition of the same measurement will necessarily confirm the result obtained previously,* $P(k|\hat{\rho}') = 1$*. It also implies that, in general, the order of measurements matters.*

The ensemble average over many such (independent) measurements approaches the expectation value

$$\sum_k P(k|\hat{\rho})F_k \to \mathtt{Tr}\{\hat{\rho}\hat{F}\} =< F > . \tag{3.298}$$

Definition 3.82 Joint measurements *Commuting operators have the same set of eigenfunctions, cf. Definition 3.18. Their respective measurements do not disturb each other, the measurement values "coexist," as would be taken for granted in the classical regime.*

The Born rule (together with the projection postulate) formalizes the fundamental stochastic nature of observational quantum mechanics. The status of this rule has given rise to some debate [Assis (2011); Saunders (2004)]: Is this rule derivable, and if so, in what sense? Based on the idea that the "abstract" quantum theory does not yet contain any aspect of observation, the Born rule (as a fundamental feature of measurement) cannot already be contained in that theory; it comes as an independent measurement postulate. Nevertheless, it is remarkable that part of this rule seems to add just a new interpretation of concepts already existing in the abstract theory. The discontinuous jump, though, is in conflict with the continuous unitary time evolution.

Remark 3.16 Gleason's theorem *Gleason choose to investigate which kind of probability measures are definable on the lattice of projection operators on a Hilbert space. For Hilbert space dimension* $d \geq 3$*, he found that the Born rule follows [Gleason (1957)].*

---

[a]This is an idealization. A known pure state could always be unitarily transformed into the ground state of the system, thermodynamically the zero temperature state. However, according to the third law of thermodynamics such a state should not be exactly reachable.

The theorem implies that, if one wants to introduce a probability measure at all, this measure has to be continuous, there can be no probability function restricted to the values 1 and 0 only. In other words, states are not dispersion free. As a hidden variable theory claims just this, that is, that a spin-component, say, is either up or not up, such theories face fundamental problems.

Definition 3.83 Measurement model *Elements of observation are measurements executed on some quantum object A (relating to some observable $\hat{F}(A)$). Typical measurement models (cf. [Ballentine (1998)]) consist of an apparatus M, which correlates the eigenvalues $F_k(A)$—and thus the states $|f^{(k)}(A) >$—with some indicator variable $x(M)$, for example, a position on a screen (which is considered a classical parameter). The measurement closes with the registration of an event marked by $x(M)$, an irreversible act, from which we infer k (measurement logic).*[a]

Typically, such (independent) measurements are repeated, leading to varying results k. If the state of A before each measurement is known to be $\hat{\rho}(A) = |\psi(A) > < \psi(A)$, the counting frequencies c_k should approach

$$c_k \rightarrow P(k|\hat{\rho}(A)) = \mathrm{Tr}\{\hat{P}^{(kk)}\hat{\rho}(A)\} = | < f^{(k)}(A)|\psi(A) > |^2 . \tag{3.299}$$

In this way, the Born rule can be tested.

Theorem 3.1 Consistency requirement *It is not possible to construct a measurement device, which would measure incompatible (noncommuting) operators "at the same time." This operational no-go-theorem gives a rather stringent and logic interpretation of the notion of "incompatibility," originally introduced by Bohr as a part of the Copenhagen interpretation.*[b] *In fact, such constructive impossibilities are needed to avoid violations of quantum theory in actual experimental situations. Their origin are not always obvious, though. Often they are related to symmetry.*

---

[a] If one puts aside the classical limit with respect to the apparatus, what remains is the built-up of correlations between object A and some object B, the roles of which are interchangeable, cf. [Rovelli (1996a)].

[b] Bohr is said to have been influenced by psychology; this is taken by some researchers to motivate a kind of retransfer of quantum analog ideas into psychology and cognitive science, cf. Section 9.2.5.

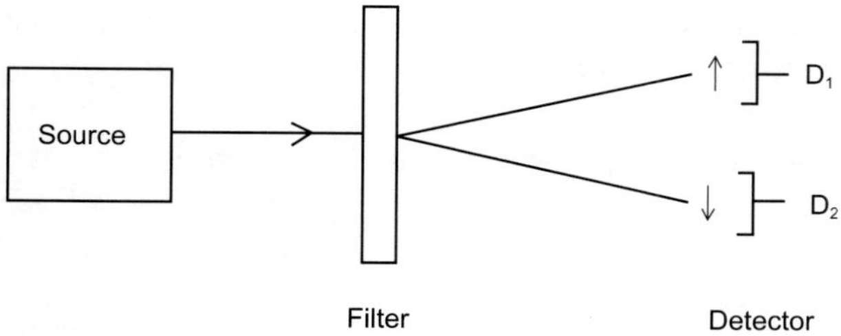

Figure 3.7 Stern–Gerlach experiment. D_1, D_2 detectors.

A notorious example is the so-called "arrow-paradox" going back to the Greek philosopher Zeno: Here it is claimed that position and change of position (linear momentum) cannot exist simultaneously. In classical mechanics, this criticism is considered obsolete, based on the assumption of a continuous trajectory and the use of infinitesimal calculus. In quantum physics, position and momentum indeed are incompatible: The momentum as a wavepattern defined everywhere is inherently nonlocal and thus cannot coexist with a selected position in real space.

Example 3.24 Stern–Gerlach apparatus *An even more transparent example is the Stern–Gerlach apparatus for the measurement of spin orientation, $\hat{\sigma}_3$. Such a measurement requires the specification of a reference direction, which is conveniently introduced by the gradient of some external magnetic field, see Fig. 3.8. But such a gradient cannot point at two different directions at the same time. (Different gradients can at most be realized sequentially.) The spin direction controls its magnetic moment and thus the deflection of the spin center-of-mass motion within an inhomogeneous magnetic field. The deflection is exploited for filtering, see Fig. 3.7. Measurements are repeated based on a supply of fresh spins entering the apparatus.*

Remark 3.17 Heisenberg microscope *Heisenberg's thought experiment has originally been intended to "explain" the momentum-position uncertainty relation. He tried to argue that the act of observation of an electron (via photons) would perturb the particle in such a way as to reproduce the uncertainty product. This model would suggest the uncertainty to be epistemic in origin, that is, due* *to our inability to perform good enough experiments; however, such an interpretation is not in accord with the quantum formalism, cf. Proposition 3.19.*

The consistency requirement, Theorem 3.1, has a different status: It guarantees that there is no operational procedure available by which one may violate the uncertainty relations known to follow from the formalism.

Example 3.25 Interference *For simplicity, we start by considering a two-dimensional Hilbert space. Let the system be prepared in the superposition state* $|\psi>= a_1|e^{(1)}> + a_2|e^{(2)}>$. *The observable* $\hat{F}$ *is going to be measured with the eigenstates* $|f^{(j)}>; j = 1, 2$ *and eigenvalues* F_j. *Then the conditional probability to find the eigenvalue* F_j *is according to the Born rule given by*

$$\begin{aligned} P(j||\psi><\psi|) &= <f^{(j)}|\psi><\psi|f^{(j)}> \\ &= |a_1|^2|<f^{(j)}|e^{(1)}>|^2 + |a_2|^2|<f^{(j)}|e^{(2)}>|^2 + \\ &+ a_1^* a_2 <f^{(j)}|e^{(2)}><e^{(1)}|f^{(j)}> + c.c. \quad (3.300) \\ &\equiv P(j|1) + P(j|2) + I(j|12). \quad (3.301) \end{aligned}$$

The third term, $I(j|12)$, *describes interference. Of course, it holds that*

$$\sum_j P(j||\psi><\psi|) = 1. \quad (3.302)$$

This trivial fact makes clear that the two interference terms $I(j|12)$, $j = 1, 2$ *have to sum up to zero: a positive (constructive) term is necessarily counterbalanced by a negative (destructive) term. Interference is nothing but a redistribution of probabilities with respect to the possible measurement outcomes j.*

The generalization to higher dimension (multiple superpositions) is straight forward:

$$|\psi> = \sum_{k=1}^{d} a_k|e^{(k)}>, \quad (3.303)$$

$$\begin{aligned} P(j||\psi><\psi|) &= \sum_k |a_j|^2|<f^{(j)}|e^{(k)}>|^2 \\ &+ \sum_{k\neq l} I(j|kl). \quad (3.304) \end{aligned}$$

The total interference term for given j is composed of pairwise contributions [Sinha (2010)].

The standard experiment to demonstrate interference is the famous optical double-slit experiment. The prepared state could be a single-photon emanating from a source via two different slits (1,2), and the measurement observable $\hat{F}$ *would be the spatial position* $\hat{\vec{r}}$ *on a screen behind. In this case, the detection variable would be continuous,* $|f^{(j)}> \rightarrow |\vec{r}>$*. Depending on position we have areas of constructive and destructive interference. Multi-slit generalizations are common practice; three-slit variants have been tested in the search for deviations from the Born rule [Sinha (2010)].*

Remark 3.18 Violation of total probability? *Interference terms such as* $I(j|12)$ *are sometimes claimed to spoil the "additivity of probabilities," namely, if the terms* $P(j|1)$ *and* $P(j|2)$ *were taken to describe mutually exclusive (and exhaustive) events. But such an interpretation is erroneous:*

(i) Never is there any loss of probability, only a redistribution on the possible outcomes $j = 1,2$*, see Eq. (3.302).*
(ii) Kolmogorov probabilities must not be combined (added, multiplied, etc.) if referring to different conditions (needed to fix the underlying statistical experiment, see Remark (2.1). Obviously, the two paths (e.g., in the case of an optical double-slit) could be isolated only by applying different experimental settings, that is, with one slit open and then(!) the other slit open. Such a scenario would be equivalent to a state prepared as an incoherent mixture. Indeed, the interference is easily shown to disappear in this case. Similarly, a classical ball would randomly pass through one of the slits, even if both are offered. The superposition-state in a strict sense is based on a different conditioning.

Proposition 3.18 Indeterminacy of experimental results *Let* $|j>$ *be the eigenbasis of* $\hat{\rho}$*. Under the conditions as stated above, the probability to measure* F_α *in this state is given by*

$$P(\alpha|\hat{\rho}) = \mathtt{Tr}\{\hat{P}^{(\alpha\alpha)}\hat{\rho}\} = \sum_j \rho_{jj}| < f^{(\alpha)}|j > |^2, \qquad (3.305)$$

and the respective measurement uncertainty can be characterized by

$$\boxed{S_F = -\sum_\alpha P(\alpha)\ln P(\alpha) \geq S_{vN} = -\sum_j \rho_{jj}\ln\rho_{jj}\,.} \qquad (3.306)$$

Proof. We distinguish two cases: If $[\hat{F}, \hat{\rho}] = 0$, the eigenvectors of $\hat{F}$ can be chosen as the eigenvectors of $\hat{\rho}$ and thus $P(\alpha) = P(j) = \rho_{jj}$. As a consequence $S_F = S_{vN}$.

If $\hat{F}$ is complete, but the above commutator is not zero, there is no joint set of eigenfunctions. Observing that

$$\sum_\alpha | < f^{(\alpha)}|j > |^2 = \sum_j | < f^{(\alpha)}|j > |^2 = 1\,, \tag{3.307}$$

we find for

$$\begin{aligned} S_F - S_{vN} &= \sum_j \rho_{jj} \ln \rho_{jj} - \sum_\alpha P(\alpha) \ln P(\alpha) \\ &= \sum_{j,\alpha} | < f^{(\alpha)}|j > |^2 \rho_{jj} \ln (\rho_{jj}/P(\alpha))\,. \end{aligned} \tag{3.308}$$

Based on the expansion

$$\ln x \geq 1 - \frac{1}{x} \tag{3.309}$$

with $x = \rho_{jj}/P(\alpha)$ we obtain

$$\rho_{jj} \ln \rho_{jj}/P(\alpha) \geq \rho_{jj} - P(\alpha)\,. \tag{3.310}$$

and the proposition follows,

$$S_F - S_{vN} \geq \sum_{j,\alpha} | < f^{(\alpha)}|j > |^2 (\rho_{jj} - P(\alpha)) = 0\,. \tag{3.311}$$

In the last step, we have used Eq. (3.307), again, and $\sum_j \rho_{jj} = \sum_\alpha P(\alpha) = 1$. This result relates to Klein's inequality (cf. Eq. (3.167)). □

S_F can be calculated once the pertinent observable, $\hat{F}$, and the state is given. Its interpretation as a measurement uncertainty rests on the measurement postulates as specified by the Born rule.

Under special conditions, the entropy originally associated with measurement uncertainty has also been given a different meaning:

Definition 3.84 Mixing entropy *Let $\hat{F} = \hat{H}$; the respective eigenstates are $|e^{(\alpha)} >$ and thus $P(\alpha) = \rho_{\alpha\alpha}$. Then*

$$S_H = -\sum_\alpha^d \rho_{\alpha\alpha} \ln \rho_{\alpha\alpha} \geq S_{vN}\,. \tag{3.312}$$

This measurement uncertainty can alternatively be interpreted to result from the state $\hat{\rho} \rightarrow \hat{\sigma}$ in energy representation with the off-diagonal matrix elements being suppressed (or declared to be inaccessible),

$$\hat{\sigma} \equiv \sum_{\alpha} P(\alpha)|e^{(\alpha)} >< e^{(\alpha)}| \,. \tag{3.313}$$

Contrary to the von Neumann entropy, this mixing entropy [Hughston (1993)] will typically change under unitary transformation. It could thus be tempting (though misguided) to suggest the mixing entropy as the thermodynamic entropy.

This entropy should not be mistaken for the classical variant describing the mixing of different species of particles.

Proposition 3.19 Uncertainty products *A simple measure for the uncertainty of $\hat{F}$ in state $\hat{\rho}$ is given by the variance, cf. Definition 2.7,*

$$(\delta F)^2 = \mathtt{Tr}\{(\Delta\hat{F})^2\} = \mathtt{Tr}\{\hat{F}^2\} - (\mathtt{Tr}\{\hat{F}\})^2 \geq 0\,, \tag{3.314}$$

where

$$\Delta\hat{F} \equiv \hat{F} - < F > \,. \tag{3.315}$$

For noncommuting operators $\hat{F}$, $\hat{G}$, there is a lower bound for the respective uncertainty product:

$$\boxed{\delta F\,\delta G \geq |\frac{1}{2i}\mathtt{Tr}\{\hat{\rho}[\hat{F},\hat{G}]\}|\,.} \tag{3.316}$$

This statement is sometimes called Robertson–Schrödinger inequality. The lower bound is, in general, state dependent.

Proof. Following [Ballentine (1998)] we start from

$$\mathtt{Tr}\{\hat{\rho}\hat{T}\hat{T}^{\dagger}\} = \sum_{j} \rho_{jj} < \hat{T}^{\dagger}e^{(j)}|\hat{T}^{\dagger}e^{(j)} > \geq 0\,. \tag{3.317}$$

Here we have applied the eigen-representation of $\hat{\rho}$ and the positive definiteness of the norm. We identify (with some real parameter g to be fixed below)

$$\hat{T} = \Delta\hat{F} + ig\Delta\hat{G}\,, \tag{3.318}$$

$$\hat{T}^{\dagger} = \Delta\hat{F} - ig\Delta\hat{G}\,, \tag{3.319}$$

$$\hat{T}\hat{T}^{\dagger} = (\Delta\hat{F})^2 + g^2(\Delta\hat{G})^2 - ig[\Delta\hat{F},\Delta\hat{G}]\,. \tag{3.320}$$

Choosing now

$$g = -\frac{\mathrm{Tr}\{\hat{\rho}\hat{C}\}}{\mathrm{Tr}\{\hat{\rho}\Delta\hat{G}\}}, \tag{3.321}$$

where

$$[\Delta\hat{F}, \Delta\hat{G}] \equiv 2i\hat{C}, \tag{3.322}$$

we immediately get

$$\mathrm{Tr}\{\hat{\rho}\hat{T}\hat{T}^{\dagger}\} = \mathrm{Tr}\{\hat{\rho}\Delta\hat{F}\}\mathrm{Tr}\{\hat{\rho}\Delta\hat{G}\} - (\mathrm{Tr}\{\hat{\rho}\hat{C}\})^2 \geq 0, \tag{3.323}$$

$$(\delta F)^2(\delta G)^2 \geq (\mathrm{Tr}\{\hat{\rho}\hat{C}\})^2. \tag{3.324}$$

Taking the square root confirms the proposition. □

Example 3.26 Heisenberg uncertainty relation *Based on the fundamental commutator, Eq. (3.49),*

$$[\hat{p}_i, \hat{r}_j] = \frac{\hbar}{i}\hat{1}\delta_{ij}, \tag{3.325}$$

we obtain the state-independent lower bound for the respective uncertainty product

$$\boxed{\delta r_j \delta p_j \geq \frac{\hbar}{2}.} \tag{3.326}$$

The bound is special insofar, as it does not depend on the state $\hat{\rho}$ the system is in—a consequence of the commutator being proportional to the unit operator.

Definition 3.85 Minimum uncertainty states *It is interesting to note that for certain Hamilton models sets of states can be constructed which are all characterized by just that minimum uncertainty product. A well-known example are the so-called coherent states for the harmonic oscillator, cf. Definition 3.36. These states are parametrized by the complex parameter α and can be written as certain superpositions of the eigenstates $|n>$:*

$$|\alpha >= \exp(-|\alpha|^2/2)\sum_n \frac{\alpha^n}{(n!)^{1/2}}|n>, \tag{3.327}$$

These states are normalized but not orthogonal; they have neither sharp energy nor sharp position or momentum. But one finds, independent of α,

$$(\delta x)^2 = \frac{\hbar}{2m_0\omega}, \tag{3.328}$$

$$(\delta p_x)^2 = \frac{\hbar m_0\omega}{2}, \tag{3.329}$$

which confirms the minimum uncertainty product. Contrary to the stationary energy eigenstates $|n>$, the expectation value of position with respect to $|\alpha>$ exhibits a coherent motion,

$$< x > (t) = |\alpha| \left(\frac{2\hbar}{m_0 \omega} \right)^{1/2} \cos(\omega t - \phi), \tag{3.330}$$

just like its classical counterpart. Also the expectation value $< p_x >$ (t) oscillates. For sufficiently large α, the uncertainty (with respect to the total amplitude of the oscillation) can be neglected, the states become most classical.

Remark 3.19 Standard quantum limit (SQL) *The lower bound of uncertainty for a given observable and a given model is called standard quantum limit. It is a fundamental constraint for the measurement accuracy. Equation (3.328) is an example for $\hat{x}$ in the case of the harmonic oscillator.*

The concept of measurement, PVM, can be generalized based on the

Definition 3.86 Positive operator-valued measure (POVM) *A set of linear operators $\hat{E}_j$ are called POVM, if with*

$$P(j) = \mathrm{Tr}\{\hat{\rho}\hat{E}_j\} \geq 0, \tag{3.331}$$

$$\sum_j P(j) = 1, \tag{3.332}$$

the operators $\{\hat{E}_j\}$ fulfill the condition

$$\sum_j \hat{E}_j = \hat{1}. \tag{3.333}$$

Definition 3.87 POVM measurement *To obtain the state of a system after measurement, one has to find measurement operators $\{\hat{M}_j\}$ such that*

$$\hat{M}_j^\dagger \hat{M}_j = \hat{E}_j. \tag{3.334}$$

These measurement operators are not uniquely defined by $\hat{E}_j$, and they need not be orthogonal. Their number can be larger than the dimension of Hilbert space. Their effect is

$$\hat{\rho}' = \frac{\hat{M}_j \hat{\rho} \hat{M}_j^\dagger}{\mathrm{Tr}\{\hat{M}_j \hat{\rho} \hat{M}_j^\dagger\}}. \tag{3.335}$$

The projective measurements discussed so far are a special subclass of POVM measurements with $\hat{E}_j = (\hat{P}^{(jj)})^\dagger \hat{P}^{(jj)} = \hat{P}^{(jj)}$.

Proposition 3.20 Neumark's dilation theorem *A POVM measurement can be lifted to a projection-valued measure in a larger dimensional Hilbert space. This is usually done by means of an auxiliary subsystem ("ancilla"), cf. [Peres (1990)].*

Primarily it is a matter of convenience regarding which measurement variant, PVM or POVM, to choose.

Remark 3.20 Measurement as state preparation *The result of a measurement is a modified, new state. This may alternatively be viewed as a preparation step. In a "complete" measurement, the new state is pure; otherwise the state will be "mixed," that is, represented by a density operator with nonzero von Neumann entropy. However, the latter case does not necessarily mean that the system will actually be in a pure state, which we simply do not know due to missing details of the preparation procedure (subjective ignorance). Such a quasi-classical interpretation would hold for inhomogeneous pure-state ensembles, that is, for identical particles having been exposed to the same complete measurement procedure—leading to different but well-specified single-particle results. The entropy can then be interpreted in the sense of Shannon.*

Complete local measurements described by the local projector $\hat{P}^{kk}(A)$ *allows the preparation of uncorrelated states (product states): The state of A after measurement is pure and the updated total system state has to be a product of this state and the state of the rest. Local measurements thus destroy the web of correlations of "everything with everything," which interactions would tend to induce. They constitute a simple but drastic "reset" operation.*

3.4.2 *Global and Local Measurements*

Definition 3.88 Bipartite systems: Global measurements *For composite systems the measurement postulate applies, as before. However, there can be "local" or "cluster"- measurements now. Bell measurements refer to two-spin systems and are based on joint measurements of the type* $\hat{K}_{jk}(A, B) = \hat{\sigma}_j(A) \otimes \hat{\sigma}_k(B)$. *For* $j = k$ *they project on maximally entangled states.*

Definition 3.89 Entanglement swapping *We refer to the example given in Definition 3.65. Here a pair of Bell states in (AB)(CD) has been prepared. We immediately see that projecting (BC) on one of the Bell states 1, 2, 3, or 4 will imply the pair (AD) to be found in the same Bell state. This projection occurs as a result of a so-called Bell measurement on (BC). Note that pair (AC) has never interacted in the past, entanglement has been swapped [Zukowski (1993)].*

Definition 3.90 Locality *A physical theory (applied to a composite system) is local, if the measurement on one subsystem does not influence any other subsystem, which is space-like separated in the sense of relativity theory.*

Definition 3.91 Bipartite systems: Local measurements *A local measurement performed on subsystem B can be described by* $\hat{1}(A) \otimes \hat{P}^{(kk)}(B)$*. After such a measurement (cf. Eq. (3.297)), we have*

$$\hat{\rho}'(B) = \hat{P}^{(kk)}(B)\hat{\rho}(B)\hat{P}^{(kk)}(B)/P(k;B)\,, \tag{3.336}$$

$$P(k;B) = \mathrm{Tr}_B\{\hat{P}^{(kk)}(B)\hat{\rho}(B)\}\,. \tag{3.337}$$

Proposition 3.21 Co-jump and nonlocality *But the state of subsystem A will, in general, be modified, too. To see this, we rewrite the initial state as*

$$\hat{\rho}(A,B) = \hat{\rho}(A) \otimes \hat{\rho}(B) + \hat{C}(A,B)\,. \tag{3.338}$$

The first term is the product of the respective reduced density operators, the second term defines the correlation with

$$\mathrm{Tr}_\nu\{\hat{C}(1,2)\} = \mathrm{Tr}\{\hat{C}(1,2)\} = 0 \quad (\nu = A, B)\,. \tag{3.339}$$

Then, after tracing out subsystem B, we obtain after measurement (of B)

$$\boxed{\begin{aligned}\hat{\rho}'(A|k) &= \frac{\mathrm{Tr}_B\{(\hat{1}(A) \otimes \hat{P}^{(kk)}(B))\hat{\rho}(A,B)\}}{P(k;B)} \\ &= \hat{\rho}(A) + \frac{\mathrm{Tr}_B\{(\hat{1}(A) \otimes \hat{P}^{(kk)}(B))\hat{C}(A,B)\}}{P(k;B)}\,.\end{aligned}} \tag{3.340}$$

The modified state $\hat{\rho}'(A|k)$ *has been called conditional density operator [Belokurov (2002); Zurek (2003)]; in so doing one explicitly refers to the history of its preparation.*

We realize that the state of subsystem A stays unchanged only, if $\hat{C}(A, B) = 0$, that is, if the initial total state has been a product state. In general, the state of subsystem A is conditioned by the result of the measurement on B, there is a co-jump [Granzow (1998)]. Such effects (already recognized by Schrödinger in 1935) are exploited by the so-called "steering" experiments [Wittmann (2011)].

Remark 3.21 Ensemble average *Such a co-jump is "instantaneous"; it underlines the nonlocal character of quantum mechanics. However, using Eq. (3.339) one easily convinces oneself that after averaging over the measurement results on subsystem B, the state for subsystem A is $\hat{\rho}'(A) = \hat{\rho}(A)$, that is, does not change on ensemble average, there is no longer a nonlocal effect.*

Definition 3.92 Measurement-based quantum conditional entropy *We define ensemble averaging over the conditional density operators $\hat{\rho}'(A|k)$ as*

$$S_{vN}(A|\{\hat{P}^{kk}(B)\}) \equiv \sum_k P(k; B) S_{vN}(\hat{\rho}'(A|k)) . \tag{3.341}$$

This conditional entropy should be seen as an alternative definition to Eq. (3.220); it depends on the measurement basis.

Definition 3.93 Quantum discord *In the quantum domain, these two definitions do not necessarily coincide; their difference can be used to define a quantumness measure [Zurek (2003b)].*

We first note that the quantum mutual information, Definition 3.62, can be rewritten as

$$I_{vN}(AB; A, B) = S_{vN}(A) - S_{vN}(A|B) . \tag{3.342}$$

Here we have employed the original definition of the quantum conditional entropy, Eq. (3.220). Alternatively,

$$I_{vN}(AB; A, B; \{\hat{P}^{kk}\}) = S_{vN}(A) - S_{vN}(A|\{\hat{P}^{kk}(B)\}) . \tag{3.343}$$

The quantum discord is now based on a kind of quantum game, cf. Section 5.5.5, in which the player has to select a (complete) measurement set $\{\hat{P}^{kk}(B)\}$, by which to infer the difference

$$\delta I_{vN}(\{\hat{P}^{kk}\}) = I_{vN}(AB; A, B) - I_{vN}(AB; A, B; \{\hat{P}^{kk}\}) \geq 0 . \tag{3.344}$$

This value function has to be minimized over all possible measurement sets. The result is the quantum discord; it is a weaker condition than entanglement. In general, it is not easy to calculate. By construction, the discord is not a "physical" quantity (there is no discord minimization principle installed in nature) and can thus not provide a direct "physical explanation." Nevertheless, the discord has proven to be useful for special purposes (see, e.g., [Dillenschneider (2009)]).

Entanglement implies correlation, correlation means (global) information—which easily comes in conflict with local information. This is intuitively seen in the so-called which-path experiments.

Example 3.27 Which-path experiments *We consider a simplified version of a double-slit experiment. Based on photons such experiments are conveniently used to demonstrate the appearance of interference on a screen positioned behind the slits. Interference occurs, if the two states (from slit 1 and from slit 2) are coherently superimposed. If coherence is gone (mixed state), the interference pattern on the screen also disappears.*

The same type of experiment can also be performed with atoms. The advantage of atoms is that in addition to the translational motion (center-of-mass observable (A)), there are additional internal degrees of freedom (B). Together they form a tensor Hilbert space (AB). Such an observable (B) can be designed to "feel" the difference between slit 1 and 2, that is, providing "which-path information." But this amounts to a built-up of correlation between (A) and (B) and should thus destroy coherence of the local observables.

Let the original state for the individual atom be

$$|\psi(A)>=\frac{1}{\sqrt{2}}(|1(A)>+|2(A)>). \tag{3.345}$$

Here, $|j(A)>$ is the state emanating from slit j. The probability p to detect the atom at x on the screen is with $\hat{\rho}(A)=|\psi(A)><\psi(A)|$

$$p(x)=\mathrm{Tr}\{|x><x|\hat{\rho}(A)\}=0.5(|<x|1>|^2+|<x|2>|^2+ \\ +2\Re(<x|2><1|x>)). \tag{3.346}$$

The last term is the interference term[a], cf. Example 3.25.

[a]This equation has the form $P(x|\hat{\rho})=P(x|1)P(1)+P(x|2)P(2)+$ interference term. It shows that for the coherent state $\hat{\rho}$, the probability P cannot be decomposed into the P's for the two substates 1 and 2. This is not in conflict with the Kolmogorov axioms. $\hat{\rho}$ and $|1><1|$ are different conditions!

Now we mark the two paths,

$$|\psi(A)> \otimes |1(B)> \rightarrow \frac{1}{\sqrt{2}}(|1(A)1(B)> + |2(A)2(B)>). \quad (3.347)$$

This may be done by means of CNOT gate, cf. Example 3.23; subsystem (B) *then functions as the control. For this entangled state, the local state is maximally mixed,* $\hat{\rho}(A) = 0.5\hat{1}$. *Then the pattern on the screen becomes*

$$p(x) = \mathrm{Tr}\{|x><x|\hat{\rho}(A)\} = 0.5(|<x|1>|^2 + |<x|2>|^2). \quad (3.348)$$

The interference is gone. An experimental realization can be found in Ref. [Dürr (1998)]. Because of the correlations between A and B, the latter can be considered an observer system (O) *in the sense of relational quantum mechanics—even though there is no irreversible act of registration yet.*

Indeed, exactly the same effect (with respect to the interference pattern) would have been obtained, if a series of independent "real" measurements had been performed. In this sense, the two scenarios are equivalent.

Remark 3.22 Relational quantum mechanics *Seen from observational quantum mechanics, what is the position of relational quantum mechanics, cf. Section 3.2.8?*

Contextual statements (properties F of subsystem S with respect to subsystem O) are of the form: Given O to be in state α, *what can be said about F? The Born rule (Definition 3.80), re-formulated as the conditional probability* $P(F\,|\,O)$, *then requires additional attention.*

With respect to an external and detached observer P, *the two subsystems* $\{O, S\}$ *would simply remain correlated (outside view). From a participatory inside view, "contextual facts" would appear as a part of a momentary state description (just like in classical physics). Superficially, this statement resembles the notorious quantum jump scenario as postulated for measurements. However, the jump can be considered a "fact" only with respect to this context. One should note that in this picture no irreversible act has to occur; as the correlation changes, so does the information O can have about S. Facts depend on time.*

It is obvious that different choices of reference systems O (different views or windows on S) will typically be mutually exclusive. In that sense, there can be "conflicting realities," and there is no absolute right or wrong. While it is possible to compare different views, the process of comparison requires a physical interaction, which, again, is quantum mechanical in nature. And as the correlations may be time dependent, the "facts" can also fade away (cf. Section 3.2.7).[a] *Contrary to erasure [Landauer (1961)], this type of "forgetting" would not be related to dissipation.*

Relational quantum mechanics thus presents an interesting approach to the concept of meaning of a "fact" in terms of correlation (cf. [Mermin (1998)]: Correlation is not a local property, it cannot be "within" O. But in quantum mechanics, correlations have a strange local impact (local uncertainty). In this way, as Rovelli argues (cf. Section 3.2.8), the observer O is able to get information about S."

The "objectification," usually associated with the macrodomain [Ghirardi (2009)], is assumed to take place generically, though in a somewhat weakened sense: In the (participatory) inside view, the correlation between S and observer O appears as an event, which actualizes the respective quantum property of S. The exact time of this event cannot be "known" by O; it would be accessible from the third person perspective only, that is, from the respective correlation between S and O. The latter might be a classical (macroscopic) subsystem, as envisioned by Bohr, but, as well, could consist of the most elementary quantum system like a single spin. In general, this correlation is time dependent; if it amounts to less than 100% at a given time, this means that the event (with respect to O) would have "occurred" with a finite probability only [Rovelli (1998)]. The concrete result *cannot be inferred in this way.*

Example 3.28 Local properties in context *Consider a pair of interacting quantum subsystems A, B evolving between time t_1 and*

---

[a] Because the existence of correlations imply local entropy, it has been argued that "entropy-decreasing transformations cannot leave any trace of their having happened" [Maccone (2009)]. This is not convincing: Local entropy can easily be made to move from subsystem to subsystem. On the one hand, we can thus register entropy-decreasing processes in properly chosen subsystems; and an arrow of time can be shown to appear in embedded quantum systems—without reference to observation/registration.

time t_2 according to

$$|\Psi(A, B)>_{t_1} = (\alpha|1(A)> +\beta|2(A)>) \otimes |0(B)> , \qquad (3.349)$$

$$|\Psi(A, B)>_{t_2} = \alpha|1(A)> \otimes|1(B)> +\beta|2(A)> \otimes|2(B)> . \qquad (3.350)$$

This unitary evolution would be the appropriate description from an outside perspective, that is, with respect to a noninteracting (i.e., detached) observer P, who knows the initial state and the total dynamics. Because of entanglement local properties become ill-defined.

Following Rovelli [Rovelli (1996b)], we may, instead, interpret B as an (inside) participatory observer O (reference) to distinguish between states $|1(A)>$, $|2(A)>$: Then O describes his effect on A as the "collapse"

$$\alpha|1(A)> +\beta|2(A)> \rightarrow |1(A)> . \qquad (3.351)$$

(Or $|2(A)>$, the concrete result is unpredictable.) There is no introspection, $O = B$ cannot know his own state. A similar argument would hold after interchanging the roles of A and B. It is therefore futile to contemplate the "reality" of those events; facts are relational and probabilistic.

Example 3.29 Optical phase *A quite similar debate has been going on in quantum optics concerning the question: Should laser light (subsystem A) be described correctly as a coherent state (cf. Definition 3.36) or as an incoherent mixture?*

Microscopically the lasing process requires a field–matter interaction, which (like any quantum interaction) tends toward entanglement and thus to local incoherence (mixed state) [Mollmer (1997)]. This reasoning would challenge the physical nature of coherence; in particular, it would imply that, for example, a laser field A (routinely described by coherent states) had to be considered incoherent.

Clearly, the definition of a phase (like any other observable) requires an external reference, so the appropriate answer may well be context-dependent, as argued by [Bartlett (2006)]: If the amplifying medium B is taken as the reference, which the detached observer O_1 includes in his quantum description, local coherence, that is, the phase of A, is entirely absent due to entanglement between field A and subsystem B.

On the other hand, for the observer $O_2 = B$ the state of A with a phase relative to B (with given local properties) could well be present.

3.4.3 *Measurement Sequences*

Definition 3.94 Realistic theory *A physical theory is called "realistic," if for any of its observables concrete values exist independent of measurement.*

Definition 3.95 Local realism and hidden variables *If quantum mechanics is taken to be realistic, one has to claim that all eigenvalues $F_j^{(m)}$ of any observable $\hat{F}^{(m)}$ "coexist" (just like in the classical domain). However, as local measurements of incompatible observables $\hat{F}^{(m)}$, $\hat{F}^{(n)}$ with $[\hat{F}^{(m)}, \hat{F}^{(n)}] \neq 0$ perturb each other—they are not jointly "knowable." As these concrete values are not contained in the quantum mechanical state, the latter would have to count as "incomplete." The completion of the description could formally be achieved by the introduction of the so-called "hidden variables."*

According to this point of view, quantum uncertainty would essentially become subjective. Whether, indeed, all those values may exist or not is tested by Bell's inequality.

Proposition 3.22 Bell's inequality *We consider two spins, $\nu = A, B$, each known to be oriented in the x–z plane,*

$$\hat{F}_\theta(\nu) \equiv \hat{\sigma}_1(\nu) \sin\theta + \hat{\sigma}_3(\nu) \cos\theta \,. \tag{3.352}$$

For spin A, we allow two orientations, θ_1, θ_3, likewise for spin B, θ_2, θ_4. According to local realism, all values for any of the four angle combinations exist, $m_\theta(\nu) = \pm 1$. Together they form datasets of the form $\{\pm 1, \pm 1, \pm 1, \pm 1\}$. There are $2^4 = 16$ such combinations, which will be numbered by $\alpha = 1, 2, \ldots 16$. Also the following products must have a definite value:

$$\begin{aligned} Z_1 &= m_{\theta_1}(A) m_{\theta_2}(B) = \pm 1 \\ Z_2 &= m_{\theta_1}(A) m_{\theta_4}(B) = \pm 1 \\ Z_3 &= m_{\theta_3}(A) m_{\theta_2}(B) = \pm 1 \\ Z_4 &= m_{\theta_3}(A) m_{\theta_4}(B) = \pm 1 \end{aligned}$$

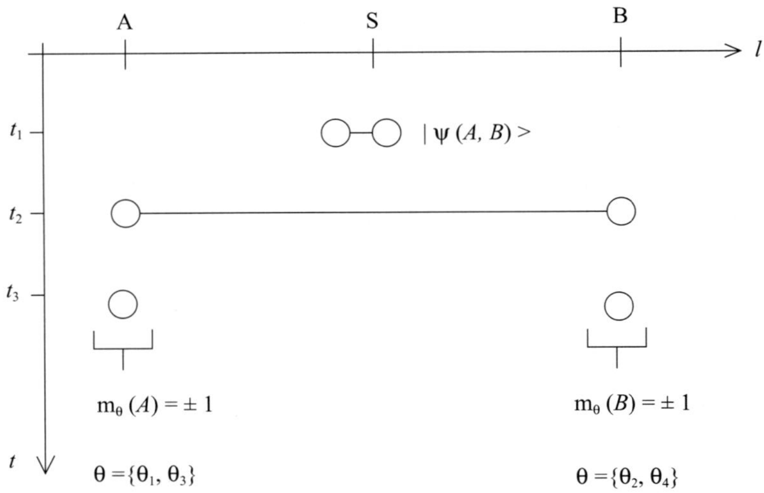

Figure 3.8 Bell measurement on spin pair (A,B). Preparation in state $|\psi(A, B)>$ at position S at time t_1, spatial separation t_2, local measurements at t_3. For details see text.

But for any such dataset α, these Z_j-values (signs) are not independent; this is conveniently expressed by the following condition

$$s(\alpha) = Z_1 + Z_3 + Z_4 - Z_2 = \pm 2\,, \tag{3.353}$$

valid for any α. (If the product results could be independently chosen, s would be -4, -2, 0, 2, or 4.) Now, the actual α in any single run is taken to be determined by a hidden variable, which, by definition, is unknown to us. The lack of knowledge can be modeled by an ensemble of independent spin pairs specified by a distribution function $f(\alpha)$. Independent of how this distribution looks like: Averaging over the ensemble must result in an $\bar{s}$ necessarily between the lower and the upper bound, that is,

$$\boxed{|\bar{s}^{cl}| = |C(\theta_1, \theta_2) + C(\theta_3, \theta_2) + C(\theta_3, \theta_4) - C(\theta_1, \theta_4)| \leq 2\,,} \tag{3.354}$$

where $C(\theta_1, \theta_2) = \overline{Z}_1$, $C(\theta_1, \theta_4) = \overline{Z}_2$, etc. The classical bound is $|\bar{s}^{cl}|_{max} = 2$.

But what does quantum mechanics tell us about the correlations?

Proposition 3.23 Spin–spin correlation function in x-z-plane *For the pure state* $|\psi(A,B)>$, *the correlation function between the pair of operators* $\hat{F}_{\theta_1}(A)$, $\hat{F}_{\theta_2}(B)$ *according to Eq. (3.352) reads (see Definition (3.59))*

$$C(\theta_1,\theta_2) = <\psi|\hat{F}_{\theta_1}(A)\otimes\hat{F}_{\theta_2}(B)|\psi> \tag{3.355}$$

$$\begin{aligned} &= C(\hat{\sigma}_1(A),\hat{\sigma}_1(B))\sin\theta_1\sin\theta_2 \\ &\quad +C(\hat{\sigma}_3(A),\hat{\sigma}_3(B))\cos\theta_1\cos\theta_2 \\ &\quad +C(\hat{\sigma}_1(A),\hat{\sigma}_3(B))\sin\theta_1\cos\theta_2 \\ &\quad +C(\hat{\sigma}_3(A),\hat{\sigma}_1(B))\cos\theta_1\sin\theta_2\,. \end{aligned} \tag{3.356}$$

Taking for $|\psi(A,B)>$ *the EPR state we have, according to Eq. (3.201),* $C(\hat{\sigma}_i(A),\hat{\sigma}_j(B)) = -\delta_{ij}$, *so that*

$$\boxed{C(\theta_1,\theta_2) = -\cos(\theta_2-\theta_1)\,.} \tag{3.357}$$

Under this condition, the correlation function is normalized like the correlation coefficient, Definition (2.10): Depending on the angles θ_1, θ_2, the correlation changes from anticorrelation ($C=-1$) via uncorrelated ($C=0$) to strict correlation ($C=1$).

Proposition 3.24 Maximal violation of Bell's inequality *We now identify the correlation functions in Bell's inequality with the respective quantum mechanical results. Choosing* $\theta_1-\theta_2=\theta_2-\theta_3=\theta_3-\theta_4=\phi=\pi/4$, *one thus finds*

$$|\bar{s}(\phi=\pi/4)| = |3\cos\phi-\cos(3\phi)| = 2\sqrt{2} > 2\,. \tag{3.358}$$

Such a violation of Bell's inequality cannot be understood on the level of local realistic models. This fact constitutes a "no-go-theorem." The new upper limit, $|\bar{s}^{qm}|_{max} = 2\sqrt{2}$, *is typical for "standard" quantum theory and also known as Tsirelson's bound.*

Tsirelson's bound can be violated by toy theories that show stronger nonlocal correlations than those allowed by quantum theory [Janotta (2011)].

Remark 3.23 Kochen–Specker theorem *Bell's theorem is supplemented and strengthened by the Kochen–Specker theorem [Kochen (1967)]: This theorem essentially demonstrates that we arrive at a paradox (within the rules of quantum mechanics) if we assume that all possible experimentally answerable questions that can* *be asked about a given quantum system have definite yes/no answers. The full proof of this theorem is technically rather demanding.*

This theorem formalizes the well-known fact that the answer to a certain question typically depends on the "context," that is, the questions having been asked before. This is known to happen under the condition of noncommutativity of the observables under consideration.

Remark 3.24 "Free will theorem" *Conway and Kochen [Conway (2009)] have summarized these strange quantum features by arguing that the spins A,B must have their own "free will" to choose how to respond to the measurements—if the experimenters were free to choose how to measure and if there is no faster-than-light communication.*

It should be clear that the notion of free will in this context is misleading, to say the least. Free will combines two distinct ideas, first that your choices are free from any external constraints, and second, that your choices are "willed by you" [Aaronson (2013)]. If anything, only the first part is applicable here.

As, in general, measurements do not commute, the conjunction of events have to observe their sequential order. (The only way to implement incompatible measurements is via separation in time, cf. Theorem 3.1.) This requirement can be formalized:

Definition 3.96 Quantum histories *A "history," h_i, is defined by*

$$\hat{C}(h_i) \equiv \hat{T} \prod_{k=1}^{n_k} \hat{P}_{i,k}(t_{i,k}) \,. \tag{3.359}$$

Here $\hat{T}$ denotes the time-order operator, $\hat{P}_{i,k}$ define propositions (e.g., projection operators) at time $t_{i,k}$. Then, based on the Born rule, the probability of history h_i is

$$P(h_i|\hat{\rho}) = \mathrm{Tr}\{\hat{C}(h_i)\hat{\rho}\hat{C}^\dagger(h_i)\} \,. \tag{3.360}$$

When all projectors commute, the history burns down to a simple conjunction of projective events. For a consistent set of histories $\{h_i, i = 1, 2, \dots\}$ defined by

$$\mathrm{Tr}\{\hat{C}(h_i)\hat{\rho}\hat{C}^\dagger(h_j)\} = 0 \,, \tag{3.361}$$

(for all $i \neq j$) the probabilities $P(h_i|\hat{\rho})$ fulfill the basic Kolmogorov axioms, that is, conventional probability theory works [Dowker (1995)]. Outside this regime it usually does not. A simple example would be the inconsistent set consisting of (1) measuring $\hat{\sigma}_1$ and (2) measuring $\hat{\sigma}_2$ of a spin system.

Remark 3.25 Discretized U/R-dynamics *The combination of (unitary, deterministic) quantum gates and (nonunitary, stochastic) measurement interactions allows to built up a discrete sequence of actions, which would carry a given system along a trajectory in state space. This strictly sequential control sequence constitutes a (partly random) walk. It has the advantage that the two types of dynamics, unitary (U) and nonunitary (R) do not occur "at the same time," that is, do not perturb each other.*

Remark 3.26 "Bang-bang" measurements *In a simplified scheme, the measurement interactions are taken to be "instantaneous," at least on the slower time scale of unitary motion. One can thus study unitary dynamics, for example, subject to periodic measurements, which, inter alia, destroy correlations. Interesting phenomena include the so-called Zeno or anti-Zeno effect [Alvarez (2010)].*

Sequential measurement projections give rise to quantum histories h_i characterized by a probability $P(h_i|\hat{\rho})$, see Eq. (3.360). If it is not the measurement result that counts but merely the back effect of quantum measurement itself, observation can be taken as an open loop control. A remarkable example for such a control is the Zeno effect [Antoniou (2001); Misra (1977)]. This effect is based on rapid periodic measurements and leads, under appropriate conditions, to a freezing of the eigendynamics of the system.

Proposition 3.25 Quantum Zeno effect *Let us start from the Born rule, Definition 3.80, taking*

$$P_k = \mathrm{Tr}\{\hat{P}_{kk}\hat{\rho}\} \qquad (3.362)$$

for the (initial) pure state $\hat{\rho}(t = 0) = |1><1|$. The survival probability under the eigendynamics generated by the time-independent $\hat{H}$ is then given by

$$P_{surv}(\Delta t) = |<1|\exp(-i\hat{H}\Delta t/\hbar)|1>|^2 \equiv |p(\Delta t)|^2 . \qquad (3.363)$$

$p(\Delta t)$ is the probability amplitude. Now include the effect of N equally spaced measurement projections (on the original state $|1>$) within the time integral $[0, \Delta t]$. Then we have

$$p_N(\Delta t) = < 1| \left(|1> < 1| \exp\left(-i\hat{H}\,\Delta t/\hbar N\right)\right)^N |1> . \qquad (3.364)$$

By considering $N = 1, 2, \ldots$ one easily convinces oneself that

$$\boxed{p_N(\Delta t) = p^N(\Delta t/N)\,.} \qquad (3.365)$$

The asymptotic limit $N \to \infty$ can conveniently be formulated based on $\tau = \Delta t/N$ as follows [Antoniou (2001)]:

$$\begin{aligned}
\lim_{\tau\to 0} p_N(\Delta t) &= \lim_{\tau\to 0} p(\tau)^{\Delta t/\tau} \\
&= \left(\lim_{\tau\to 0}\left(1 - \frac{(1-p(\tau))/\tau}{1/\tau}\right)^{1/\tau}\right)^{\Delta t} \\
&= \begin{cases} 0 & \text{for} \quad p'(0) = \infty \\ \exp(-c\Delta t) & \text{for} \quad p'(0) = -c \\ 1 & \text{for} \quad p'(0) = 0\,. \end{cases}
\end{aligned} \qquad (3.366)$$

Here, $p'(0)$ means first derivative of the function $p(\Delta t)$ taken at $\Delta t = 0$. Different models of $p(\Delta t)$ thus show different asymptotic behavior. The Zeno effect obtains for

$$p(\Delta t) = 1 - (\Delta t)^\alpha \quad \text{for all} \quad \alpha > 1\,. \qquad (3.367)$$

Exponential decay, on the other hand, $p(\Delta t) = \exp(-c\Delta t)$, cannot be stopped by the Zeno effect: case $p'(0) = -c$ in Eq. (3.366).

Remark 3.27 Proliferation of Zeno effects *Besides this quantum Zeno effect, a kind of "anti-Zeno effect" has also been analyzed, implying* speed-up *due to measurements [Kofman (1996)]. This further underlines the fact that special conditions have to be satisfied for the Zeno effect to occur. Despite these severe limitations, the quantum Zeno effect has also enjoyed some popularity in certain philosophical circles: H. Stapp [Stapp (2007)] claims that the "quantum mind" may hold the brain in a superposition of states using this very effect. Needless to say that there is no evidence whatsoever, cf. Section 9.2.4.*

3.4.4 *Continuous Measurements*

Remark 3.28 Continuous measurements and modified eigendynamics *Eventually, the eigendynamics and the interactions with the environment (leading to measurements) may be taken to be continuous. The coexistence of both types of evolution then leads to an effective dynamical law. One approximate way to deal with such a rather complex situation is based on "stochastic unraveling," which, in turn, depends on master equations.*

Proposition 3.26 Lindblad master equation *The most general Markovian master equation that is trace-preserving and and completely positive can be written as [Lindblad (1976)]*

$$\boxed{\begin{aligned} \tfrac{d}{dt}\hat{\rho} &= \hat{\mathcal{L}}(\hat{\rho}), \\ \hat{\mathcal{L}} &= \hat{\mathcal{L}}_{coh} + \hat{\mathcal{L}}_{inc}^{(D)} + \hat{\mathcal{L}}_{inc}^{(OD)}. \end{aligned}} \tag{3.368}$$

It holds that $\mathrm{Tr}\hat{\mathcal{L}} = 0$. *The first ("coherent") term reproduces the von Neumann–Liouville equation*

$$\hat{\mathcal{L}}_{coh}(\hat{\rho}) = -\frac{i}{\hbar}[\hat{H}, \hat{\rho}]. \tag{3.369}$$

The following two (incoherent) terms

$$\hat{\mathcal{L}}_{inc}^{(D)}(\hat{\rho}) = \sum_q W_q \hat{G}_q \hat{\rho} \hat{G}_q^\dagger, \tag{3.370}$$

$$\hat{\mathcal{L}}_{inc}^{(OD)}(\hat{\rho}) = -\frac{1}{2}\sum_q W_q (\hat{\rho}\hat{G}_q^\dagger \hat{G}_q + \hat{G}_q^\dagger \hat{G}_q \hat{\rho}), \tag{3.371}$$

involve non-negative parameters W_q *associated with the so-called environment operators* $\hat{G}_q$. *While acting within the d-dimensional Hilbert space of that system, they describe the influence of the environment; they are constrained by*

$$\mathrm{Tr}\{\hat{G}_q\} = 0, \tag{3.372}$$

$$\mathrm{Tr}\{\hat{G}_q \hat{G}_r^\dagger\} = \delta_{qr}. \tag{3.373}$$

The master equation is a closed, effective evolution equation for the density operator of the open system.

The inverse parameters $(W_q)^{-1}$ *have dimension of time and can thus often be interpreted as decay times.*

Remark 3.29 Violation of conservation laws *The unconditional constants of motion, Proposition 3.16, are based on unitary evolution; they no longer hold under the Lindblad master equation. This becomes quite intuitive from stochastic unraveling: An original pure state becomes a mixture, the entropy increases, the purity goes down.*

Example 3.30 Energy balance *Let us assume the Hamiltonian to be explicitly time independent. Then*

$$\frac{d}{dt} < \hat{H} >= \mathtt{Tr}\left\{\hat{H}\frac{d}{dt}\hat{\rho}\right\} = \mathtt{Tr}\{\hat{H}\,\hat{\mathcal{L}}_{inc}\} \equiv \dot{Q} \tag{3.374}$$

This energy flow is due to the incoherent term only and interpreted as a heat current.

Remark 3.30 Distinguishability of states and information flow *Distance measures (cf. Definition 3.44) are of paramount importance for the distinguishability of states. Strictly speaking, only orthogonal states can uniquely be distinguished by measurement: (Note that the eigenfunctions of any nondegenerate observable are orthogonal, cf. Definition 3.16). Under nonunitary evolution (resulting, e.g., from a Lindblad master equation for open systems), two original distinguishable states will become more and more indistinguishable. This can be interpreted as a loss of information; the information may be said to have flown into the environment. (The environment has been traced out and is no longer accessible, though.)*

Remark 3.31 Non-Markovian generalization *For the Lindblad equation to be Markovian, the Lindblad operator $\mathcal{L}$ has to be explicitly independent of time (memoryless). Generalizations lead to time-dependent environment operators $\hat{G}_q$ and rates W_q. Such cases may be encountered for externally driven systems, cf. Section 6.2.2. If some W_q become negative for some time, the master equation is non-Markovian [Romero (2012)]—despite in Lindblad form.*[a]

Remark 3.32 Heisenberg picture *Correspondingly, the open dynamics for the Heisenberg operator $\hat{F}_H(t)$ reads, cf.*

[a] As noted by [van Kampen (1981)], the need to go beyond Markovian treatments can often be avoided by "clever physical insight."

[Feldmann (2003)]:

$$\boxed{\frac{d}{dt}\hat{F}_H = \hat{\mathcal{L}}^*(\hat{F}_H) + \left(\frac{\partial \hat{F}_H}{\partial t}\right)_{expl}} \; . \qquad (3.375)$$

This equation generalizes the Heisenberg equation of motion, Eq. (3.277).

Example 3.31 Single transition channel *Consider quantum system with energy spectrum $E_j > E_i$ for $j > i$. Then the environment operator $\hat{G} = \hat{P}^{(ij)}$ with $\hat{G}^\dagger = \hat{P}^{(ji)}$ and $\hat{G}^\dagger\hat{G} = \hat{P}^{(jj)}$ describes a relaxation process with characteristic time $\tau = (W_{ij})^{-1}$ and is specified by*

$$\hat{\mathcal{L}}_{inc}(\hat{\rho}) = \frac{1}{2\tau}\left(2\hat{P}^{(ij)}\hat{\rho}\hat{P}^{(ji)} - \hat{\rho}\hat{P}^{(jj)} - \hat{P}^{(jj)}\hat{\rho}\right) . \qquad (3.376)$$

Because of the first part (deriving from $\hat{\mathcal{L}}_{inc}^{(D)}$), the lower level i gains in population, the upper level j looses. This happens within the diagonal matrix elements (D) of the density operator and characterized by the life-time τ. The last two parts originating from $\hat{\mathcal{L}}_{inc}^{(OD)}$ influence off-diagonal matrix elements (OD) of the density operator, characterized by the "decoherence time" $\tau_{dec} = 2\tau$.

There are damping channels that contribute to dephasing only; in fact, dephasing is often much faster than population decay. In the classical, incoherent limit, there would be nothing to decohere, the latter processes would not contribute. Note the different origin of any such times and the correlation time defined following Eq. (3.282).

Remark 3.33 Incoherent driving *The Lindblad form can also be used to model incoherent driving by means of a thermal field. Pertinent examples could be a photon or a phonon field. For a single transition channel, we would have to add to $\hat{G}_1 = \hat{P}^{(ij)}$ a gain term $\hat{G}_2 = \hat{P}^{(ji)}$, each with appropriate prefactors. As before, $E_j > E_i$ for $j > i$.*

Proposition 3.27 Effective non-Hermitian Hamiltonian *Motivated by the different roles played out by the two incoherent terms in the Lindblad operator, we realize that with the non-Hermitian Hamiltonian*

$$\hat{H}_{eff} = \hat{H} + \Delta H_{eff} \, , \qquad (3.377)$$

where

$$\Delta \hat{H} \equiv -\frac{i\hbar}{2}\sum_q W_q \hat{G}_q^\dagger \hat{G}_q\,, \tag{3.378}$$

we can re-write

$$\frac{d}{dt}\hat{\rho} = -\frac{i}{\hbar}\left(\hat{H}_{eff}\hat{\rho} - \hat{\rho}\hat{H}_{eff}^\dagger\right) + \hat{\mathcal{L}}_{inc}^{(D)}(\hat{\rho})\,. \tag{3.379}$$

Note that the dynamics based on the non-Hermitian Hamiltonian alone (i.e., without the last term) would no longer be trace preserving.

Proposition 3.28 Stochastic unraveling *Consider a single transition channel as given above; W_{ij}^{-1} defines a time scale. Let $\mathtt{Tr}\{\hat{\rho}(t)\} = 1$ for some initial time t. Then, putting aside $\hat{\mathcal{L}}_{inc}^{(D)}(\hat{\rho})$, we have, after the infinitesimal time $dt \ll W_{ij}^{-1}$,*

$$\mathtt{Tr}\{\hat{\rho}(t)\} = 1 - dp\,, \tag{3.380}$$

$$dp = W_{ij}\mathtt{Tr}\{\hat{G}^\dagger\hat{G}\hat{\rho}(t)\}dt > 0\,. \tag{3.381}$$

Stochastic unraveling [Dalibard (1992)] is then based on the following rule (algorithm):

$$\hat{\rho}(t+dt) = \begin{cases} \dfrac{\hat{\rho}(t+dt)}{1-dp} & \text{with probability} \quad 1-dp \\ \dfrac{\hat{\mathcal{L}}_{inc}^{(D)}(\hat{\rho}(t))dt}{dp} & \text{with probability} \quad dp \end{cases}\,. \tag{3.382}$$

This algorithm incorporates the quantum jump postulate. It generates a discontinuous pure state trajectory. The method can easily be generalized to include several damping channels.

Remark 3.34 Interpretation of the stochastic trajectories *The ensemble average over many such realizations (pure state trajectories) can be shown to reproduce the solution of the respective master equation. In that sense, stochastic unraveling can be used as a convenient method to numerically solve the master equation.*

However, it is tempting to associate any such discontinuous jump with a measurement event, thus making up a whole measurement record. Knowledge of the type of transitions and the waiting times in between would allow to reconstruct what "really happens" within the open system. We thus would get a remarkable illustration of observational quantum mechanics.

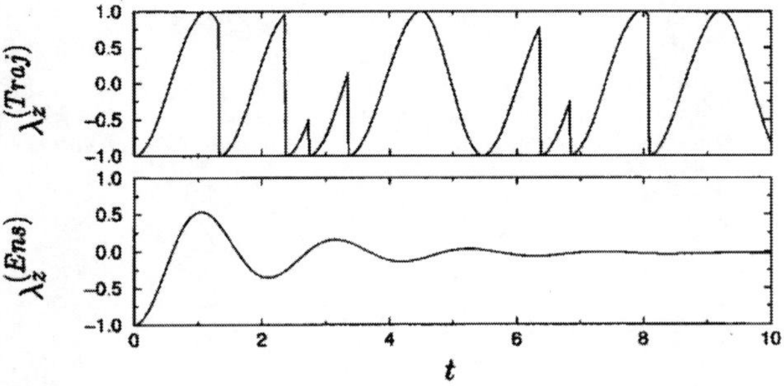

Figure 3.9 Open two-level system: Stochastic unraveling. Parameters: $g = 3$, $\delta = 0$; arbitrary time-units. Shown is the expectation value of $\sigma_3 = \lambda_z$. Lower section: Ensemble average.

While such an approach has become quite popular, for example, in quantum optics, there are some caveat to remember: The master equation only contains the impact of the environment, not the dynamics of the environment itself; but measurement records primarily derive from those external state changes. And the relation between the underlying master equation and the stochastic model is anything but unique: There are many stochastic unravelings reproducing the same ensemble average. Nevertheless, that kind of quantum trajectories do support intuition.

Example 3.32 Driven spin system under observation *The driven two-level system according to Eq. (3.248) can conveniently be described in terms of the Pauli operators, cf. Example 3.3,*

$$\hat{H} = \frac{\hbar}{2} \left(\delta \hat{\sigma}_3 + g \hat{\sigma}_1 \right) , \tag{3.383}$$

where $\hbar\delta = E_2 - E_1 - \hbar\omega$ is the detuning. Assume a single damping channel with $\hat{G} = \hat{P}_{12}$. Then a numerical simulation of Eq. (3.382) is shown in Fig. 3.9 [Granzow (1999)].

When averaged over many such pure state trajectories, we recover the ensemble result as would be obtained directly from a solution of the Lindblad master equation.

Chapter 4

Thermodynamics

Jonas Burgert: Untitled, 2008, based on photograph by Rainer Iglar
[Reprinted with permission from Schreiber and Wipplinger (2011)].

Quantum Thermodynamic Processes: Energy and Information Flow at the Nanoscale
Günter Mahler

ISBN 978-981-4463-73-7 (Hardcover), 978-981-4463-74-4 (eBook)
www.panstanford.com

Summary: Contrary to what its name may suggest, equilibrium thermodynamics does not contain the notion of time; there are no equations of motion.

Thermodynamics is a control theory in that it allows to specify, to what extent certain state changes can be executed on the basis of given resources. This is an operational approach and requires modeling of the pertinent observer/agent.

Thermodynamics is a powerful effective theory: It requires the introduction of an appropriate (macroscopic) level of description. This description is closed, that is, does not require a microscopic input.

Thermodynamics is based on rather few basic assumptions but gives rise to a wealth of predictions, applicable to almost any branch of physics, chemistry, and biology.

Nevertheless, thermodynamics is considered "reducible" to an underlying theory (classical mechanics or, more recently, quantum mechanics). In fact, quantum features enter the statistical analysis right from the start. Such a microscopic foundation is indispensable for the study of conditions as well as generalizations of the emergent thermodynamic theory.

Information turns out to be a natural aspect of thermodynamics, the pertinent link being provided by the concept of entropy. There are two different sources: In the context of quantum measurement information retrieval perturbs the thermal state under consideration. Its respective entropy changes; but this change will be transient, the state returns to equilibrium—if the system stays embedded in a thermalizing environment. The entropy can also change during a thermodynamic process (i.e., while remaining in momentary equilibrium). Under certain conditions such changes can even be measured via the heat being exchanged.

Eventually, surplus information might be considered a resource by which to go beyond the limits of standard thermodynamics. However, as shown by the analysis of Maxwell's demon, such attempts are seriously constrained. There are costs involved.

4.1 Historical Remarks

Thermodynamics started as the science underlying heat and work. While these terms had been widely discussed since the 17th century, it was not before the middle of the 19th century that the unifying concept, the notion of energy, had been clarified. The first and second law of thermodynamics emerged around 1850, primarily due to the studies by Rudolf Clausius and William Thomson. These laws introduce the internal energy and entropy as thermodynamic state functions. The optimization of heat engines had been a strong practical motivation behind all these exciting developments.

Finally, with the appearance of atomic and molecular theories thermodynamics was given a statistical interpretation: In a sense it thus became reducible to mechanics. The rather abstract thermodynamic entropy could be re-expressed in terms of the statistical distribution function over the accessible micro-states. Correspondingly, heat and work had to be re-interpreted as essentially mechanical energy forms subject only to different means of control. This completes thermodynamics as a kind of unification theory.

While in conventional equilibrium thermodynamics time does not show up explicitly (there are no equations of motion), the notion of "change" is formalized in terms of the so-called thermodynamic processes.

What can be done, what cannot be done—under given constraints and/or resources: These features constitute a control theory. A successful theory of that type should thus be largely immune against further developments: This is one reason why A. Einstein could argue—without much qualification—that thermodynamics "is the only physical theory of universal nature concerning which I am convinced that ... it will never be overthrown." [Schilpp (1969)].

This does not mean, though, that thermodynamics would not see new and fascinating developments. These developments deal with foundational issues [Gemmer (2009)], the quest for ultra-small quantum thermodynamic machines (see, e.g., [Linden (2010)]), and non-equilibrium aspects [Garrido (2011)], to name a few.

While it is now (almost) generally accepted that thermodynamics should be considered a "reducible" theory, that is, reducible to a more fundamental basis theory, axiomatic approaches can still be found [Lieb (2002)]. However, these appear to be inapplicable in the context of (small) quantum systems, which will be of prime interest here.

4.2 The Big Questions

With thermodynamics viewed as an effective (higher level) description the question appears, whether "we have a choice": Putting aside technical difficulties, could we just switch between this and lower (more detailed) levels at will? Would it make sense? And could this be beneficial? Tentative answers cannot be given without explicit reference to the observer/agent.

4.2.1 *What are Classical Foundations of Thermodynamics?*

Consider a (closed) classical n-particle system. Its micro-state can be visualized as a single point $\{\vec{r}_\nu, \vec{p}_\nu\}$, $\nu = 1, 2, \ldots N$ in an abstract $6N$-dimensional space, the so-called Γ-space. The Hamiltonian motion of this point generates a trajectory. The quasi-ergodic hypothesis postulates that this trajectory in Γ-space would eventually pass arbitrarily close to any point on the so-called energy surface. (This surface comprises all points in Γ-space with a given total energy.) As a consequence the long time average of any observable would equal the so-called ensemble average, the latter being the average over all those accessible micro-states.

However, ergodicity has been proven for very few model systems only; furthermore, the idea that all thermal measurements would be long-time averages in the above sense is anything but convincing.

The Gibbs ensemble theory tries to bring in another form of coarse-graining: The idea is that a given macroscopic observation will be consistent with a very large number of micro-states. This uncertainty follows from the assumed limited control available to the observer/agent, it is thus operational in origin. Those "possible" micro-states form the appropriate ensemble; it can be visualized as a certain distribution of points in Γ-space.

Now it is tempting to associate this uncertainty, the volume of occupied points, with the entropy. Unfortunately, this volume turns out to be a constant of motion under conventional Hamiltonian dynamics. So, in this classical picture another coarse-graining has to come for rescue: The Γ-space is divided into cells, on which a coarse-grained entropy can be defined. While this entropy, indeed, may increase as a function of time, it depends on the artificially introduced cell size.

4.2.2 *What are Quantum Foundations of Thermodynamics?*

The term "typicality" has been coined by Lebowitz [Lebowitz (1993)]. It is not far away from the Gibbsian ensemble concept. The idea is that very many micro-states may—for some functions defined on the micro-state space—yield very similar results. These results are thus "typical" for that region in state space. On the other hand side, from knowing those results the various micro-states appear to be indistinguishable. One may thus say that all those states are "occupied at the same time." Quite obviously this has an operational touch.

In the quantum domain the micro-state space will be the respective Hilbert space. While typicality applies to a wide range of observables, it turns out that it does not apply for the state as a whole. The situation changes for partitioned systems, though: There may be typicality of the subsystem state (i.e., its reduced density matrix) given a large enough environment. The evolution from low toward high local entropies can be expected.

The grand picture is this: Assuming the total system to be in a pure state, the local entropy becomes a measure for the quantum entanglement between the system of interest and its quantum embedding. The typicality of high local entropy thus amounts to saying that among all possible pure states of the total system most local states are highly entangled [Lubkin (1978)]. In this sense entanglement is not special but typical. And thermal equilibrium is typical—as a consequence of our inability to isolate. The local state relaxes toward a stationary equilibrium state, while the total state continues to evolve unitarily. This is an interesting answer to the old problem: How to reconcile fundamental reversible motion with irreversible behavior, which so often dominates actual phenomena. At the end of the day it is a question of perspective. The intuitive and nice feature of quantum thermodynamics is the fact that the emergence of thermal behavior can easily be studied based on exactly solvable quantum models [Borowski (2003)].

In summary: Systems are not thermal as such—they become thermal in response to their embedding. Thermal behavior is contextual. And irreversibility is only "apparent" (based on the context): The total system dynamics remains unitary.

The typicality picture helps to clarify also the tension between subjective ignorance and fundamental uncertainties: We may not know the initial state of the system of interest. But even if we did, it would not make any difference: The final state is entangled, and locally looks almost the same, irrespective of its initial preparation. It is an attractor state, which represents fundamental uncertainties. A measurement on this local state would imply a new initial state, at least some perturbation; however, this modified state would rapidly relax back toward the previous local equilibrium. The measurement (as a subjective gain of information) is thus "useless" with respect to predicting the stable properties. The influence of the environment wins.

4.2.3 *What is the Origin of Irreversibility?*

Fundamental evolution equations are typically invariant under time inversion. This is in stark conflict with our everyday experience in the macroscopic world. An arrow of time is said to exist if one can reach some state B from A but not A from B. Such instants of irreversibility dominate thermodynamic processes, which "spontaneously" occur only if the respective entropy change is non-negative: as a consequence very many phenomena (like the spontaneous fixing of a broken glass) are never observed. Any attempt to derive thermodynamics from an underlying reversible theory thus faces a severe problem (see, e.g., [Gemmer (2009)]). This is highlighted by the so-called Loschmidt paradox [Loschmidt (1876)]: By an (instantaneous) reversal of all velocities of a classical gas, for example, one should be able to get back to the initial state. Boltzmann's reaction to this assertion has essentially been operational: "Then try to do it" [Gorin (2006)]. This argument indicates that irreversibility may not just be "out there" but rather a consequence of the limited control by the respective observer/agent.

How to incorporate such limitations into the model of the observer? One basic "trick" is to give him access to some kind of limited resources only, so that he has to settle for a so-called coarse-grained description and incomplete control. Appropriate observer interfaces can be modeled in various ways.

L. Boltzmann around 1866 came up with such a model by introducing cells for the state space of classical particles each characterized by its position and velocity (so-called μ-space). Each cell would be numbered by an average position $\vec{q}$ and an average velocity $\vec{v}$, that is, six coordinates. He then introduced a function $f(\vec{q}, \vec{v}, t)$ as the number of particles per cell at time t. The evolution of f is then controlled by an "effective equation", meanwhile known as the Boltzmann equation. This equation has to account for the scattering of particles (resulting in many-particle correlations). An autonomous equation for f can thus be reached only by employing approximations.

In what sense is quantum mechanics able to account for irreversibility? It has been shown [Waldherr (2010a)] that the arrow of time in a quantum system subject to unitary dynamics is not an absolute property of the system as such but rather a consequence of two externally imposed conditions: incomplete control (while preparing the Loschmidt echo) and local observation (focusing on a small part of the total system only). The emergence of the respective time asymmetry can be demonstrated by reducing the size of the observed part with respect to the total system. The "world as a whole" has no arrow of time.

4.2.4 *What are Thermodynamic Processes?*

According to quantum thermodynamics (cf. Section 4.2.2) there is, in principle, an effective equation of motion for the embedded subsystem, but the marginal dynamics can be neglected in proper limits. Equilibrium thermodynamics is essentially a static theory (cf. Section 4.2.7).

In order to induce a dynamics at all (in a well-defined fashion), macro parameters have to be changed from the outside. A pertinent example is the change of volume V of an ideal gas, say, under isothermal conditions, that is, temperature T being constant. The resulting process defines a trajectory in the control space spanned by the macro variables V and T. While time does not show up here explicitly, one has to make sure that the actual process is carried out slowly enough in order to guarantee thermal equilibrium at each instant.

From the point of view of the system the process is non-autonomous; the rules, by which the parameter dynamics is implemented, are not part of the thermodynamic theory. These rules can be taken to define an external "agent".

It is clear, though, that the actions of the agent have to be subject to the laws of physics. The agent is not a demon operating from outside the physical world.

The change of volume of a gas can be realized by means of a movable piston. Because of the availability of this mechanical device the volume is a very convenient control parameter for the agent. The same holds for the temperature, a thermal control parameter: Coupling the system to a thermal reservoir of some given temperature an isothermal process can easily be implemented. This is because in equilibrium the system and reservoir are on the same temperature (zeroth law of thermodynamics).

Such control features can be applied even in the quantum domain (cf. Section 5.1). The analogue to the gas in a cylinder with movable piston is a single particle in a box of varying size V. Decreasing the size (i.e., modifying the particle potential) will increase the energy of the quantized levels. For fixed occupation of levels the average energy of the particle will thus increase: This represents a positive pressure p. (p is the negative partial derivative of the average energy with respect to volume V.)

But where does the changing volume V come from? It could be re-interpreted as an effective dynamics—in turn based on quantum theory. From a more detailed point of view the external agent can thus be included in the theoretical model as another quantum system. We thus attempt to replace the non-autonomous control by pertinent design requirements (cf. [Schröder (2010); Abah (2012)]). Limitations encountered in such preliminary models have to do with the fact that interactions in quantum mechanics typically lead to entanglement, that is, the effective local subsystem dynamics looses coherence. Mechanical driving, however, requires the persistence of coherence on time scales very long compared to any observation time, that is, the validity of a classical limit, cf. Section 3.3.8.

4.2.5 *What is the Difference between Work and Heat?*

An important aspect of thermodynamic processes is their energetic impact. For the first law of thermodynamics work has to be distinguished from heat: This law states that the internal energy of a system can change in two ways: by adding work or by adding heat. Work and heat thus constitute two different process types rather than two different forms of energy: There is no way to uniquely decompose a given total internal energy into work and heat. Consequently, neither work [Talkner (2007)] nor heat can be an observable, that is, in the quantum context there are no operators for these terms. If there were, we could calculate their respective energy contributions for any given state—in contradiction to the above statement.

The two different processes are related to the typical means of control available in the thermodynamic setting. As the internal energy is the average energy of the system under consideration for a given thermal state, there are two fundamental ways for change: Change the system or change the state.

The change of the thermal state is related with heat. The change of the system requires to modify certain parameters that specify the system; in principle, such modifications might affect anything, from the particle mass to the particle–particle interaction strength. But a simple and well-established parameter (because easily accessible) would be the volume of a gas. Indeed, to compress a gas, say, requires work.

The concept of work and heat, like the other concepts of thermodynamics, are based on thermal equilibrium. Furthermore, introducing changes means to couple the system to an appropriate physical environment. These couplings must be weak in order to guarantee additivity of the various subsystem energies involved.

Work is a concept originally defined within classical mechanics. It characterizes the change of the total energy of some mechanical system due to a process, during which an external force is applied. A simple example would be the action of the gravitational force on a falling body. Such entirely mechanical models may serve as work reservoirs, see Definition 4.8.

Models which can only accept heat are called heat reservoirs. General (thermal) systems have access to both types of energy change.

Observation is a central theme not only in quantum mechanics but also in thermodynamics. Not being an observable does not mean that work or heat could not be measured at all. However, the respective strategy has to be indirect: In quantum mechanics the measurement of work can be reduced to two separate energy measurements, one at the beginning and one at the end of the respective process. Of course, for this strategy to hold one has to make sure that no heat has been exchanged in parallel.

Such quantum processes under observation have been studied in the context of the so-called quantum Jarzynski relation. The results are measurement-induced fluctuations [Mukamel (2003)]. These fluctuations are quantum mechanical in origin, but, nevertheless, fulfill the same relations as derived within a classical context (cf. Section 4.2.7).

4.2.6 *What is the Role of Information in Thermodynamics?*

Can one switch between different levels of description? A key concept appears to be information. However, for information to have a say, it should be connected to physics, it must become operational.

"It's about this man who is a very good man, who dies and goes up to heaven and he meets an angel there, and the angel has a big bowl filled with hard spheres. And the guy asks: Do you know every position and velocity? The angel says yes. Then the guy asks, what is the entropy of this system, the thermodynamic entropy?"

Obviously, the angel has everything he needs to answer all kind of questions. Does this mean that "his entropy" would have to be zero?

"Sure, you can answer any questions but . . . it is still true that if you start with one part hot and the other part cold, heat will go from the hot to the cold, and as you know, entropy describes that evolution of going from hot to cold but it doesn't change at all that angel knows everything" [Lebowitz (2011)].

This little tale illustrates the delicate role information plays in thermodynamics. As far as the heat exchange is concerned, the angel would not make use of his knowledge; it remains abstract and entirely disconnected from the actual scene. This situation would change, however, if the angel had means to translate his knowledge into directed physical interventions (cf. Section 1.3), that is, if he became an "active agent." In this way he might be able to extract more work than would be possible otherwise.

However, if the angel and thus his actions were subject to physical laws, such interventions can easily turn out to be rather costly in terms of resources needed. Such limitations have thoroughly been discussed in the context of Maxwell's demon. Physical consistency then shows that—contrary to naive expectation—the second law of thermodynamics is not violated (see, e.g., [Raizen (2009)]). For this to hold the concept of entropy and thus of equilibrium thermodynamics must still be applicable.

It is well known that the statistical entropy can formally be identified with the Shannon entropy, which supports an entirely information-theoretical interpretation. An extreme position has been taken by Jaynes [Jaynes (1957)]: His basic idea is that entropy presents a measure for missing information about the "actual" micro-state of the thermal system in question. The whole approach can thus be interpreted as a method to rationally deal with incomplete knowledge. While interesting in its own right, it is not a physical theory at all.

Furthermore, this subjective ignorance idea would be compatible only with the classical picture, according to which a given system, however complicated, always has to be in some well-defined, if (subjectively) unknown state. This picture is not in accord with the quantum approach (cf. Section 4.2.2), where the equilibrium state would be mixed and stationary as a consequence of its (weak) interaction with a large quantum environment. "Knowing the state" would then require appropriate observations/measurements, which, according to quantum mechanics, will in general change the state under observation.

In fact, complete knowledge would imply a pure state rather than the original mixed (thermal) state.[a] Such a pure state, though, would not be stable (unless the thermal state was already pure, that is, of zero temperature): Due to the environment the state will rapidly relax back to its initial thermal state, the information gained from the measurements is useless and lost again. Given certain models for system and environment, the thermodynamic description becomes mandatory.

Nevertheless, there are fairly stable routes to information change: this is by means of an entropy change induced by a thermodynamic process (within a thermodynamic machine, say). Under appropriate conditions this entropy change can directly be related to extractable work [Alicki (2004)]. One may thus talk about the "work value of information." This is an effective description, though: The entropy, interpreted as missing information, is changed via a change of its macroscopic control parameters, not by explicit information retrieval. It has the advantage that this change of information immediately becomes operational—contrary to the angel's surplus knowledge discussed above. And it is optimal in terms of expenditure, cf. Maxwell's demon (Theorem 4.19).

4.2.7 *What is the Nature of Thermal Fluctuations?*

According to quantum thermodynamics the observer considers a partition of the total system into the system of interest and its environment. The eternal motion of the total state then leads to marginal temporal fluctuations in the quasi-thermal state of the subsystem of interest. There are various ways to describe these fluctuations. A simple parameter is the ground state occupation

[a]As knowledge requires measurement and measurement tends to change the state in question, a known and an unknown state must, in general, differ. Because of this feedback, knowledge (information) must physically be accounted for to guarantee consistency [Dahlsten (2011)].

probability; its variance can be shown to disappear proportional to $1/\sqrt{n^E}$, where n^E is the number of accessible states in the environment [Gemmer (2009)]. An alternative measure would be the local (reduced) entropy. This means that for small environments the entropy should fluctuate around its equilibrium value. For large enough environments any such "autonomous state fluctuations" can be neglected, though. In the following, we constrain ourselves to such stationary equilibrium states in a strict sense. Note that, beyond the choice of partition, the theoretical description of the system does not yet include any concrete physical interaction with the observer.

In the underlying statistical theory, it is the time-independent distribution function which characterizes the thermal state. The macro-variables are interpreted as average values (expectation values) over this distribution function, and thus cannot fluctuate—in agreement with the assumptions of equilibrium thermodynamics. Likewise the distribution function can be used to calculate higher moments of those macro-variables and thus uncertainty measures like the variance. But what does this uncertainty mean with respect to actual observations, with respect to concrete thermal phenomena? We are back to measurement problems. Here we restrict ourselves to the quantum description.

Measurement results of quantum observables like energy or angular momentum can be predicted as usual (cf. Section 3.2.3): we expect respective eigenvalues to occur with a probability that can uniquely be calculated from the state. (This procedure does not depend on the fact that we have a thermal state; it could as well be a pure state or anything else.) If such measurements are repeated on the same initial state (i.e., after thermalization), the results are not the same but fluctuate. These measurement-induced fluctuations cannot be avoided; they confirm the underlying quantum uncertainties. The time average over a longer and longer sequence of such tests tends to reproduce the thermal expectation value.

The intensive thermal parameters such as pressure or temperature are introduced as derivatives of thermodynamic potentials; they are not observables in the sense of quantum mechanics, there are no associated operators. Measurements are thus indirect at most. Should we still expect fluctuations?

Temperature measurements have to be based on an estimation scheme, the input of which are quantum measurements in a strict sense. One way to do so is to exploit the fact that there is a unique relationship between average energy and temperature. Replacing the unknown average energy by the single energy measurement result, an "estimated" temperature obtains. For not too small quantum objects prepared in a state with not too low temperature T this estimated temperature T_m will fluctuate according to $(\Delta T)^2 = T^2/C$, where $\Delta T = T_m - T$ and C is the heat capacity [Jahnke (2011a)]. C is extensive so that the importance of those fluctuations rapidly goes down as the system size increases. It is remarkable that projective quantum measurement theory reproduces an old classical result [Landau (1980)], but now with a clear operational meaning. There are no fluctuations, "when nobody looks": T does not fluctuate, only T_m—eventually due to the quantum uncertainty of the energy E! The so-called "thermal fluctuations" (i.e., fluctuations of thermal variables) are thus traced back to (measurement-induced) quantum fluctuations.

This expression for the temperature fluctuation is a pertinent example for a fluctuation–dissipation theorem, where the heat capacity C as a thermal response function characterizes dissipation. In fact, the (classical) temperature fluctuation formula can formally be derived from an effective linear response, just as is done to reach the famous Nyquist formula [Balatsky (2003)].

Besides the above individual measurements (local in time) one can also consider periodically repeated measurements as a kind of "permanent supervision." Because of the feedback associated with each quantum measurement step (information retrieval) the resulting quantum thermodynamics under observation will, in general, differ from the isolated case. Fortunately, there are situations in which both versions coincide, though in a very specific sense [Jahnke (2011b)]: The time average of the measured data approaches the ensemble average as obtained in the unobserved case. Once again, quantum mechanics confirms and reinterprets a controversial classical concept, in this case the ergodic theorem. For more details see Section 5.5.

4.2.8 *Beyond Thermodynamics: Are There Fundamental Limits?*

Thermal behavior has been recognized as contextual; non-thermal behavior should thus also result from appropriate contexts. Different contexts often mean different levels of description. When these are at our disposal or even become mandatory, would it necessarily improve our abilities to perform tasks?

There should be scenarios, which show deviations from thermal behavior on various scales. Such deviations can be classified by the gradual breakdown of certain thermal properties. There could be dependencies on initial conditions, increased fluctuations, violation of specific fundamental relations between thermal variables, etc. In the nano-domain specific preparation strategies can be designed to realize stationary non-thermal states [Ji (2011)]. For a single spin embedded in spin environments, for example, the relaxation to non-thermal states has been investigated in Ref. [Schmidt (2005)]. Here, specific spatial structures have been taken to map onto specific coupling pattern between system and environment. These pattern are shown to exclude simple analysis in terms of standard thermal predictions.

The dynamical response of closed systems after a quench has found considerable interest recently: Here not all observables can be expected to equilibrate. In a theoretical investigation [Gogolin (2011)] it has been shown that even the condition of non-integrability is no guarantee for thermalization. The interest in such kinds of quench dynamics has mainly been triggered by experiments with cold atomic gases.

Additional information might be used to break out from the pure thermodynamic level to the underlying, more detailed one. More information should increase control: "... you can do more things if you know more" [Lebowitz (2011)]. With respect to thermodynamic machines the conversion of heat into work might thus be improved—motivating the question: What is the work value of information? Could the second law be violated? Maxwell's demon in thermodynamics (and its exorcism) presents an interesting example: In fact, the demon could be seen to represent a computer-controlled agent. It has been shown that system and demon together cannot violate the second law in terms of total cost [Raizen (2009)].

When it comes to a real "quantum control" far beyond the thermal standards, quantum computation [Nielsen (2000)] should count as a pertinent example, in fact, as some remote "promised land." The well-known challenge is decoherence, that is, the need for almost perfect isolation. Right now, nobody knows whether there are hard fundamental constraints prohibiting large-scale quantum information processing; what one can say, though, is that any such implementation would have to become very costly in terms of the underlying control machinery. *Beating typicality cannot be easy.*

4.3 The Formalism

Thermodynamics is a "closed" macro-theory. Nevertheless, on a foundational level its basic concepts (like equilibrium state, entropy) have a macro- and a micro-formulation. Obviously these two descriptions are strictly interrelated—despite the conceptual closedness of the former.

Macro-descriptions are coarse-grained, they provide much less "information" than the (full) micro-description. A strong link to information theory should thus be expected. Questions about the possible "use" of (additional) information to improve on thermodynamic functioning have to be addressed; such problems are often introduced via so-called "demons," like Maxwell's demon. However, information (i.e., the reduction of the lack of information) is not free of charge. Its cost has to enter any complete thermodynamic modeling.

4.3.1 *Micro-Description*

Remark 4.1 Macro versus micro *The notion of a macro-state is relative and cannot be defined without reference to an underlying micro-description. As such we can take classical mechanics (the traditional view) or quantum mechanics. The following discussion will be based on quantum mechanics as the appropriate micro-theory.*

Quantum theory shares with thermodynamics a "natural" concept of entropy, cf. (3.46). In general, however, these two concepts do not coincide.

Basic to the analysis of a physical scenario on any level of description is the distinction between system, state, and dynamics.

Definition 4.1 Micro-description $\{\hat{H}, \hat{\rho}\}$ *For a full quantum description a (closed) system is defined via its respective Hamilton operator $\hat{H}$, the state via the corresponding density operator $\hat{\rho}$, and the dynamics via the Liouville–von Neumann equation (cf. Section 3.3).*

Micro-states underlying thermodynamics are stationary:

Definition 4.2 Stationary states: Parametrization *From the Liouville–von Neumann equation, Eq. (3.270), we have*

$$\frac{d\hat{\rho}}{dt} = -\frac{1}{i\hbar}\left[\hat{\rho}, \hat{H}\right] . \tag{4.1}$$

A stationary state as a constant of motion should thus commute with the Hamiltonian, cf. Proposition 3.14. If $\{\hat{G}_k\}$, $k = 1, 2, \cdots$ are a complete set of commuting operators including $\hat{H}$, cf. Definition 3.18, a stationary state $\hat{\rho}$ must be an operator function of the form

$$\hat{\rho} = f(\hat{G}_k) . \tag{4.2}$$

Stationary states can thus be defined only relative to a given model $\hat{H}$.

In the space of stationary states there is, by definition, no eigendynamics. Changes—if they occur at all—are thus non-autonomous and must be imposed from the outside.

Definition 4.3 Equilibrium states *Equilibrium states $\hat{\rho}_{equ}$ are stationary states, which maximize the respective von Neumann entropy (cf. Definition 3.46) subject to certain so-called macro-constraints (cf. (Section 4.3.2)).*

$$S_{vN}(\hat{\rho} = \hat{\rho}_{equ}) = \quad max! \tag{4.3}$$

In the following we will mainly be concerned with such states.

Definition 4.4 Internal energy *The internal energy is a functional, defined as the expectation value of the underlying Hamiltonian $\hat{H}$ with respect to a micro-state taken from the subset of equilibrium-states,*

$$\boxed{U(\hat{H}, \hat{\rho}_{equ}) = \mathrm{Tr}\{\hat{H}\,\hat{\rho}_{equ}\} .} \tag{4.4}$$

While such an expectation value can be defined for any micro-state, it has the meaning of an internal energy for $\hat{\rho} = \hat{\rho}_{equ}$ only.

Definition 4.5 Mechanical control, *Hamilton operators always contain certain (classical) model parameters. It is assumed that at least one of those, γ, can be controlled from the outside:*

$$\hat{H} = \hat{H}(\gamma) = \sum_i E_i(\gamma)|e^{(i)}(\gamma)><e^{(i)}(\gamma)|\,. \tag{4.5}$$

The parameter γ will be taken to be dimensionless. In general, the eigenfunctions for different parameters are not orthogonal. Consider, for example, their real-space representation. Obviously, for $\gamma_1 \neq \gamma_2$

$$<e^{(i)}(\gamma_1)|e^{(j)}(\gamma_2)>=\int dx <x|e^{(i)}(\gamma_1)>^*<x|e^{(j)}(\gamma_2)>\neq \delta_{ij}\,. \tag{4.6}$$

If all the $\hat{H}(\gamma)$ are defined in the same Hilbert space, a given eigenfunction for one parameter can be expanded in terms of the eigenfunctions of any other parameter. (This does not work, e.g., for the particle in a box, cf. Example 6.8.)

Definition 4.6 Spectral deformation *For the spectrum it is convenient to restrict oneself to relations of the type [Birjukov (2008)]*

$$\boxed{E_i(\gamma) = g(\gamma)\epsilon_i\,,} \tag{4.7}$$

where g is some (dimensionless) monotonous scaling function and the ϵ_i are to be regarded as a set of characteristic constants (in units of energy). In this way, the ratio between any two transition frequencies, ω_{ij}/ω_{kl}, where $\omega_{ij} = (E_i - E_j)/\hbar$, is invariant under a change of γ. The "motion" of each energy level can easily be followed, a non-degenerate spectrum remains non-degenerate.

Example 4.1 Mechanical control functions *Simple model Hamiltonians giving rise to such functions $g(\gamma)$ are*

- *Spin in a magnetic field B (in given direction): The spectrum is $E_\pm = \pm(B/B_0)\hbar\Gamma B_0/2 \equiv \gamma\epsilon_\pm$. Here, Γ is the gyromagnetic ratio and B_0 some standard unit magnetic field. We thus have*

$$B/B_0 = \gamma \mapsto g(\gamma) = \gamma\,. \tag{4.8}$$

- *Non-interacting network of N spins in a magnetic field B:*

$$E_j = j\hbar\Gamma B_0(B/B_0) \equiv \epsilon_j\gamma \quad j = 1, 2, \ldots N\,, \tag{4.9}$$

$$B/B_0 = \gamma \mapsto g(\gamma) = \gamma\,. \tag{4.10}$$

Except for the lowest and highest state these levels are degenerate.

- *Linear, harmonic oscillator:* $E_n = \hbar\gamma\omega_0(n + 1/2) \equiv \gamma\epsilon_n$, $n = 1, 1, 2, \ldots$

$$(\omega/\omega_0) = \gamma \mapsto g(\gamma) = \gamma\,. \tag{4.11}$$

For a mechanical oscillator we have $\omega_0 = \sqrt{(k_0/m_0)}$*, where* m_0 *is the mass of the particle,* k_0 *the spring constant, both defining some standard reference unit. Instead of* ω *we could alternatively use* $\gamma = k/k_0$ *with* $g(\gamma) = \gamma^{1/2}$ *or* $\gamma = m/m_0$ *with* $g(\gamma) = \gamma^{-1/2}$ *as the external control.*

- *Particle in an D-dimensional box of volume* $V = L^D$*: Independent of D the spectrum is* $E_n(L) = \hbar^2\pi^2n^2/(2mL_0^2)(L_0/L)^2 \equiv \epsilon_n(1/\gamma^2)$, $n = 1, 2, \ldots$ *Here, m is the mass of the particle and* L_0 *some standard unit length. We thus have*

$$\boxed{(V/V_0) = \gamma \mapsto g(\gamma) = \gamma^{-2/D}\,.} \tag{4.12}$$

This micro-model is close to the macro-control pertaining to a many-particle gas in volume V, cf. Definition 4.23.

- *Ultra-relativistic Dirac particle in a one-dimensional box: Based on the Compton wavelength*

$$\lambda_C \equiv \frac{2\pi\hbar}{mc}\,, \tag{4.13}$$

the spectrum for a particle of mass m confined to a box of length L is found to be [Muñoz (2012)]

$$E_n^{Dirac} = mc^2\left(\sqrt{1 + (n\lambda_C/2L)^2} - 1\right) \equiv \epsilon_n\,, \tag{4.14}$$

with $n = 1, 2, \ldots$*. In the ultra-relativistic limit,* $\lambda_C \to \infty$*, on finds*

$$E_n^{Dirac} = \frac{n\pi\hbar c L_0}{(L/L_0)} \equiv \epsilon_n\gamma\,, \tag{4.15}$$

where

$$(L/L_0) = \gamma \mapsto g(\gamma) = \gamma\,. \tag{4.16}$$

This limit can be of interest also for graphene systems.

Note that the change of any Hamilton parameter and any combinations thereof—even including coupling constants within a complicated multipartite system, say—would allow to influence the spectrum, though not necessarily of the form Eq. (4.7). (For an example see Section 7.5.3.)

Definition 4.7 Partial quantum control *Let the Hamiltonian of the control system be split into two parts, cf. [Feldmann (2003)]:*

$$\hat{H}(\gamma) = g(\gamma)\hat{H}_c + \hat{H}_{nc}, \tag{4.17}$$

$$[\hat{H}_c, \hat{H}_{nc}] \equiv \hat{C}. \tag{4.18}$$

$g(\gamma)\hat{H}_c$ is the "controlled part" and via $g(\gamma)$ explicitly time dependent; taken by itself it is an instant of spectral deformation with parameter-independent eigenstates. $\hat{H}_{nc}$ is the "uncontrolled" part, $\hat{C}$ the commutator between the two. The controlled part might describe two non-interacting spins in an external field, the uncontrolled part their mutual interaction, cf. Example 3.9. In general, such combined models tend to violate the condition of spectral deformation.

Note that for $\hat{C} \neq 0$ the total Hamiltonian at different γ does no longer commute, in general.

Definition 4.8 Work reservoir *A simple one-particle model (momentum $\hat{p}_x$, position $\hat{x}$) subject to mechanical driving is*

$$\hat{H}(\gamma) = \hat{H}_0(\hat{p}_x, \hat{x}) - F_x(\gamma)\hat{x}, \tag{4.19}$$

with $F_x(\gamma)$ denoting the external force; its time dependence must be given. Also this model is non-autonomous.

Such a system functions as a "work reservoir": it only accepts (mechanical) work as an input to change its internal energy, $\Delta U =< \hat{H}(\gamma_2) > - < \hat{H}(\gamma_1) >$.

Remark 4.2 Mechanical closure *The time-dependent control, $g(\gamma(t))$, constitutes an effective theory; it must eventually derive from an enlarged quantum system (with a time-independent Hamiltonian) in some appropriate quasi-classical limit. Examples of autonomous extensions are discussed in Section 7.5.2.*

4.3.2 *Equilibrium and Non-Equilibrium on the Micro-Level*

Theorem 4.1 Second law of thermodynamics: Irreversibility *The entropy S of a closed system can only increase (or stay constant). In equilibrium the state reaches maximum entropy (under given constraints).*

This property defines the notion of irreversibility: One can reach state $\hat{\rho}_b$ from $\hat{\rho}_a$ only if $S(b) \geq S(a)$. This does not mean, however, that this specific transition could actually be observed[a]*; the second law is a statement on the asymptotically reached state, not about the actual path.*

The equilibrium states are special stationary states. Their standard form can be derived from the maximum entropy principle (MEP), cf. Example 2.4. Here we follow Ref. [Mahler (1998)].

Proposition 4.1 MEP *Let the operators $\hat{G}_k$ specify some appropriate level of observation; they are taken to commute with $\hat{H}$. Subject to the macro-constraints ($k = 1, 2, \ldots b$)*

$$G^{(k)} = \mathrm{Tr}\{\hat{G}_k \hat{\rho}\} \tag{4.20}$$

and the normalization condition

$$1 = \mathrm{Tr}\{\hat{\rho}\}, \tag{4.21}$$

the von Neumann entropy $S_{vN}(\hat{\rho}) = -\mathrm{Tr}\{\hat{\rho} \ln \hat{\rho}\}$ is maximal for $\hat{\rho} = \hat{\rho}_{equ}$, where

$$\boxed{\begin{aligned} \hat{\rho}_{equ} &= \frac{1}{Z} \exp\left\{-\sum_{k=1}^{b} \beta_k \hat{G}_k\right\}, \\ Z(\beta_k) &= \mathrm{Tr}\left\{\exp\left(-\textstyle\sum_k \beta_k \hat{G}_k\right)\right\}. \end{aligned}} \tag{4.22}$$

Here Z is the partition sum, and the β_k are Lagrange multipliers determined by the maximum principle and the constraints.

[a]A classical example has been given in [Mead (1977)]: Here state (a) and (b) are both non-equilibrium states. In order to avoid ending up in (b) the authors suggest an additional condition termed "mixing character." But such a "sharpening" of the second law is not needed. The true equilibrium state is simply that of maximum entropy.

Proof. Let $|\rho_m>$ denote the eigenfunctions of the trial operator $\hat{\rho}$ with $\hat{\rho}|\rho_m>=\rho_m|\rho_m>$ and $|k_m>$ denote the eigenfunctions of $\hat{\rho}_{equ}$ with $\hat{\rho}_{equ}|k_m>=k_m|k_m>$. Then

$$\mathtt{Tr}\{\hat{\rho}\ln\hat{\rho}\}-\mathtt{Tr}\{\hat{\rho}\ln\hat{\rho}_{equ}\}=\sum_m \rho_m\left(\ln\rho_m-<\rho_m|\ln\hat{\rho}_{equ}|\rho_m>\right). \tag{4.23}$$

Completeness of the basis functions $|k_n>$, that is, $\hat{1}=\sum_k|k_n><k_n|$, allows to rewrite

$$\begin{aligned}<\rho_m|\ln\hat{\rho}_{equ}|\rho_m> &= \sum_{n,n'}<\rho_m|k_n><k_n|\ln\hat{\rho}_{equ}|k_{n'}><k_{n'}|\rho_m> \\ &= \sum_n|<\rho_m|k_n>|^2\ln k_n. \end{aligned} \tag{4.24}$$

As $\sum_n|<\rho_m|k_n>|^2=1$ for any m, we obtain

$$\begin{aligned}\mathtt{Tr}\{\hat{\rho}\ln\hat{\rho}\}-\mathtt{Tr}\{\hat{\rho}\ln\hat{\rho}_{equ}\} &= \sum_{m,n}|<\rho_m|k_n>|^2[\rho_m(\ln\rho_m-\ln k_n)] \\ &\geq \sum_{m,n}|<\rho_m|k_n>|^2[\rho_m-k_n]=0 \end{aligned} \tag{4.25}$$

In the last step we have used the inequality

$$\ln x\geq 1-\frac{1}{x} \tag{4.26}$$

for $x=\rho_m/k_n$ and, again, the above completeness relation.

In the final step we conclude from the definition of $\hat{\rho}_{equ}$ that

$$\ln\hat{\rho}_{equ}=-\sum_k\beta_k\hat{G}_k-\ln Z \tag{4.27}$$

and thus

$$\mathtt{Tr}\{\hat{\rho}_{equ}\ln\hat{\rho}_{equ}\}=-\sum_k\{\beta_k\mathtt{Tr}(\hat{\rho}_{equ}\hat{G}_k\})-\ln Z. \tag{4.28}$$

According to Eq. (4.9) $\hat{\rho}_{equ}$ in the trace terms can be replaced by $\hat{\rho}$, so that

$$\mathtt{Tr}\{\hat{\rho}_{equ}\ln\rho_{equ}\}=\mathtt{Tr}\{\hat{\rho}\ln\rho_{equ}\}. \tag{4.29}$$

Putting this into Eq. (4.25) we finally get

$$S_{vN}(\hat{\rho})\leq S_{vN}(\hat{\rho}_{equ}). \tag{4.30}$$

The trial states $\hat{\rho}\neq\hat{\rho}_{equ}$ are non-equilibrium states . □

The energy representation is the joint eigenrepresentation of all the $\hat{G}_k$ and $\hat{\rho}_{equ}$. In this representation we find

$$\rho_\nu = < \nu|\hat{\rho}_{equ}|\nu > = \frac{1}{Z}\exp\left(-\sum_k \beta_k G_\nu^{(k)}\right), \tag{4.31}$$

$$Z = \sum_\nu \exp\left(-\sum_k \beta_k G_\nu^{(k)}\right). \tag{4.32}$$

Here we have used the abbreviation

$$G_\nu^{(k)} = < \nu|\hat{G}_k|\nu > . \tag{4.33}$$

The constraints (expectation values) can be rewritten as

$$G^{(k)} = \sum_\nu \rho_\nu G_\nu^{(k)} = -\frac{1}{Z}\frac{\partial Z}{\partial \beta_k}. \tag{4.34}$$

They implicitly specify the parameters β_k. One easily convinces oneself that

$$S_{vN} = \ln Z + \sum_k \beta_k G^{(k)}. \tag{4.35}$$

Theorem 4.2 *T* ensemble or canonical ensemble *Here, the only relevant observable is assumed to be $\hat{G}^{(1)} = \hat{H}$, there are no additional constants of motion. The resulting equilibrium state is then given by*

$$\boxed{\hat{\rho}_{can} = \frac{1}{Z_{can}}\exp\left(-\beta_1 \hat{H}\right).} \tag{4.36}$$

Written in the energy eigenbasis the probability for state ν is

$$P(\nu) = < \nu|\hat{\rho}_{can}|\nu > = \frac{1}{Z_{can}}\exp\left(-\beta_1 E_\nu\right). \tag{4.37}$$

With $G^{(1)} = U = \sum_\nu \rho_{\nu\nu} E_\nu$ and $\beta_1 = \beta$ we thus obtain

$$\begin{aligned} S_{vN} &= \ln Z_{can} + \beta U \,, \\ F &\equiv U - S_{vN}/\beta = -\frac{1}{\beta} Z_{can}\,. \end{aligned} \tag{4.38}$$

Microscopically, this so-called free energy F (cf. Section 4.3.6) can be seen as a function of $\hat{\rho}_{can}$ and $\hat{H}$. The Lagrange multiplier β will have to be related to phenomenological thermodynamics (see Eq. (4.71)).

Proposition 4.2 Internal energy and partition sum *It follows from*

$$Z_{can} = \mathtt{Tr}\{\exp(-\beta\hat{H}) \tag{4.39}$$

that

$$-\frac{1}{Z_{can}}\frac{\partial Z_{can}}{\partial\beta} =< \hat{H} >= U\,. \tag{4.40}$$

This relation can be re-written as

$$U = -\frac{\partial}{\partial\beta}\ln Z_{can}\,. \tag{4.41}$$

The internal energy U *—like the partition sum—is taken here as a function of* T, V. *Along similar lines one obtains for the variance*

$$\sigma_U^2 = \frac{\partial^2}{\partial\beta^2}\ln Z_{can} = -\frac{\partial}{\partial\beta}U\,. \tag{4.42}$$

Example 4.2 Internal energy of harmonic oscillator *Based on the eigenvalue spectrum Eq. (3.95) the partition sum is*

$$Z_{can} = \exp(-\beta\hbar\omega/2)\sum_\nu \exp(-\beta E_\nu) \tag{4.43}$$

The second factor is easily evaluated. We get

$$Z_{can} = \frac{\exp(-\beta\hbar\omega/2)}{1-\exp(-\beta\hbar\omega)} = \frac{1}{\exp(\beta\hbar\omega/2)-\exp(-\beta\hbar\omega/2)} \tag{4.44}$$

With the help of Eq. (4.41) and using $d/dx\ln(\exp x - \exp(-x)) = (\exp x + \exp(-x))/(\exp x - \exp(-x)) = \coth x$ *we finally obtain*

$$\boxed{U = \frac{1}{2}\hbar\omega\coth(\beta\hbar\omega/2)} \tag{4.45}$$

Its high-temperature (classical) limit is $U \approx k_B T$.

Remark 4.3 Emergence of equilibrium states *The maximum principle does not explain, why and how the equilibrium states are actually reached. According to quantum thermodynamics most pure states of a bipartite system—subject to some appropriate constraints—are equivalent in so far as the smaller subsystem is found to be in a specific thermal equilibrium state. These features are thus "typical" cf. Section 5.1.1).*

According to Eq. (3.166) the relative entropy of $\hat{\rho}_1$ with respect to $\hat{\rho}_0$ is given by

$$S_{vN}(\hat{\rho}_1 \| \hat{\rho}_0) = -\mathtt{Tr}(\hat{\rho}_1 \ln \hat{\rho}_0) + \mathtt{Tr}(\hat{\rho}_1 \ln \hat{\rho}_1) . \tag{4.46}$$

Proposition 4.3 Minimum principle for generalized free energy *For a given Hamiltonian model $\hat{H}$ the incremental free energy, that is, the (generalized) free energy (cf. Def. (4.29) defined for some arbitrary (non-equilibrium) state $\hat{\rho}_1$ and some given β,*

$$\mathcal{F}(\hat{\rho}_1) = U(\hat{\rho}_1) - S_{vN}(\hat{\rho}_1)/\beta \tag{4.47}$$

can be interpreted as the relative entropy of $\hat{\rho}_1$ relative to that of the respective canonical equilibrium state $\hat{\rho}_0 = \hat{\rho}_{can}$ with that β [Donald (1987)]:

$$\boxed{\mathcal{F}(\hat{\rho}_1) - F(\hat{\rho}_{can}) = \frac{1}{\beta} S_{vN}(\hat{\rho}_1 \| \hat{\rho}_{can}) .} \tag{4.48}$$

This term is minimized for $\hat{\rho}_1 \rightarrow \hat{\rho}_{can}$.

Proof. For the canonical equilibrium state, Eq. (4.36), we have, according to Eq. (4.38),

$$F(\hat{\rho}_{can}) = -\frac{1}{\beta} \ln Z_{can} . \tag{4.49}$$

The free energy difference thus is

$$\mathcal{F}(\hat{\rho}_1) - F(\hat{\rho}_{can}) = \frac{1}{\beta}\mathtt{Tr}\{\hat{\rho}_1 \ln \hat{\rho}_1\} + \mathtt{Tr}\{\hat{H}\hat{\rho}_1\} + \frac{1}{\beta} \ln Z_{can} . \tag{4.50}$$

Observing that

$$\ln \hat{\rho}_{can} = -\beta \hat{H} - \ln Z_{can} , \tag{4.51}$$

$$\mathtt{Tr}\{\hat{\rho}_1 \ln \hat{\rho}_{can}\} = -\beta \mathtt{Tr}\{\hat{\rho}_1 \hat{H}\} - \ln Z_{can} , \tag{4.52}$$

we immediately confirm the proposition. □

From Klein's inequality, Eq. (3.167), and Eq. (4.48) we conclude that $\mathcal{F}(\hat{\rho}_1)$ is minimized by the thermal state $\hat{\rho}_{can}$.

Theorem 4.3 T–μ or grand canonical ensemble *The relevant observables are total energy, $\hat{G}^{(1)} = \hat{H}$, and particle number operator*

$\hat{G}^{(3)} = \hat{N}$. *From the MEP we immediately get, using the two Lagrange multipliers* $\beta_1 = \beta$ *and* $\beta_3 = -\beta\mu$:

$$\boxed{P(\nu) = \rho_{\nu\nu} = \frac{1}{Z}\exp\left(-\beta E_\nu + \beta\mu N_\nu\right).} \tag{4.53}$$

μ *is the so-called chemical potential. We thus obtain with the expectation value* $G^{(1)} = U$ *and* $G^{(3)} = N$

$$S_{vN} = \ln Z + \beta U - \beta\mu N\,, \tag{4.54}$$

$$J \equiv U - S_{vN} + \beta U - \beta\mu N = -\frac{1}{\beta}\ln Z\,. \tag{4.55}$$

J *is the so-called grand canonical potential.*

Remark 4.4 Multi-component systems *The grand canonical ensemble can easily be generalized to various particle species:* $\hat{G}^{(3)} = \hat{N}_1, \hat{G}^{(4)} = \hat{N}_2, \ldots$. *For each particle type we get a corresponding chemical potential, cf. Eq. (4.146).*

Example 4.3 Ideal Bose gas *We consider a system of* N *identical particles described by the Hamiltonian* $\hat{H}(m)$, $m = 1, 2, \ldots, N$. *The energy spectrum of each particle is* ϵ_j, $j = 1, 2, \ldots$. *The micro-state is taken to be*

$$\nu \equiv \{n_1^{(\nu)}, n_2^{(\nu)}, \ldots n_j^{(\nu)}, \ldots\}\,. \tag{4.56}$$

Here, the $n_j^{(\nu)}$ *are the number of particles in the single-particle energy state* ϵ_j *for the micro-state* ν. *We thus have*

$$E_\nu = \sum_j \epsilon_j n_j^{(\nu)}\,, \tag{4.57}$$

$$N_\nu = \sum_j n_j^{(\nu)}\,. \tag{4.58}$$

It is convenient to fix the average $\overline{N^{(\nu)}}$ *(macro constraint) rather than to require* $N^{(\nu)} = N$ *(micro-constraint). With the two macro-constraints,* $G^{(1)} = \hat{H}, G^{(3)} = \hat{N}$, *the pertinent equilibrium state is the grand canonical state. The particles do not interact; it is thus not surprising to find that the micro-state distribution factorizes into single-particle terms,* $P(\nu) = P_1(n_1)P_1(n_2)\ldots$, *with*

$$P_1(n_j) = \frac{1}{Z^{(j)}}\exp\left(-\beta(\epsilon_j - \mu)n_j\right). \tag{4.59}$$

The average single-particle state occupation thus is

$$\boxed{\overline{n_j} = \sum_{n_j} n_j P_1(n_j) = \frac{1}{\exp\left(\beta(\epsilon_j - \mu)\right) - 1},} \tag{4.60}$$

and the corresponding result for an ideal gas of Fermions

$$\overline{n_j} = \frac{1}{\exp\left(\beta(\epsilon_j - \mu)\right) + 1}. \tag{4.61}$$

Theorem 4.4 *T–p* ensemble *The relevant observables are* $\hat{G}^{(1)} = \hat{H}$*, and the mechanical "volume operator"* $\hat{G}^{(2)} = \hat{V}$*. In particular,* $\hat{V}$ *could relate to the position* $\hat{x}$ *of a confining wall (piston), so that* $V = < x > A$*, when* A *is the fixed area of the wall and* V *the volume, cf. Section 5.5.2. In this scenario the volume as an expectation value would be "uncertain" [Diu (1989)], an idea that should sound quite familiar within quantum thermodynamics (cf. Section 5.5.3), but less so within the classical domain. Not surprisingly, this ensemble has seen much less applications.*

From the MEP, see Section 4.3.2, we thus obtain in the energy-eigenbasis

$$\tilde{P}(\nu) = \tilde{\rho}_{\nu\nu} = \frac{1}{\tilde{Z}} \exp\left(-\beta E_\nu(V_\nu) + \beta p V_\nu\right). \tag{4.62}$$

Here we have identified $\beta_2 = -\beta p$*, where* p *is the pressure, cf. Eq. (4.84). Formally this result is very similar to the grand canonical ensemble.*

Remark 4.5 Adaptive behavior and entropy maximization *An interesting shortcut method to adaption has been suggested in [Wissner-Gross (2013)]: Just as a non-equilibrium physical system tends to evolve toward higher-entropy macro-states (thus increasing its range of accessible micro-states, cf. MEP), a "path-based causal entropic force" might suffice to produce complex adaptive (intelligent) behavior (thus maximizing the overall diversity of accessible future paths). Unfortunately, the physical basis for that force is unclear yet—as is its relation to living systems.*

4.3.3 From Micro to Macro: Particle in a Box and Ideal Classical Gas

The microscopic standard derivation of the ideal gas law is via the canonical partition sum Z_{can}, from which one obtains the free energy F. The macroscopic limit here means the limit of large particle numbers, cf. Section 3.3.8.

Definition 4.9 Thermodynamic limit *The thermodynamic limit is traditionally seen as the large N-limit, where N is the number of particles of a many-particle system. As a consequence of the central limit theorem of statistics, the ratio of the size of the fluctuations over the mean value scale like $1/\sqrt{N}$. In this sense thermodynamics—neglecting fluctuations—becomes appropriate.*

Here we want to avoid this macro-limit. We choose a different route, namely via the large L-limit for the quantum particle in a D-dimensional box of edge length L. In this limit the respective energy spectrum (of a single particle!) becomes quasi-continuous.

Proposition 4.4 Continuum limit for quantum particle in a box *Given the particle to be in a canonical state its internal energy approaches for large $V = L^D$*

$$U = D/(2\beta) \tag{4.63}$$

per particle [Jahnke (2008)].

Proof. We consider a micro-model, for which

$$E_j(\gamma) = g(\gamma)\epsilon_j\,. \tag{4.64}$$

This is a spectral deformation in the sense of Eq. (4.7) with $\gamma = (V/V_0)$ and $g(\gamma) = \gamma^{(-2/D)}$, see Eq. (4.12). Let α parametrize the canonical equilibrium state, Eq. (4.49), in energy basis,

$$\rho_j(\alpha) = \frac{1}{Z_{can}} \exp\left(-\alpha\epsilon_j\right), \tag{4.65}$$

$$\alpha = g(\gamma)\beta\,. \tag{4.66}$$

The internal energy is then given by

$$U(\gamma, \alpha) = g(\gamma)\sum_j \epsilon_j\rho_j(\alpha) \equiv g(\gamma)h(\alpha)\,. \tag{4.67}$$

For large reference volume V_0 the spectrum ϵ_j becomes quasi-continuous and we approximate

$$h(\alpha) = \frac{\sum_j \epsilon_j \exp(-\alpha\epsilon_j)}{\sum_j \exp(-\alpha\epsilon_j)} \rightarrow \frac{\int_0^\infty \epsilon\rho(\epsilon)\exp(-\epsilon\alpha)d\epsilon}{\int_0^\infty \rho(\epsilon)\exp(-\epsilon\alpha)d\epsilon} = \frac{N1}{N2}. \quad (4.68)$$

For the box potential the density of states is found to be [Liboff (2002)]

$$\rho(\epsilon) \approx \epsilon^{D/2-1}, \quad (4.69)$$

so that both integrals above can be expressed via the Γ-function with argument $(D/2)$, namely $N1 = D \cdot \Gamma(D/2)/(2\alpha^{D/2+1})$ and $N2 = D \cdot \Gamma(D/2)/\alpha^{D/2}$. This leads to the simple result

$$h(\alpha) = \frac{D}{2\alpha} \quad (4.70)$$

and with Eq. (4.66) the proposition is confirmed. □

Definition 4.10 Absolute temperature *Identifying*

$$\boxed{\beta \equiv 1/(k_B T),} \quad (4.71)$$

where k_B is the Boltzmann constant and T the absolute temperature, the result, Eq. (4.63), coincides with that of the classical ideal gas,

$$U = Dk_B T/2, \quad (4.72)$$

if D is identified with f, the number of degrees of freedom per particle.

While the approach can easily be generalized to $N \geq 2$ non-interacting particles, inter-particle correlations are not included, there are no corrections due to Fermi or Bose statistics, see Example 4.3. As a result, the model becomes inapplicable in the low-temperature/high-density limit. This constraint can conveniently be expressed by reference to the respective thermal de Broglie wavelength [Yan (2000)]:

Definition 4.11 Thermal de Broglie wavelength

$$\lambda_{dB} = \frac{2\pi\hbar}{\sqrt{2\pi m k_B T}}, \quad (4.73)$$

where m is the particle mass.

Definition 4.12 Classical limit *Quantum effects are negligible as long as*

$$\tau = \lambda_{dB}/R \ll 1, \tag{4.74}$$

that is, if the average particle distance $R \leq V^{(1/D)}$ is large compared to the Broglie wavelength. For a single particle R is infinite.

The differential form of Eq. (4.67) is

$$dU(\alpha, \gamma) = \left(\frac{\partial U}{\partial \alpha}\right)_\gamma d\alpha + \left(\frac{\partial U}{\partial \gamma}\right)_\alpha d\gamma. \tag{4.75}$$

We now consider both terms separately. We rewrite the first term as

$$\left(\frac{\partial U}{\partial \alpha}\right)_\gamma d\alpha = \left(\frac{\partial U}{\partial \alpha}\right)_\gamma \left(\frac{dS_{vN}}{d\alpha}\right)^{-1} dS_{vN} = \left(\frac{\partial U}{\partial S_{vN}}\right)_\gamma dS_{vN}. \tag{4.76}$$

Observing

$$S_{vN}(\alpha) = -\sum_j \ln \rho_j(\alpha)\rho_j(\alpha), \tag{4.77}$$

we find

$$\left(\frac{\partial U}{\partial S_{vN}}\right)_\gamma = = -g(\gamma)\frac{\sum_j \epsilon_j d\rho_j/d\alpha}{\sum_j \ln \rho_j d\rho_j/d\alpha}$$
$$= \frac{g(\gamma)}{\alpha}. \tag{4.78}$$

Using Eq. (4.66) we recover the standard temperature definition:

Definition 4.13 Absolute temperature

$$k_B T = \left(\frac{\partial U}{\partial S_{vN}}\right)_\gamma = \frac{g(\gamma)}{\alpha}. \tag{4.79}$$

Definition 4.14 Heat *We identify*

$$\left(\frac{\partial U}{\partial \alpha}\right)_\gamma d\alpha = k_B T\, dS_{vN} \equiv đQ(\alpha, \gamma). \tag{4.80}$$

The partial derivative of the second term is

$$\left(\frac{\partial U}{\partial \gamma}\right)_\alpha = \frac{dg}{d\gamma}h(\alpha). \tag{4.81}$$

With $dg/d\gamma = -2\gamma^{-(2/D+1)}/D = -2g/(D\gamma)$ and Eq. (4.70) one obtains

$$\boxed{-\left(\frac{\partial U}{\partial \gamma}\right)_\alpha \gamma = \frac{g(\gamma)}{\alpha} = \frac{1}{\beta} = k_B T\,.} \tag{4.82}$$

This is the ideal gas law (for one particle)

$$\tilde{p}\gamma = k_B T \tag{4.83}$$

provided we identify $\tilde{p}$ as follows:

Definition 4.15 Generalized pressure *$\tilde{p}$ as the (negative) conjugate variable to γ, that is, with*

$$\boxed{\tilde{p} = -\left(\frac{\partial U}{\partial \gamma}\right)_\alpha} \tag{4.84}$$

and $\gamma = V/V_0$ as a dimensionless volume, cf. Eq. (4.12). $\tilde{p}$ is positive and has the dimension of an energy.

It follows that

Definition 4.16 Work

$$d_V U = \left(\frac{\partial U}{\partial \gamma}\right)_\alpha d\gamma = -\tilde{p}d\gamma = -\tilde{p}V_0 dV = -pdV \equiv đW(\alpha, \gamma)\,. \tag{4.85}$$

Putting these results back into Eq. (4.75), we get the first law of thermodynamics,

$$\boxed{dU(\alpha, \gamma) = đQ(\alpha, \gamma) + đW(\alpha, \gamma)\,.} \tag{4.86}$$

This result is exact, though model dependent: We have taken the particle in a box given in Example 4.1, which is close to the classical idea of an ideal gas within a given volume. However, the other Hamilton models would also work, provided we adjust the definition of the generalized pressure; $dg/d\gamma$ may be positive.

The present outcome suggests that typical micro-definitions of U as a function of $\hat{\rho}$ and $\hat{H}$ support an effective higher-level description parametrized in terms of some appropriate control parameters $\{\alpha, \gamma\}$ or $\{S_{vN}, V\}$.

The classical limit in the above example is not essential. Also note that we do not presuppose the thermodynamic limit; quantum thermodynamics, see Chapter 5, will be seen to be applicable down even to single particles, single spins, provided the respective quantum system is embedded by an appropriate quantum environment.

4.3.4 Macro-Description

The thermodynamic macro-description need not be based on the simple Hamilton models discussed so far; its range of applicability is much broader, details can be left open. It is an example of an effective theory, cf. Section 1.1. Its basic definitions are as follows:

Definition 4.17 Macro-description $\{Z_\nu\}$ *The thermodynamic macro-description is defined by a set of macro-variables* Z_ν, $\nu = 0, 1, 2, \cdots, m-1$. *Their total number m is small compared with the number of micro-variables. The choice of the* Z_ν *is operational: their selection reflects accessibility from the outside.*

Definition 4.18 State function *There are functional relations of the type,* $Z_1 = Z_1(Z_0, Z_2, Z_3, \cdots Z_{m-1})$. Z_1 *defines the system in the thermodynamical sense and is considered a state function of the m–1 appropriately chosen independent variables constituting the macro-state* $\vec{Z}$. *The description is "closed," that is, any thermodynamic variable of interest can be found from the state function and nothing else.*

A finite variation of macro-state variables will generally change also the dependent variable Z_1. *For a state function this change does not depend on the respective path in the m–1-dimensional macro-state space, but only on the end points. This means that*

$$\oint dZ_1 = 0 \tag{4.87}$$

along any closed circuit in state space.

Definition 4.19 Extensive/intensive variables *The* Z_ν *may be classified as extensive (i.e., scaling proportional to system size) or intensive (independent of system size).*

Definition 4.20 Equilibrium-state *Those macro-states* $\vec{Z}$ *refer to a subset of micro-states, the so-called equilibrium states,* $\hat{\rho}_{equ}$, *cf. Definition 4.3 directly connected with the notion of equilibrium is the property of stationarity. Stationarity means that closed macro-systems approach a state* $\hat{\rho}_{equ}$, *which then does no longer change spontaneously.*

Definition 4.21 Closed macro-system *"Closed" macroscopically means that no energy or particle exchange is allowed with the* *outside. This definition does not exclude interactions between a quantum system and a quantum environment that would lead to entanglement—even without the exchange of energy. Such a possibility is indispensable for the concept of quantum thermodynamics, see Proposition 5.5.*

We now start to associate the macro-variables with concrete physical entities. These identifications also specify a connection between macro- and micro-description as considered before.

Definition 4.22 Internal energy *In the standard "energy representation," we identify*

$$Z_1 \equiv U \,. \tag{4.88}$$

U is the internal energy; it is an extensive state function.

Theorem 4.5 Second law of thermodynamics: Entropy as state function *The micro-states $\hat{\rho}_{equ}$ enter the macro-description via a single macro-variable only, the thermodynamic entropy,*

$$Z_0 \equiv S \,. \tag{4.89}$$

This entropy is, in turn, an extensive state function, $S = S(U\,; Z_2, Z_3, \cdots)$.

Theorem 4.6 Equivalence of entropies I: Thermo-quantum *The thermodynamic entropy S for the macro-system U is identified with the von Neumann entropy for the underlying micro-states, $\hat{\rho}_{equ}$:*

$$S/k_B = S_{vN}(\hat{\rho}_{eq}) \,. \tag{4.90}$$

S_{vN} fulfills the thermodynamic relations for the subset of thermal equilibrium states only. This identification serves as a further corner stone for the macro–micro relationship.

Definition 4.23 Mechanical macro-control *In the underlying micro-description the "mechanical" control variable γ is understood to refer to a parametrization of the respective Hamiltonian. According to Eq. (4.85) the associated work is*

$$d_V U = \left(\frac{\partial U}{\partial \gamma}\right)_\alpha d\gamma \tag{4.91}$$

with $V = \gamma V_0$. Here we generalize to the control $d_{Z_2} U(S, Z_2) = \xi_2 dZ_2$, where Z_2 need not even be mechanical in the strict sense.

Instead, the parameters are phenomenological in that they refer to a coarse-grained description. But they still control the work done on the respective thermal system (macroscopic body).

Here are some standard examples [Römer (2005)].

(i) *Change of volume V (of a macroscopic gas, say): Rather than being directly related to a Hamilton parameter, the volume is here a phenomenological concept, cf. Section 2.2.7.*

$$d_V U = -p dV \,, \tag{4.92}$$

where $p = -\xi_2$ is the pressure.

(ii) *Change of position of macroscopic body against external force $\vec{F}$:*

$$d_x U = -d\vec{x}\vec{F} \tag{4.93}$$

(iii) *Change of polarization $\vec{P}$: Exchange of energy is here between two (classical) fields: a phenomenological "matter" field $\vec{P}$, and $\vec{E} \to \xi_2$, the electric field (acting as an external force on $\vec{P}$). The effective interaction is described by the Hamilton density $\vec{P} \cdot \vec{E}$.*

$$d_P U = \vec{E} d\vec{P} \,. \tag{4.94}$$

(iv) *Change of magnetization $\vec{M}$: Exchange of energy is between $\vec{M}$, a phenomenological field, and $\vec{B} \to \xi_2$, the external magnetic field.*

$$d_M U = \vec{B} d\vec{M} \,. \tag{4.95}$$

Note that Z_2 as well as ξ_2 are numbers that refer to the thermodynamic system as a whole. However, under (local) non-equilibrium conditions these numbers become fields, cf. Remark 4.7. This is typical for the electromagnetic models considered above.

Definition 4.24 Chemical control *Let us allow for a change of particle number $Z_3 = N$:*

$$d_N U = \mu dN \,. \tag{4.96}$$

$\mu = \xi_3$ is the chemical potential. Given that the total particle number is conserved, N can only change via exchange with the environment, $dN + dN^{env} = 0$. This kind of control can be generalized to account

for several species $N_1, N_2, \ldots$; typical applications include chemical reactions, for which individual particle number conservation does no longer apply.

Those various contributions could be combined to establish a multidimensional control space $\{Z_2, Z_3, \ldots\}$. A special two-dimensional space will briefly be discussed in the context of biological motors, Section 7.5.4. As our standard reference we will use the model characterized by $Z_2 = V$, $\xi_2 = -p$.

Theorem 4.7 First law of thermodynamics *Energy can only change via transfer from the outside: there is no internal energy production. The two types of variables, $Z_0 = S$ and Z_2, provide a natural splitting of this transfer in terms of heat and work :*

$$\boxed{\begin{aligned} dU &= \left(\frac{\partial U}{\partial S}\right)_{Z_2} dS + \left(\frac{\partial U}{\partial Z_2}\right)_S dZ_2 \\ &= đQ + đW\,. \end{aligned}} \tag{4.97}$$

This equation can be interpreted as an energy conservation law: Heat Q and work W are forms of energy, which both serve as a source of total energy change.

Remark 4.6 Operational aspects *There is no way to split a given total internal energy U of a thermal system uniquely into heat and work. The decomposition has to refer to special processes; these can easily be realized by coupling the system either to a heat or to a work reservoir. Such a reservoir can accept energy of one type only.*

While the internal energy U is a state function and thus zero along any closed path in the macro-state space, this does not hold for the heat and work terms separately: This is why we use the notation $đQ$ rather than dQ and $đA$ instead of dA.

Heat and work are energy forms but no state functions. They both depend on the path, not only on the final points reached in state space; this is essential for the operation of cyclic thermodynamic machines.

The respective partial derivatives are identified by

$$T(S, Z_2) = \left(\frac{\partial U}{\partial S}\right)_{Z_2}, \tag{4.98}$$

$$\xi_2(S, Z_2) = \left(\frac{\partial U}{\partial Z_2}\right)_S. \tag{4.99}$$

T is conjugate to S and denotes the absolute temperature, ξ_2 is conjugate to Z_2. Being defined as partial derivatives (ratios) of extensive terms, these conjugate variables must be intensive. The expression for the heat can be rewritten as

$$đQ = T dS, \tag{4.100}$$

the expression for the work as

$$đW = \xi_2 dZ_2, \tag{4.101}$$

and the balance equation, Eq. (4.97), in the so-called

Definition 4.25 Gibbs fundamental form

$$\boxed{dU = T \cdot dS + \xi_2 dZ_2 + \xi_3 dZ_3 .} \tag{4.102}$$

This means for the entropy

$$dS = \frac{1}{T} dU - \frac{\xi_2}{T} dZ_2 + \frac{\xi_3}{T} dZ_3 \tag{4.103}$$

and thus

$$\frac{1}{T(U, Z_2, Z_3)} = \left(\frac{\partial S}{\partial U}\right)_{Z_2, Z_3}, \tag{4.104}$$

$$\frac{\xi_2(U, Z_2, Z_3)}{T} = -\left(\frac{\partial S}{\partial Z_2}\right)_{U, Z_3}, \tag{4.105}$$

$$\frac{\xi_3(U, Z_2, Z_3)}{T} = \left(\frac{\partial S}{\partial Z_3}\right)_{U, Z_2}. \tag{4.106}$$

It is understood that the underlying micro-states always belong to the set of thermal equilibrium states, $\hat{\rho}_{equ}$ for $\hat{H}$.

4.3.5 *Equilibrium and Non-Equilibrium on the Macro-Level*

Theorem 4.8 Second law of thermodynamics: Entropy balance *The entropy S (as defined in Theorem 4.5) can change via two different contributions:*

$$dS = \delta_{ex} S + \delta_{in} S \,. \tag{4.107}$$

The "external production" $\delta_{ex} S$ is given by

$$\delta_{ex} S = \frac{đQ}{T} \,. \tag{4.108}$$

For the energy such an external source (transfer) would be the only one allowed (cf. theorem 4.7).

The additional "internal entropy production" $\delta_{in} S$ reflects the nature (amount of irreversibility) of the respective transfer or internal relaxation. For a closed macro-system, cf. Definition 4.21, it holds that:

$$\delta_{in} S \geq 0 \,. \tag{4.109}$$

Equivalently, the entropy production rate is positive or zero.

In equilibrium S approaches its maximum value subject to the macro-constraints given.

In general, we thus have

$$\boxed{dS \geq \frac{đQ}{T}} \,, \tag{4.110}$$

which is the Clausius formulation of the second law.

Obviously, the maximum principle can only become operative once we include non-equilibrium in some form or other. But how could this be possible on the macro-level, for which the underlying micro-state, whether equilibrium or not, does not show up explicitly? A well-known way out of this quandary is to introduce multiple macro-variables.

Definition 4.26 Macro-non-equilibrium *Consider a spatial partitioning of a many-particle system, cf. Fig. 4.1. Each part is in a separate thermal state, that is, there are multiple macro-variables, one complete set for each part. For a two-part system (A,B), for example, we would thus start from the system description*

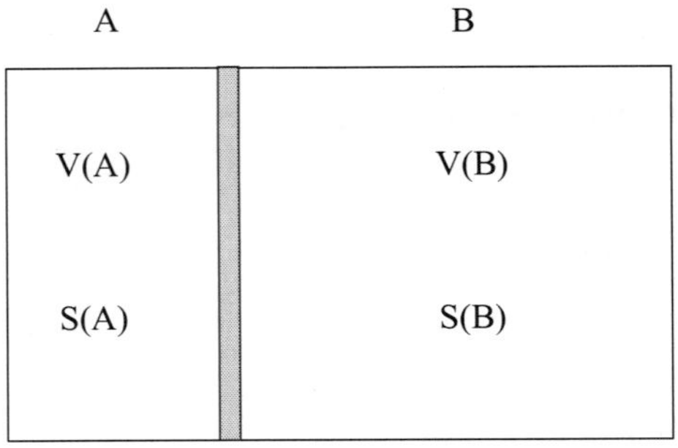

Figure 4.1 Macro-non-equilibrium. Partition via "impenetrable wall."

$U(S(A), S(B); V(A), V(B))$, that is, the number of independent state variables has doubled. Nevertheless, this state would be a stationary equilibrium state, as long as the exchange of volume and entropy was suppressed (via some appropriate "impenetrable wall"). After removing the wall the state represents a non-equilibrium state, which should thus undergo an irreversible relaxation to the global equilibrium (with but two independent state variables left). Macro-non-equilibrium underlies thermodynamic machines.

Alternatively, the wall may be replaced by membranes with selective transfer characteristics, for example, allow for heat but no particle exchange or let pass particles of a special type only. For a generalization in the nano-domain see Remark 8.1.

Remark 4.7 Thermodynamic fields *Multiple macro-variables can be seen as a special case of thermodynamic fields, based on a quasi-continuous partitioning in real space. Pertinent examples would be temperature fields, particle-density fields, or (electromagnetic) polarization fields. Thermodynamic fields based on local equilibria are constitutive ingredients of irreversible thermodynamics.*

Theorem 4.9 Thermal contact: Zeroth law *Let us consider a bipartite macro-system (A, B) with weak mutual interaction, that is, the contribution of the interaction to the total energy can be neglected. There is no energy or entropy exchange with the outside, that is, $dU = 0$ and $\delta S_{ex} = 0$. The two subsystems are allowed to exchange heat only, no work, that is, $V(A) = const.$, $V(B) = const.$ Then*

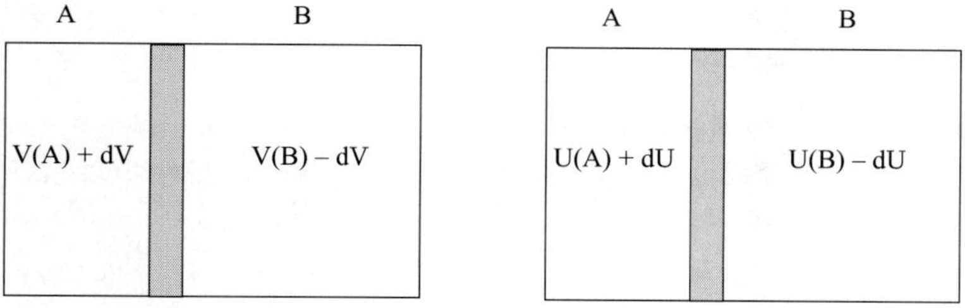

Figure 4.2 Contact equilibria: Volume exchange (left) and heat exchange (right).

$$dU = dU(A) + dU(B) = đQ(A) + đQ(B) = 0\,, \tag{4.111}$$

$$dS = \delta S_{in} = dS(A) + dS(B) = \left(\frac{1}{T(A)} - \frac{1}{T(B)}\right) đQ(A) \geq 0\,. \tag{4.112}$$

According to the second law equilibrium is characterized by maximum entropy and thus $dS = 0$. This condition implies that $T(A) = T(B)$, and the process of heat transfer $đQ(A)$ will have to fade away. This is the zeroth law and an example for a contact equilibrium, cf. Fig. 4.2.

Theorem 4.10 Mechanical contact *Under the same conditions as above the two subsystems are now allowed to exchange volume only, while $U(A) = const.$, $U(B) = const.$ Then*

$$dV = dV(A) + dV(B) = 0\,, \tag{4.113}$$

$$\begin{aligned} dS &= dS(A) + dS(B) \\ &= \left(\frac{\partial S(A)}{\partial V(A)}\right)_{U(A)} dV(A) + \left(\frac{\partial S(B)}{\partial V(B)}\right)_{U(B)} dV(B) \geq 0\,, \end{aligned}$$

$$dS = \left(\frac{p(A)}{T(A)} - \frac{p(B)}{T(B)}\right) dV(A) \geq 0\,. \tag{4.114}$$

Here contact equilibrium means that for $T(A) = T(B)$ the local pressures must be the same.

Theorem 4.11 Chemical contact *Likewise we may allow for particle exchange only, that is,*

$$dN = dN(A) + dN(B) = 0\,, \tag{4.115}$$

$$dS = \left(\frac{\partial S(A)}{\partial N(A)}\right)_{U(A)} dN(A) + \left(\frac{\partial S(B)}{\partial N(B)}\right)_{U(B)} dN(B) \geq 0\,. \tag{4.116}$$

In contact equilibrium the local chemical potentials, $\mu = T\partial S/\partial N$, must be the same.

Remark 4.8 Transitivity *Thermal equilibrium is said to establish a "transitive relationship" between thermal systems in contact: If A is in equilibrium with B and B in equilibrium with C, then A is in equilibrium with C.*

In any case, $S(A)$ and $S(B)$ will have readjusted in order to maximize $S = S(A) + S(B)$. More general: Under the condition of weak coupling and for subsystems of the same type (i.e., with the same set of independent macro-variables), the extensive variables become additive, the respective intensive variables approach each other. During the irreversible relaxation process and for a heat transfer $dQ(A) > 0$ we have $T(A) < T(B)$, that is, the heat goes from the warmer to the colder part, as expected. Likewise, if for $T(A) = T(B)$ we originally have $p(A) > p(B)$, the volume of A increases. The reverse processes cannot be observed. The entropy balance further restricts what kind of macro-state changes could be considered possible.

Remark 4.9 Thermodynamic process on the macro-level *Contact equilibria of this sort allow to change some intensive state variable in a prescribed way: Operationally one has to think of a series of environments (baths) of slightly different temperatures, say, with which the system is sequentially brought in contact. In this way the respective environment establishes a time-dependent control on the macro-system under consideration. Such a time-dependent control underlies, for example, the process cycles of thermodynamic machines.*

Definition 4.27 Substitute process *There have been various attempts to apply thermodynamical reasoning even outside the immediate range of applicability. A systematic way to do this is via substitute processes.*

In quantum thermodynamics stable thermodynamic states are obtained via appropriate embedding. Changing these states would thus be impossible unless the embedding is changed or, at most, entirely decoupled. Thermal states can then be manipulated via additional tools like thermal coupling to auxiliary environments, (addressable) auxiliary state transformations, (addressable) measurement projections.

Eventually one can also think of process control on the micro-level. Examples will be discussed in the context of quantum thermodynamics.

Example 4.4 Sudden thermal contact *Given the exchanged heat, ΔQ, what is the entropy production, ΔS, if two thermal bodies A, B with (fixed) different temperatures $T_A > T_B$ are brought in contact? As this is a highly non-equilibrium situation, thermodynamics does not seem to be of any help. Nevertheless, (reversible) substitute processes come for rescue. Simply assume that the heat ΔQ from A is reversibly transferred to an auxiliary bath of infinitesimally smaller temperature, T_A^-; the same heat is transferred on to B from a second auxiliary bath of slightly higher temperature, T_B^+. Then the total entropy change is*

$$\Delta S = \Delta Q \left(\frac{1}{T_B} - \frac{1}{T_A} \right) . \tag{4.117}$$

The above result can be applied to infer the entropy production due to the contact between sun (high-temperature bath) and earth (low-temperature bath): Given an average earth temperature of $T_E = 290$ K, the solar constant $1400 W/(m^2)$, an albedo of 0,3, and the earth's cross-sectional area, one obtains the power $dQ/dt = 1,2 \times 10^{17} W$ and thus $dS/dt \approx dQ/dt(1/T_E) = 4,0 \times 10^{14} W/K$.

Theorem 4.12 Measurability of entropy change *Let us consider a system A at temperature $T(A)$ interacting with the environmental system B specified by the fixed temperature $T_B = T(A) + \delta T$; the positive infinitesimal temperature difference $\delta T > 0$ is required to drive the heat transfer, while the total state is infinitesimally close to equilibrium. The process is thus slow ("quasi-static") and reversible, so $\delta_{in} S = 0$.*

Then, after a long enough process time, that is, long enough to get a significant heat transfer, we end up with some $\Delta Q(A) = -\Delta Q(B)$. From the measurement of the transferred heat (and temperature) we can infer the entropy change of subsystem A (and of subsystem B),

$$\boxed{\Delta S(A) = \Delta S_{ex}(A) = \frac{\Delta Q(A)}{T(B)} .} \tag{4.118}$$

In this way the entropy becomes measurable on the macro-level and for a single thermal object.

Definition 4.28 Cyclic thermodynamic machine *A thermodynamic system can be forced (via external control) to follow a path in macro-state space. This process is non-autonomous, that is, not derivable from the eigendynamics of the system. (Such an eigendynamics cannot exist in the space of equilibrium states.)*

A machine should operate without being (irreversibly) changed during the process of operation. Any state function of a thermodynamic machine must thus return to its initial value after the machine has undergone a cycle in macro-state space.

Of course, this does not exclude non-zero inputs/outputs, for example, the flow of energy through this device. Typically, the machine function is characterized by its impact on the outside world.

Remark 4.10 Non-equilibrium and irreversibility *Deviations from the simple model of spectral deformation, Definition 4.6, typically lead to irreversibility: Any adiabatic process will leave the quantum system in a non-equilibrium state (even if it started from a thermal one); contact with a thermal bath later on will inevitably induce irreversible relaxation behavior. As a consequence, nanoscopic machines of that general type cannot reach the ideal efficiency limits.*

Corollary 4.1 Second law of thermodynamics: Kelvin–Planck *The second law (cf. Theorem 4.5 and 4.8) in the Kelvin–Planck formulation constitutes a "no-go theorem":*

It is impossible that a machine, operating in a cycle, produces no effect other than the conversion of heat from a single bath into an equal amount of work.

The condition of "no other effect" is important here. Note that the machine forbidden here would be in accord with the first law of thermodynamics, the conservation of energy.

Thermal equilibrium—like mechanical equilibrium—comes equipped with appropriate stability criteria. These allow to differentiate between equilibrium and non-equilibrium.

Theorem 4.13 Thermodynamic stability *The MEP does not only give rise to the contact equilibrium, but also leads to additional* *relations, called stability conditions. As an example for a macro non-equilibrium consider the partitioned system (A, B) described by the entropy state function $S(A, B)$,*

$$S(A, B) = S(A) + S(B) = S(U(A), U(B), V(A), V(B)), \quad (4.119)$$

constrained by

$$U(A, B) = U(A) + U(B), \quad (4.120)$$

$$V(A, B) = V(A) + V(B). \quad (4.121)$$

Now we allow energy exchange between the two subsystems:

$$dU(A, B) = dU(A) + dU(B) = 0. \quad (4.122)$$

The change of entropy can then be expanded as

$$dS(A, B) = \left(\frac{\partial S(A)}{\partial U(A)}\right)_{V(A)} dU(A) + \left(\frac{\partial S(B)}{\partial U(B)}\right)_{V(B)} dU(B) + \frac{1}{2}\left(\frac{\partial^2 S(A)}{\partial U(A)^2}\right)_{V(A)} dU(A)^2 + \frac{1}{2}\left(\frac{\partial^2 S(B)}{\partial U(B)^2}\right)_{V(B)} dU(B)^2 + \ldots. \quad (4.123)$$

The maximum principle for $S(A, B)$ implies that the first derivative be zero, which leads to the contact equilibrium. In addition, the second derivative should be negative, leading (with $dU(A)^2 = dU(B)^2 > 0$) to the stability condition

$$\left(\frac{\partial^2 S(A)}{\partial U(A)^2}\right)_{V(A)} + \left(\frac{\partial^2 S(B)}{\partial U(B)^2}\right)_{V(B)} < 0. \quad (4.124)$$

As we can partition the subsystem A, say, again, this time in two identical parts (a) so that $S(A) = 2S(a); U(A) = 2U(a)$, the inequality must hold also for subsystem A separately,

$$2\left(\frac{\partial^2 S(a)}{\partial U(a)^2}\right)_{V(a)} = \left(\frac{\partial^2 S(A)}{\partial U(A)^2}\right)_{V(A)} < 0. \quad (4.125)$$

Based on $(\partial S/\partial U) = 1/T$ the stability condition can be re-written as:

$$\boxed{\left(\frac{\partial T}{\partial U}\right)_V \equiv \frac{1}{C_V} > 0.} \quad (4.126)$$

C_V is the specific heat for constant volume. It must be positive, as confirmed by Eq. (5.275). (Violations have been discussed for the

microcanonical ensemble, though: For example, at different energies different parts of phase space may be accessible.)

If, instead, we allow for "volume exchange," we get the stability condition

$$\left(\frac{\partial^2 S}{\partial V^2}\right)_U < 0\,, \tag{4.127}$$

which, based on Eq. (4.105), can be re-written as

$$\boxed{\frac{\partial}{\partial V}\left(\frac{p}{T}\right)_U < 0\,.} \tag{4.128}$$

Theorem 4.14 Third law *The third law further constraints the possible thermal properties in the limit of the temperature going to zero. A formal definition reads:*

$$\lim_{T\to 0} S(T, X) = 0 \quad \textit{independent of}\, X\,. \tag{4.129}$$

The limiting behavior for the entropy S is universal ("absolute") in that it is independent of the various constraints X. As a simple example consider

$$đQ = TdS = T\left(\frac{\partial S}{\partial T}\right)_V dT \equiv C_V dT\,, \tag{4.130}$$

where C_V is the heat capacity under constant volume V. With $dS = C_V dT/T$ the entropy can be expressed as

$$S(T, V) = \int_0^T \frac{C_V dT'}{T'}\,, \tag{4.131}$$

which goes to zero for $T \to 0$, if

$$\lim_{T\to 0} C_V = 0 \tag{4.132}$$

$$\to C_V \sim T^\delta \quad \delta \geq 1\,. \tag{4.133}$$

This functional requirement for the heat capacity $T \to 0$ severely limits our ability to reach zero temperature (see Section 5.4 and Section 8.3.1.) The operational consequences of the third law can qualitatively be stated as follows:

For low temperatures some tasks get very difficult to carry out (i.e., become "physically hard problems," cf. Definition 5.18)—due to limiting resources such as energy, time, and/or (memory-) space, cf. Corollary 5.1.

Remark 4.11 Third law for non-ergodic systems *The standard example for a "violation" of the third law in classical thermostatistics is the expected behavior of systems with frozen disorder like glasses. This is interpreted as being due to non-ergodicity, that is, in the process of cooling the system tends to be captured in long-living non-equilibrium states. This behavior may be interpreted to violate Theorem 4.14.*

But in a more general context, this limitation becomes an instant of a physically hard problem: As a consequence of non-ergodic behavior there is a broad distribution of relaxation times; the accessible state space is hard to specify and control in advance. For any given "reasonable" waiting time the approach to zero entropy (zero temperature state) cannot be guaranteed—not knowing what should count as the pertinent state space.[a]

(Partly) integrable systems are also non-ergodic, but with a well-defined accessible region—once we know the constants of motion, cf. Definition 5.4.

4.3.6 *Thermodynamic Potentials*

Thermodynamic systems can be specified by the state functions $U(S, V)$, alternatively by so-called thermodynamic potentials. These are defined as specific Legendre transformations of the former.

Definition 4.29 Free energy *The free energy F is the Legendre transformation of the internal energy state function $U(S, V)$, by which we replace $x = S$ by $z = (\partial U/\partial S) = T$. The transformation is as follows (cf. definition 3.23):*

$$\mathcal{L}U(S, V) = U(S(T, V), V) - \left(\frac{\partial U}{\partial S}\right) S = F(T, V). \qquad (4.134)$$

The differential form for F is

$$dF = dU - d(TS) = -SdT - pdV, \qquad (4.135)$$

$$\left(\frac{\partial F}{\partial T}\right)_V = -S(T, V) \qquad \left(\frac{\partial F}{\partial V}\right)_T = -p(T, V). \quad (4.136)$$

[a] Albert Einstein already complained that in order to use the Boltzmann principle one had to know the number of accessible states, which is impossible without a complete molecular mechanical theory [Pais (1982)].

Note that—contrary to $S(U, V)$—the derivative $S(T, V)$ is not a state function. What is the potential advantage for using $F(T, V)$ instead of $U(S, V)$? One pertinent application is a reformulation of the second law under isothermal conditions (constant temperature T):

From the Clausius inequality it follows that

$$đ W_{dis} = T dS - đ Q \geq 0 . \tag{4.137}$$

$đ W_{dis}$ is the dissipated work [Ritort (2003)], as will become clear immediately. We first note that

$$dF = dU - T dS - S dT \tag{4.138}$$

$$= đ W + (đ Q - T dS) - S dT = đ W - đ W_{dis} - S dT . \tag{4.139}$$

Without dissipation and under isothermal conditions dF can directly be identified with the work, $đ W$, retrievable from a system in contact with a heat bath T. In the presence of dissipation the actual work is smaller by the dissipated work, $đ W_{dis} \geq 0$, that is,

$$\boxed{dF - đ W \leq 0 .} \tag{4.140}$$

For $dV = 0$ and $dT > 0$ we conclude from Eq. (4.135) that $dF < 0$. $đ W$ is thus also negative, that is, lost by the system:

$$0 > đ W = dF + (S dT + đ W_{dis}) \geq dF . \tag{4.141}$$

The modulus of the extracted work is bounded from above by the change of the system free energy.

Definition 4.30 Enthalpy *A further example for thermodynamic potentials is the enthalpy, replacing variable V by p,*

$$\mathcal{L}U(S, V) = U(S, V(p, S)) - \left(\frac{\partial U}{\partial V}\right) V$$

$$= U + pV = H(S, p) , \tag{4.142}$$

$$dH = dU + d(pV) = T dS + V dp , \tag{4.143}$$

Definition 4.31 Gibbs free energy *This potential results from a two-fold Legendre transformation of U replacing S and V by T and p, respectively.*

$$\mathcal{L}U(S, V) = U(S(p, T), V(p, T)) - \left(\frac{\partial U}{\partial V}\right) V - \left(\frac{\partial U}{\partial S}\right) S$$

$$= U + pV - TS = G(T, p) . \tag{4.144}$$

The differential form is

$$dG = Vdp - SdT\,. \tag{4.145}$$

The Gibbs free energy $G(T,\, p,\, N_1,\, N_2,\, \ldots)$ *is particularly useful for the description of chemical reactions, last not least in biological systems:*

$$(dG)_{T,\,p} = \sum_{\nu} \mu_\nu dN_\nu\,. \tag{4.146}$$

μ_ν *is the chemical potential for particle type* ν.

The reaction $X + Y \rightarrow Z$, *e.g., with* $\Delta N \equiv -\Delta N_X - \Delta N_Y = \Delta N_Z$, *would be governed by*

$$\Delta\mu \equiv \mu_Z - \mu_X - \mu_Y = \frac{\Delta G}{\Delta N}\,. \tag{4.147}$$

Any of these potentials—like the state functions $U(S,\, V)$ or $S(U,\, V)$—contains everything that might be of interest thermodynamically about a given macro-system.

Theorem 4.15 Stability with respect to thermodynamic potentials *Under appropriate constraints the maximum principle for entropy can be translated into minimum principles for the state functions* U, F, H, G. *In order to see this, consider system* A *in contact with a heat reservoir* B *specified by*

$$dU(B) = T(B)dS(B) \tag{4.148}$$

and a volume reservoir C,

$$dU(C) = -p(C)dV(C)\,. \tag{4.149}$$

The total system is closed, so that

$$dU(A,\, B,\, C) = dU(A) + T(B)dS(B) - p(C)dV(C) = 0\,, \tag{4.150}$$

$$dV(A,\, B,\, C) = dV(A) + dV(C) = 0\,, \tag{4.151}$$

$$dS(A,\, B,\, C) = dS(A) + dS(B) \geq 0\,. \tag{4.152}$$

Temperature $T(B)$ *and pressure* $p(C)$ *are given constants. Making use of the last two lines we can re-write the first line as*

$$\boxed{d(U(A) - T(B)S(A) + p(C)V(A)) \leq 0\,.} \tag{4.153}$$

We thus obtain

$$dU(A) \leq 0 \quad \textit{for } V(A), S(A) \textit{ constant}, \tag{4.154}$$

$$dF(A) \equiv d(U(A) - T(A)S(A)) \\ \leq 0 \quad \textit{for } V(A), T(B) = T(A) \textit{ constant}, \tag{4.155}$$

$$dH(A) \equiv d(U(A) + p(A)V(A)) \\ \leq 0 \quad \textit{for } p(B) = p(A), S(A) \textit{ constant}, \tag{4.156}$$

$$dG(A) \equiv d(U(A) - T(A)S(A) + p(A)V(A)) \\ \leq 0 \quad \textit{for } p(B) = p(A), T(B) = T(A) \textit{ constant}. \tag{4.157}$$

Based on these conditions for subsystem A, non-equilibrium can be introduced by further partitioning of A, as has been done in Section 4.3.5. Returning to equilibrium the appropriate contact and stability conditions apply.

Remark 4.12 Notation *Internal energy U, entropy S as well as the thermodynamic potentials have a double meaning: (i) they each define a specific physical property (with a given value); (ii) they each specify a state function, but only with respect to the appropriate set of "natural" independent variables, as given above.*

Example 4.5 Sudden temperature quench *Suppose a quantum system has been prepared in a canonical state with temperature T_0. It is then brought in contact with a bath at temperature $T_1 > T_0$. What is the work ΔW the system can do while approaching its new equilibrium state?*

According to Eq. (4.141) and assuming $\Delta W_{diss} = 0$ we have for $\Delta T = T_1 - T_0 > 0$ and thus $\Delta F = F(T_1) - F(T_0) < 0$

$$0 > \Delta W = F(T_1) - F(T_0) + (T_1 - T_0)S(T_1) \tag{4.158}$$

$$\geq F(T_1) - F(T_0). \tag{4.159}$$

This result can be interpreted in terms of the quantum Renyi entropy, Eq. (3.155). Following [Baez (2011)]:

$$\boxed{\frac{\Delta W}{T_1 - T_0} = \frac{F(T_1) - F(T_0)}{T_1 - T_0} = -k_B S_\alpha(T_0),} \tag{4.160}$$

where $\alpha = T_0/T_1$ and $S_\alpha(T_0)$ is the Renyi entropy of order α of the system in its original canonical state T_0,

$$S_\alpha(T_0) = \frac{\ln(\mathrm{Tr}\hat{\rho}^\alpha)}{1-\alpha} = \frac{1}{1-\alpha}\ln \mathrm{Tr}\left(\frac{\exp(-\alpha\hat{H}/T_0)}{Z(T_0)^\alpha}\right). \tag{4.161}$$

Note that in the case of initial equilibrium, that is, $T_1 \to T_0$ and thus $\alpha \to 1$, the difference equation (4.160) reduces to the well-known partial derivative, Eq. (4.136),

$$\left(\frac{\partial F}{\partial T}\right) = -k_B S_{vN}\,. \tag{4.162}$$

We are back in standard thermodynamics.

4.3.7 *Thermodynamics and Information Theory*

The three fundamental fields, information, quantum physics, and thermodynamics, share a common concept, the concept of entropy. Information theory is characterized by the Shannon entropy, S_I, quantum physics by the von Neumann entropy, S_{vN}, and thermodynamics by the thermodynamic entropy, S. We have argued (cf. theorem 4.6) that the latter two can be identified under equilibrium conditions. How about S_I?

The role of information in physics is not that clear at all, cf. [Roederer (2005)]; reference to an observer/agent is indispensable. We approach this problem from three different aspects:

(1) Subjective ignorance.
(2) Communication metaphor: Measurements as "messages" from nature.
(3) Information as a physical resource for an agent.

Remark 4.13 Subjective ignorance: Ensemble description *The probability of an event can intuitively be related to its relative frequency of occurrence n/N, where N is the total number of independent trials, and n the number of "successful" outcomes, Eq. (2.24). All these N trials together may be taken to constitute an ensemble. While each member of this ensemble clearly has definite properties, the ensemble as a whole can be used to represent our subjective ignorance: We do not know, which is actual. When one member of the ensemble is picked at random and then identified, the average information gain is given by the Shannon entropy, S_I, cf. proposition (2.8).*

Remark 4.14 *In the subjective information language an ensemble of identical pure states would be associated with zero information, while* *the (maximally) mixed state ensemble "carries" maximum information (cf. [Plenio (2001)]). In a different parlance S is said to represent "Lack of information." These are two sides of the same token, specifying the situation after/before getting the result of the statistical experiment.*

Remark 4.15 Communication metaphor *The operational aspect of subjective ignorance is communication, see Definition 2.29, which, in turn, is based on "hypothetical measurements." (Except for simplified cases, complete measurements can hardly be carried out in full detail.) In any case, the individual measurement result would have to be considered a message from the physical system to the observer—taken from the set of all possible messages as specified by the ensemble; this ensemble is a theoretical construct.*

The Gibbs entropy and the Boltzmann entropy may be said to represent different variants of subjective ignorance about classical many-body systems.

Definition 4.32 Gibbs entropy *Consider a mono-atomic fluid of N (classical) particles. Its micro-state is described as a point in the 6N-dimensional phase space,*

$$\vec{s}(t) \equiv \{\vec{r}^{(1)}, \vec{p}^{(1)}, \vec{r}^{(2)}, \vec{p}^{(2)}, \ldots, \vec{r}^{(N)}, \vec{p}^{(N)}\}\,. \tag{4.163}$$

We define the so-called Gibbs entropy [Jaynes (1965)] as the Shannon entropy for the random "messages" $\vec{s}(t_j)$:

$$S_I(p_N) = S_I^G = -\int p_N \ln p_N ds^{6N}\,. \tag{4.164}$$

It is the Gibbs entropy, which, based on a canonical distribution, can be identified with the thermal entropy S of the N particle gas [Jaynes (1965)],

$$S_I^G = S/k_B \tag{4.165}$$

If the n-particle system is taken to be closed, its entropy is necessarily a constant of motion.

Definition 4.33 Boltzmann entropy *The single-particle probability density is defined by*

$$p_1(\vec{r}^{(1)}, \vec{p}^{(1)}; t) = \int ds^{6(N-1)} p_N(\vec{s})\,. \tag{4.166}$$

The integration is over the $6(N-1)$-dimensional phase space. p_1 represents a kind of "coarse-grained" description. Based on *the "messages"* $(\vec{r}^{(1)}, \vec{p}^{(1)})$ *the corresponding Shannon entropy is [Jaynes (1965)]*

$$S_I(p_1) = S_I^B = -N \int p_1 \ln p_1 ds^6 \, . \tag{4.167}$$

The famous H theorem by Boltzmann indicates that $dS_I^B/dt \geq 0$, in apparent agreement with the second law.[a] However, the identification $S_I^B = S/k_B$ is unjustified: The two descriptions, Gibbs and Boltzmann, are equivalent if and only if

$$p_N = p_1(\vec{r}^{(1)}, \vec{p}^{(1)}) p_1(\vec{r}^{(2)}, \vec{p}^{(2)}) \cdots p_1(\vec{r}^{(N)}, \vec{p}^{(N)}) \, , \tag{4.168}$$

that is, the many-particle distribution is reducible to the one-particle description [Jaynes (1965)]. In general, Gibbs entropy and Boltzmann entropy do not coincide. There are many more ignorance models which would share this limitation.

Definition 4.34 Canonical pure state ensemble *For quantum systems the observation plays a much more decisive role. Let us consider a quantum system prepared in the initial state* $\hat{\rho}$*. A complete measurement then projects onto a pure state* $|j>$ *with given probability. Repeat the measurement on a large number* N *of independent identical copies of this state, that is, on the total product input state* $\prod_\nu \otimes \hat{\rho}(\nu)$*. Then the set of* N *results defines the inhomogeneous pure state ensemble. Its information is given by the Shannon entropy,* S_I*, based on the probability = relative frequency* $P_j = \rho_{jj}$ *for each* $|j>$*.*

For sufficiently small quantum systems this feature can easily be made explicit: As discussed in Section 5.5, periodic supervision of a quantum system embedded in a quantum environment makes the system follow a stochastic trajectory through instantaneous pure states. The relative frequency of those outcomes will be counted, n_j. Eventually, the set of N such measurement states generates an inhomogeneous pure state ensemble, cf., the relative frequency of which, n_j/N, would asymptotically approximate the "objective" P_j according to quantum thermodynamics. The ensemble average coincides with the time average (ergodicity).

---

[a]The irreversibility has aroused much controversy (cf. Loschmidt's reversibility objection). But this feature is apparent only and should be seen as a logical consequence of the incomplete 1-particle description.

Theorem 4.16 Equivalence of entropies II: Info-quantum *The von Neumann entropy S_{vN} of a density operator $\hat{\rho}$ can be reinterpreted as the Shannon entropy S_I: The maximum number of qubits needed to store the information expressed by that very quantum state:*

$$S_I = S_{vN} \,. \tag{4.169}$$

This is an operational statement concerning the pure state ensemble (in the asymptotic limit $N \rightarrow \infty$).

Theorem 4.17 Equivalence of entropies III: Info-thermo *As indicated above, the Shannon entropy based on some ignorance model does not necessarily correspond to a thermodynamic entropy in a strict sense. The identification works, for example, for the Gibbs or the inhomogeneous pure state ensemble, Definition (4.34):*

$$S_I = S/k_B \,. \tag{4.170}$$

S is the thermodynamic entropy. In this operational sense the two entropies coincide.

Remark 4.16 Brillouin's negentropy *Brillouin [Brillouin (1995)] postulated that any information gain, ΔS_I (typically due to observation), would have to be accompanied by some process of dissipation somewhere in the physical system, leading to an entropy increase, ΔS such that*

$$\Delta S/k_B - \Delta S_I \geq 0 \,. \tag{4.171}$$

Superficially this looks quite similar as our entropy equivalence condition, Theorem 4.17. However, in the negentropy principle it is argued that the increase of information (equivalent to "negative entropy") must be paid for by an increase of "physical" (positive) entropy, thus leading to a kind of "generalized second law of thermodynamics."

The entropy equivalence, on the other hand, is based on a formal construction, the pure state ensemble. Change of information (if not merely subjective) means change of state; the measurement process itself has to be included physically. The talk about information specifies a substitute rather than an addendum.

Remark 4.17 Macro-bit *The thermodynamic entropy S has the dimension of the Boltzmann constant k_B, that is, Joule/Kelvin. Expressing the Shannon entropy in bits, we find the equivalence*

$$S[J/K] = S_I[bit]k_B \ln 2\,, \qquad (4.172)$$
$$0.956J/K \approx 10^{23} bit\,.$$

J/K is the typical order of magnitude of entropy in the macro-domain; if taken to represent our subjective ignorance, one might argue that roughly 10^{23} yes–no answers were needed to locate the actual micro-state.

However, such a search algorithm is unrealistic, not only because of the extremely large numbers. Suppose we had to find a microstate within $N = 2^{10} = 1024$ possibilities. Our ignorance is $\log_2(N) = 10$ bit. However, 10 yes–no answers would suffice only, if the search was highly structured (e.g., as a sorted array) or if the item searched for had a property that would stick out from the properties of all the other items. If, for example, the items were taken to be massive balls, the searched-for ball would have to be "heavier" than the others (taken to be of equal weight). Then a pair of balances would do to compare the total weight of 1024/2 balls against the other half. Continue with the heavier half etc. In general, such a strategy does not exist, we need on average N/2 queries.

These operational limitations render the philosophical idea of the world being characterized in terms of fundamental yes/no decisions ("Ur-theory", cf. [Lyre (1997)]) highly problematic.

Finally, we turn to the information as a potential source for improved control by some agent.

Remark 4.18 Agency *Information in physics is not detached from physics. It relates to observation and agency. Information about the state of a physical system requires measurements (physical interaction). For these to have any impact, for example, to improve control, further conditioned interactions are needed. Alternatively, the observer, O_1, can communicate the measurement results to some other agent, O_2, who, based on this information, can also try to improve control on the physical system. This potential benefit might serve as a measure for the "value" of information.*

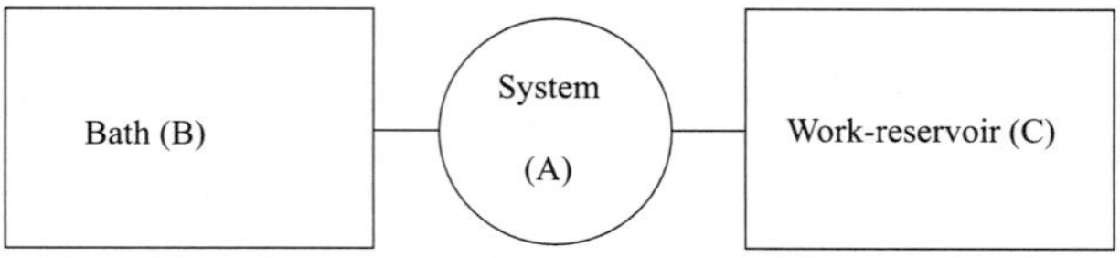

Figure 4.3 Single bath(B)–system(A)–work reservoir(C).

The scenario in Fig. 4.3 has a surprisingly large number of interesting applications, eventually relating to the Landauer principle and Maxwell's demon.

Proposition 4.5 Maximal work from a single bath $T(B)$ *Consider a d-level system A defined by* $\hat{H}(A)$*. Initially it is in a diagonal state with entropy* $S_{vN}(A) = S_{initial}(A)/k_B$*. When brought in contact with an "infinite" heat bath B at temperature* $T(B)$ *the maximal extractable work is given by*

$$\boxed{\Delta W_{max}(A) = -k_B T(B)(\ln d - S_{initial}(A)/k_B)\,.} \tag{4.173}$$

Proof. The system A is taken to act as an interconnecting node between the single heat bath B of temperature $T(B)$ and an unspecified work reservoir C. It follows that the energy exchange between A and B would be heat only, between A and C work only. The exchange process is finite. Subject to an as yet unspecified mechanical control parameter γ, it is taken to be constrained by two additional conditions (cf. [Alicki (2004)]):

- Reversibility: no internal entropy production for the total system, $\delta_{in} S = 0$, cf. Theorem 4.8.
- Conservation of energy of subsystem A: $U(A) = \text{const.}$

These conditions imply:

$$dS = dS(A) + dS(B) = 0\,, \tag{4.174}$$

$$dU(A) = đW(A) + đQ(A) = 0\,, \tag{4.175}$$

and thus

$$đW(A) = -đQ(A) = đQ(B) = T(B)dS(B) = -T(B)dS(A)\,. \tag{4.176}$$

The work transferred, $\Delta W(A)$, should thus be given by

$$\Delta W(A) = -T(B)(S_{final}(A) - S_{initial}(A))\,. \tag{4.177}$$

Heat from the heat reservoir is transferred into work for the work reservoir, a reversible heat engine process. The maximum output (negative with respect to subsystem A) for given initial state would be reached for $S_{final}(A) = S_{max}(A) = k_B \ln d$, which confirms the proposition. □

Remark 4.19 Work value of information *Obviously, if we knew the initial state of subsystem A completely, $S_I(A)/k_B = S_{initial}(A) = 0$ (i.e., zero temperature $T(A)$), the maximum work, $-k_B T(B) \ln 2$, would be gained (as exploited by Maxwell's demon, see below).*

As a possible implementation of such a transfer process we consider a micro-model based on a $d = 2$-level system (qubit) as the interconnecting node A [Alicki (2004)]. The model has two free parameters for external control.

Example 4.6 Micro-model: control space *Let the respective micro-state be given by the density matrix*

$$\rho_{equ}(A) = \begin{pmatrix} \rho_1 & 0 \\ 0 & (1-\rho_1) \end{pmatrix} . \tag{4.178}$$

in the energy eigenbasis of the corresponding Hamiltonian $\hat{H}(A)$. We take ρ_1 as a convenient control parameter for the micro-state and thus the entropy, $S(A) = S(A; \rho_1)$. The spectrum of $\hat{H}(A)$ is parametrized by the mechanical parameter, ΔE, and responds to ρ_1:

$$E_1^{(A)} = -\Delta E(1-\rho_1) , \tag{4.179}$$

$$E_2^{(A)} = \Delta E \rho_1 . \tag{4.180}$$

By construction this "spectral deformation" leaves the average energy invariant, $U(A) \equiv 0$, as required. The effective energy splitting is

$$\Delta E_{eff} = \Delta E . \tag{4.181}$$

This model would allow a reversible implementation of the work extraction process as defined in Proposition 4.5.

Example 4.7 Three-step heat engine *In this two-dimensional control space $\{\rho_1, \Delta E\}$ one-dimensional trajectories are selected as follows, cf. Fig. 4.4:*

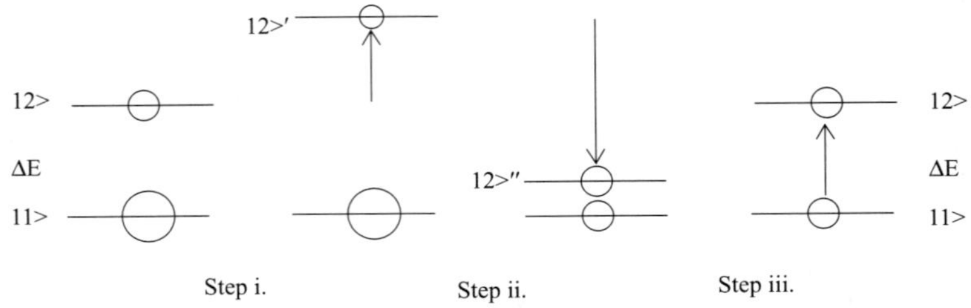

Figure 4.4 Three-step heat engine (two-level system). Circle radius indicates occupation probability of states.

(i) *Adiabatic step* $(t_0 \rightarrow t_1) : S(A) = const.$

The subsystem A is isolated and prepared in some state with ground-state probability $1 > \rho_1 = \rho_1^0 > 0.5$ *corresponding to a finite positive temperature* $T(A) \neq T(B) > 0$*. Then the mechanical parameter* ΔE *is slowly changed until the corresponding density matrix represents a thermal equilibrium state for the temperature* $T(B)$ *of the bath. There is no energy input. But note that the energy splitting is needed to allow for a definition of temperature (and thus of thermal states)!*

This pre-step makes subsystem A ready for the contact with bath B (without inducing irreversible heat exchange!); it allows us to start from an arbitrary diagonal state, not necessarily from the canonical state with the "correct" temperature $T(B)$*. (This only works if the initial entropy of subsystem A is not exactly zero.)*

(ii) *Isothermal step* $(t_1 \rightarrow t_2) : T(A) = T(B) = const.$

The subsystem A is brought in contact with the bath B and the work reservoir C. Slowly decreasing ΔE *for fixed* $T(B)$ *will reduce the effective energy splitting until we reach zero. In this limit both energy levels will exhibit the same population, independent of* $T(B)$*. We thus have reached maximum entropy* $S_{final}(A) = k_B \ln 2$*. During the whole process energy has flown in from the bath as heat, and, as the node A cannot store any energy, must have been passed on to the work reservoir as work.*

(iii) *Adiabatic step* $(t_2 \rightarrow t_3)$: $S_{final}(A) = const.$

We return the control parameter ΔE *to its original value. Again, no energy is needed. This makes the external driving cyclical, while system A has been heated up.*

This process reproduces, indeed, the heat engine result Eq. (4.173) with $d = 2$.

Example 4.8 Three-step refrigerator *The reverse process constitutes a three-step refrigerator. For* $S_{initial}(A) = k_B \ln 2$ *and* $S_{final}(A) = 0$ *one needs a work input of*

$$\Delta W_{max}(A) = k_B T(B) \ln 2 \tag{4.182}$$

Both three-step processes are reversible in the thermodynamic sense (constant total entropy!); under irreversible conditions the necessary work input (in the refrigerator case) would increase, the work output (in the heat engine case) decrease. The refrigerator result does not imply that we could actually reach zero temperature, as this would require infinite energy splitting in the node A to reach $S_{initial}(A) = 0$ for finite $T(B)$.

The refrigerator process can directly be related to the so-called erasure problem.

Definition 4.35 Erasure *Erasure of information is defined as the task to reset the initial state of a (quantum) system A to some (standard) pure state* without knowing *the initial state (i.e., maximum Shannon entropy). The erasure is (logically) irreversible; it may be seen as a kind of "game" specified by rules.*

After storage of one bit of information, the quantum register is in a pure state. The complete subjective ignorance about this state can be modeled in the sense of the Shannon entropy $S_I(A)$: the state is taken to be picked at random from an ensemble with both states being equally likely [Alicki (2004)]. For a large enough ensemble we know that the Shannon entropy approaches the thermodynamic entropy $S(A)/k_B$. But, as argued above, this equivalence does not only hold for the initial state but for the whole time-dependent evolution. This is so because the dynamics is the same, irrespective of the initial state. As a consequence the evolution cannot depend on the nature and origin of the corresponding density matrix.

Theorem 4.18 Landauer principle *The process of erasure is equivalent to the operation of a refrigerator, which cools the target (the qubit) from infinite to zero temperature: In fact, even though, objectively, we start each time from a pure state, the trajectories—when ensemble averaged over the two initial states—exactly reproduce the above thermodynamic cooling process and thus require, on average, the work input given above,*

$$\boxed{\Delta W_{max}(A) = S_{initial}(A)T(B) = k_B T(B) \ln 2\,.} \tag{4.183}$$

As the underlying refrigerator model is based on the second law, so is the Landauer principle. Basic properties of the erasure game can thus be modeled by thermodynamics.

This principle should apply also to the classical domain, if the relevant state space can appropriately be coarse-grained into two discrete alternatives of equal weight. However, due to the smallness of $\Delta W_{max}(A) = k_B T(B) \ln 2$, it can play a practically important role in the quantum domain only. (For an experimental verification see [Lambson (2011); Berut (2012)].)

Originally the principle has been derived and interpreted in a somewhat different way: the term $k_B T(B) \ln 2$ is usually described as dissipated heat[a] rather than work input. There is no contradiction, though, as should be clear from the above model analysis. The work has to leave the qubit as heat into the bath. This is one and the same process, merely seen from a different perspective. The interpretation preferred here is the effective refrigerator language.

Remark 4.20 Erasure for a d-level system *One may attempt to generalize Eq. (4.183) for a d-level system:*

$$\Delta W_{max}(A) = k_B T(B) \ln d\,. \tag{4.184}$$

However, this does not hold in general, cf. Remark 4.24. It works for a N-spin system, $d = 2^N$, though.

Remark 4.21 Erasure under partial knowledge *We may generalize the erasure game to allow for partial knowledge of the initial state of (A). Subjective knowledge is conveniently characterized by the*

---

[a]The release of heat, $đQ < 0$, is in accord with the Clausius equality, $đQ = T\,dS$, for $dS < 0$.

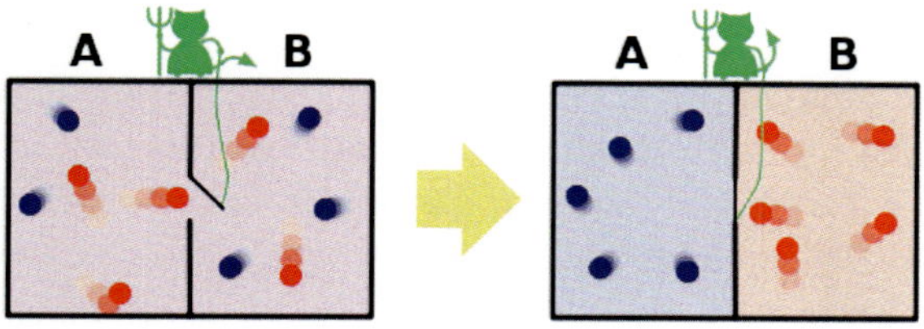

Figure 4.5 Maxwell's demon: The creature is supposed to control the shutter between two compartments. Hot molecules are preferably allowed to pass from left to right. In this way the temperature difference between the compartments would be enhanced.

Shannon entropy, $S_I(A)$ in bits. It is essential that the corresponding state interpreted as an ensemble, is still canonical. The work needed is then still given by Eq. (4.183) with Eq. (4.172),

$$\boxed{\Delta W(A) = S_I(A) k_B T(B) \ln 2\,.} \tag{4.185}$$

The work is reduced; for complete knowledge, $S_I(A) = 0$, no work-input is needed at all[a].

Remark 4.22 Repetition of process? *The above engine processes based on Example (4.6) represent terminating processes: The final state of subsystem A is changed—as desired. The process is reversible (in the ideal case), we could undo the operation—but then without any benefit.*

But in order to extract $\Delta W_{max}(A)$ once more, we would first have to re-install the initial state by some other means. An interesting option appears to be measurement. This leads to the Maxwell demon paradox.

Theorem 4.19 Maxwell's demon paradox *A classical version of Maxwell's demon is shown in Fig. 4.5 Its essential features are most easily interpreted in terms of a simple quantum model, a 2-level system (qubit) A [Lloyd (1997)] with energy splitting $\hbar\omega$. Suppose that the*

[a]This conclusion is based on special models like Example 4.6, where the spectral deformation has been taken to depend on the initial state. In general, the transition from a higher energy initial pure state, say, to the (lower energy) ground state would require transfer of energy.

spin A has been in contact with a heat bath B with $T(B) \gg \hbar\omega$ *so that its entropy is* $S(A) \approx k_B \ln 2$*. It is decoupled now, and the demon measures whether the spin is in the upper,* $| \uparrow>$*, or lower energy state,* $| \downarrow>$*. This information (equivalent to a state preparation) can easily be converted into work: If the spin is found in the high-energy state, the demon sends an electromagnetic* π*-pulse to extract the energy* $\hbar\omega$*, which will amplify the field C (work reservoir). The* π*-pulse guarantees that initial and final state are energy-eigenstates, so that the work extracted (= the difference in spin energies before and after the pulse) would not fluctuate. If in the low energy state the spin is left untouched.*[b] *Then the spin is recoupled to the bath B and the process repeated. We thus switch between minimum and maximum entropy states, as suggested by the heat engine model before. But now this can apparently be done in a cyclic way!*

On average (in the high temperature limit) the demon extracts from the single bath the energy $\hbar\omega/2$ *per cycle, without limit, in violation of the second law.*

Proposition 4.6 Resolution of paradox *The Landauer principle has been suggested as a solution for the Maxwell's demon paradox [Bennett (1982)]: The work cost for erasure compensates for the claimed work benefit by the demons's action:*

The measurement-information about the state of A has to be stored to become physically active as a condition. But for cyclic operation, the demon has to erase this information, before he can continue; the dissipated heat is $k_B T(B) \ln 2$ *and connected with an entropy increase* $k_B \ln 2$ *which at least compensates for the above reduction.*

Ironically, we find ourselves back in the same position as before: no benefit.

This conclusion could be avoided if we had an infinite set of fresh spins at our disposal rather than a single spin to recycle. Such infinite resources are not permitted, though.

Remark 4.23 Heat cost of control *One should note that concrete processes (interventions) are needed to translate information into a physical effect: Knowledge as such does not help. Landauer's principle*

---

[b]This is an idealized feedback control scenario, cf. Section 6.4. The idealization relates to the neglect of any back-action noise.

shows: Information is useful, but—unfortunately—control is not for free.

Within mechanics there appears to be no limit with respect to the use/transformation of local energies. The difference between mechanics and thermodynamics is not inherent, but depends on the control one is able to built up from the outside. One may thus attempt to transform the "heat" of a large system composed of N 2-level-subsystems into work by iterating Maxwell's demon.[a] *While this could be achieved, in principle, the underlying physical process of establishing the control would entirely spoil its potential benefit by producing heat according to the Landauer principle. (It is dangerous to talk about "we know" or some observer "knows"; information is not a resource independent of physics!)*

This limitation can be seen as a fundamental expression of the second law: To recover the energy from all those spins, even more energy would be lost to other degrees of freedom in the environment (cf. [Duncan (2005)]). In this sense the limits imposed by the second law are surprisingly robust, even beyond thermodynamics.

Remark 4.24 Information and non-equilibrium *Information* per se *(about a quantum state or anything else) has no relation to the underlying Hamilton model and thus to equilibrium thermodynamics. Incomplete (subjective) knowledge (represented as a certain pure state ensemble) will thus typically amount to a non-equilibrium state, for which the rules of thermodynamics (and thus of thermodynamic processes) do not apply. This underlines the fact that the Shannon entropy can by defined for any probability distribution, while the thermal entropy is restricted to equilibrium functions only, cf. Definition 4.33.*

As exploited above, two-level systems play a special role here, as any diagonal state in the energy basis can be interpreted as a thermal state. In general, this simple reasoning no longer works for d-level systems ($d > 2$). An exception is the n-particle Szilard engine.

Definition 4.36 *n* Particle Szilard engine *A two-level system may be visualized as a box, occupied by a single particle. There are two compartments, left L, and right R. These represent the two states*

[a] Cf. *n*-particle Szilard engine, Definition (4.36).

under consideration. We generalize to n such systems, where each subsystem is in contact with a heat bath at temperature $T(B)$. There is virtually no interaction between the subsystems. For $N = 1$ this is the standard Szilard engine.

How much work can be extracted from the bath in a single operational step? According to Eq. (4.173) we have

$$\Delta W_{max}(A) = -k_B T(B)(\ln(d) - S_{initial}(A)/k_B)\,. \tag{4.186}$$

Taking for the Hilbert space dimension $d = 2^N$ we find

$$\boxed{\Delta W_{max}(A) = -k_B T(B)\ln(2)(N - S_{initial}(A)/(k_B \ln(2)))} \tag{4.187}$$

as the maximal work retrievable from a bath B by an array A of N 2-level systems under reversible conditions. Reversibility requires the contact of the spin array with the bath to be constrained by $T(A) = T(B)$. If this condition is not fulfilled, additional preparatory steps are needed, which may influence the pertinent balance equations for energy and/or entropy.

A class of convenient initial states consists of product states with each local state being diagonal in its respective energy eigen basis. If all those local states are the same, this is a global thermal state. A spectral deformation may then be applied to reach the contact condition—without hidden costs.

But there are other classes of initial states. Their use may require pre-steps in terms of measurements and/or unitary transformations.

Example 4.9 Local measurement *Let a 2-particle Szilard engine $A = \{A1, A2\}$ be prepared in state $\hat{\rho}$ with the diagonal matrix-elements*

$$\rho_{LL,LL} = \rho_{RR;RR} = 0.5\,. \tag{4.188}$$

The matrix elements are defined in the product-state basis, that is, $\rho_{LR,LR} = < L(A1)R(A2)|\hat{\rho}|L(A1)R(A2) >$ etc. All non-diagonal matrix-elements are zero. It is a correlated non-equilibrium state, known to us. Based on this "resource" we proceed with the following process:

Step 1*: We apply a controlled-NOT operation (cf. Example 3.23), which allows to reversibly transform into the state*

$$\rho_{LL,LL} = \rho_{RL;RL} = 0.5\,,$$
$$\rho_{RR,RR} = \rho_{LR;LR} = 0\,. \qquad (4.189)$$

This is a product state: each subsystem can be considered to be in a local equilibrium—there is no global equilibrium, though.

Step 2: *This state can be used to extract the maximum work $\Delta W_{max}(A) = -k_B T(B)$ via particle A2, as given by Eq. (4.187) and suggested by Maxwell's demon; here we do not consider cyclic operation and the cost for erasure (Definition (4.35)). The final state is maximal mixed. We may argue to have found a procedure to exploit correlations [Dahlsten (2011)].*

Alternatively we may consider the following substitute process:

Step 1*: *The system A is prepared in state $\hat{\rho}$ with the diagonal matrix-elements*

$$\rho_{LL,LL} = \rho_{RR;RR} = 0.25\,,$$
$$\rho_{LR,LR} = \rho_{RL;RL} = 0.25\,. \qquad (4.190)$$

This state can be considered thermal, it has maximum entropy $S(A;t_1) = 2k_B \ln 2$.

Step 2*: *Now selectively measure particle A2 and find it in local state $|L(A2)>$. Then, after measurement, the new state is given by Eq. (4.189), the entropy has been reduced to $S(A;t_2) = k_B \ln 2$.*

Step 3*: *This state can be used to extract the maximum work $\Delta W_{max}(A) = -k_B T(B)$ via particle A2, as discussed before. We may argue that the information gain due to measurement has been exploited to extract work. In fact, both schemes, the original one and its substitute, are equivalent.*

Example 4.10 Complete measurement *We consider a n-particle Szilard engine A, prepared in the correlated state*

$$|\Psi(A)> = \frac{1}{\sqrt{2}}|LLL\ldots L> + \frac{1}{\sqrt{2}}|RRR\ldots R>\,. \qquad (4.191)$$

Step 1: *Knowing this initial state we apply an appropriate unitary transformation that generates the pure state $|LLL\ldots L>$ [Dahlsten (2011)].*

Step 2: *This product zero-entropy state is thermal and can therefore directly be used to extract work from a bath B. According to Eq. (4.187) this is the maxmimum work, $\Delta W_{max}(A) = -Nk_B T(B)$.*

Alternatively, a complete measurement could be performed (step1). Let the measurement result be the pure state* $|LLL...L>$. *This state can directly be used to extract work from a bath B, as discussed before.*

Example 4.11 Combination of elementary procedures *Let us consider a composite system AC within an environment B. As suggested by [del Rio (2011)] the properly prepared system AC may be used to extract work from B, while subsystem A is then operated in the cooling (or erasure) mode thus consuming work. Under special conditions the total work extracted remains positive: Erasure would thus deliver work! This is not so surprising as it might seem: The resources needed are "hidden" in the respective initial state. Remarkably, the overall effect can be recast into the quantum-information language based on the conditional von Neumann entropy,* $S_{vN}(A|C)$, *see Eq. (3.220):*

$$\Delta W_{max} = S_{vN}(A|C) k_B T(B) . \quad (4.192)$$

In any case, the preparation of the initial state constitutes an important resource for the subsequent functionality. However, this procedure may be subject also to subjective ignorance, that is, lack of control. Lack of control also increases the risk to fail in a given task.

Remark 4.25 Consistency between information and physics *As already stated in Section 2.2.7, information may serve as an input for effective descriptions. These are typically incomplete. In quantum mechanics and quantum thermodynamics the origin of uncertainty is objective. It would thus be inconsistent to talk about a spin in a thermal state and inquire about the effect of "knowing" its energy eigenstate. If we knew the eigenstate, the spin would be in a pure, not in a thermal state (cf. Theorem (4.19)). Likewise, any correlations as expressed by the quantum conditional entropy must be based on some appropriate physical network, not on abstract reasoning of an observer (as apparently assumed by [del Rio (2011); Dahlsten (2011)]). Of course, this caveat does not exclude the possibility to consider situations of subjective uncertainty even in the quantum domain. However, the subjective uncertainty is always classical in nature.*

Partial knowledge of a quantum state may easily imply non-equilibrium, for which there is no direct relationship to standard thermodynamics, cf. Remark 4.24.

Part II

Quantum Thermodynamic Processes

Chapter 5

Quantum Thermodynamics

Jonas Burgert: Pulse Run, 2009, based on photograph by Lepkowski Studios [Reprinted with permission from Schreiber and Wipplinger (2011)].

Quantum Thermodynamic Processes: Energy and Information Flow at the Nanoscale
Günter Mahler

ISBN 978-981-4463-73-7 (Hardcover), 978-981-4463-74-4 (eBook)
www.panstanford.com

In Chapter 4 macroscopic features of thermodynamics have been introduced, essentially based on a micro-description taken from quantum physics. Fundamental relations to information theory have already been included—another contribution to "agent-based" physics. Here we further explore the pertinent statistical consequences based on the partitioning of quantum systems.

According to quantum thermodynamics thermal behavior is not a property of a system as such; it follows from its embedding into a larger quantum environment under appropriate conditions. Rather than asking, "is the system thermal," we have to inquire, "do descriptions exist for which thermal properties follow for a selected part." There could be none, there could be many.

While quantum thermodynamics tries to give a foundation of conventional thermodynamics,[a] it allows to locate thermal behavior in quite novel contexts, for example, down to the nano-limit. But the unrestricted validity of all thermodynamic relations is not guaranteed. There could be scenarios for which we have to generalize thermal concepts, possibly to a point, where they loose their original meaning and thus their universal validity.

Quantum thermodynamics is based on entanglement. Entanglement theory is notoriously complex; to guide research there have been attempts to find analogies between entanglement and thermodynamical concepts [Horodecki (2008)]. If successful, this would mean—from the thermodynamics point of view—that certain quantum features tend to "repeat themselves" on a higher level, cf. Definition 9.3. So far evidence is not convincing, though. (See also Section 9.3.)

5.1 Zeroth Law: Equilibrium

Definition 5.1 Zeroth law *The zeroth law will be understood here to imply the existence of thermal equilibrium; on the macro-level this includes the contact equilibrium conditions, which, in turn, establish*

---

[a] One may contemplate, nevertheless, multiple origins of thermal behavior. Some researchers might find it rather unsatisfactory to accept quantum mechanics as the driving force behind the thermal aspects even of climate models, say.

equilibrium as a transitive relation between subsystems (cf. Theorem 4.14).

5.1.1 *Typicality*

What is typical? As a first attempt for clarification one may address the everyday meaning of "typical behavior" of ourselves or of our friends: Given you find out that your friend Bob is not at home and the weather is fine, you bet that he will be hiking up in the mountains; it would be very surprising to find him watching TV. That is the way you know Bob, but, of course, you cannot be sure in this particular case. . .

Clearly, one would not call a behavior typical, if it was entirely determined by the conditions we know to be satisfied. And the behavior we want to predict cannot be fully detailed: We do not dare to predict at what time Bob should be found on what position of which hiking trail.

Also physical systems may exhibit typical behavior—provided we refer to a sufficiently coarse-grained description, that is, we do not care about all details. In the context of thermodynamics the notion of typicality has been introduced by Lebowitz [Lebowitz (1993)]. Here the idea is that very many microstates may yield very similar outcomes with respect to a some observable F. In that sense nominally different microstates become indistinguishable, they form an equivalence class. Even that condition may be weakened, though. Being satisfied to know that most microstates of those accessible are members of that equivalence class would lead us to the above introduced notion of typicality. We have good reasons to expect certain properties, we can bet on that, but we cannot be sure.

These ideas can be made more quantitative (cf. [Gemmer (2009)]). Consider a set of accessible microstates $\vec{X}$ within a D-dimensional state space with $\int d^D X = \Omega \approx$ number of states. Then the average value of some function or property $h(\vec{X})$ would be given by

$$E[h] = \frac{1}{\Omega} \int h(\vec{X}) d^D X \tag{5.1}$$

and the variance

$$\Delta^2[h] = E(h - E[h])^2 = E[h^2] - E^2[h]. \tag{5.2}$$

The value $h(\vec{X})$ will almost be the same for all $\vec{X}$ if and only if

$$\Delta[h] \ll h_{max} - h_{min}. \tag{5.3}$$

Here, h_{max}, h_{min} denote the bounds of h within the accessible region, cf. Proposition 5.3. This notion of typicality can eventually be linked to dynamics. Partition the accessible microstate space into two regions, $R1$, $R2$. Let $\vec{X}$ belong to $R1$ if

$$|h(\vec{X}) - E[h]| \leq \epsilon \tag{5.4}$$

and to $R2$ otherwise. Define the size of region R by

$$\Omega(R) = \int_R d^D X. \tag{5.5}$$

If we choose ϵ such that $\epsilon \ll h_{max} - h_{min}$ typicality implies that $\Omega(R1) \gg \Omega(R2)$. Consider a time interval $t, t + dt$ in which a certain amount of microstates, $d\Omega_t(R1 \to R2)$, leave region $R1$ for region $R2$, and the microstates $d\Omega_t(R2 \to R1)$ undergo the opposite transfer. If the dynamics preserves the microstate volume (which is the case for classical as well as quantum dynamics), we must have $d\Omega_t(R1 \to R2) = d\Omega_t(R2 \to R1)$ and thus

$$\frac{d\Omega_t(R1 \to R2)}{\Omega(R1)} \ll \frac{d\Omega_t(R2 \to R1)}{\Omega(R2)}. \tag{5.6}$$

This means that the relative frequency of states that evolve from the equilibrium set $R1$ to the non-equilibrium one is much smaller than the relative frequency for the inverse transition. This observation can be taken as a preliminary justification of the second law.

5.1.2 *Scenarios Based on Partitions*

Various scenarios can be distinguished based on specific partitions of the total system into subsystems $\mu = A, B, C, \ldots$, cf. Definition 3.57. The subsystems of such a quantum network[a] are characterized

[a]Quantum networks are basic also to quantum computing, where the subsystems are treated on equal footing. In quantum thermodynamics the roles of the subsystem are typically different, the description incomplete.

by their size (Hilbert space dimension $d(\mu)$) and the strength of their mutual interaction.

Such a partition structure constitutes a kind of reference frame; it is essential for understanding the contextualize of (effective) physical phenomena. Unfortunately, there seem to be a few guidelines by which to select a partition:

Definition 5.2 Weak coupling *Consider a pair of subsystems*

$$\hat{H}(AB) = \hat{H}(A) \otimes \hat{1}(B) + \hat{1}(A) \otimes \hat{H}(B) + \lambda \hat{V}(AB) . \qquad (5.7)$$

Here, $\hat{V}$ specifies the interaction with λ being an adjustable (dimensionless) parameter. Weak coupling requires

$$\boxed{\lambda < V(AB) > \ll < H(\mu) >, \quad \mu = A, B .} \qquad (5.8)$$

$< F >$ is a short-hand notation for $\mathtt{Tr}\{\hat{F}\hat{\rho}\}$. Weak coupling thus means that the interaction energy can be neglected with respect to any of the local energies.

Weak coupling helps to define a kind of "local identity": The partition into subsystems may be said to be physically justified. We introduce the following hierarchy:

(1) Closed single system: (A).
(Here and in the following the parenthesis indicates the subsystem of interest.) In the present context the so-called eigenstate thermalization hypothesis (ETH) has been suggested [Deutsch (1991)]. According to this hypothesis thermalization happens in each individual eigenstate, but is initially hidden by coherences between them. In the course of time decoherence reveals the thermal properties, diagonal elements of specific observables take on a constant value.
While interesting in its own right [Reimann (2008)]; [Ikeda (2011)], isolated systems are unrealistic. Furthermore, thermodynamics is explicitly based on external control and partitioning. Any approach that allows to incorporate such features right from the start should thus appear more appropriate.

(2) Closed bipartite system: (AB).
A and B are treated primarily in a symmetric fashion. Mutual coupling in terms of energy exchange (or volume exchange) can be studied.

(3) Embedded system: $(A)B$.
The total system is closed, but the (much smaller) subsystem A is of prime interest only. When weakly coupled to a much larger environment B quantum thermodynamics [Gemmer (2009)] predicts that the subsystem A should—under appropriate conditions—be found in a thermal state with some temperature $T(A) = 1/(k_B\beta(A))$,
$$\hat{\rho}_{equ}(A) = \frac{1}{Z(A)} \exp\left(-\beta(A)\hat{H}(A)\right), \tag{5.9}$$
even if the total state is pure (and then remains pure). $T(A)$ reflects the degeneracy structure of the embedding system with respect to A, see Eq. (5.29). This local thermalization is a consequence of entanglement between A and B. For it to occur an interaction allowing for energy exchange (even if extremely weak) has to be present.
This thermal state of A commutes with the local Hamiltonian $\hat{H}(A)$. It is stationary even in the limit $\lambda \rightarrow 0$, that is, after the interaction has faded away. However, the thermal state is an attractor state only as long as the interaction with the environment is present. Without this coupling, any perturbation, in particular, measurements, will persistently change the state.

(4) Embedded bipartite system: $(AB)C$.
Again, the total tripartite system is closed. The subsystems A, B, C are weakly coupled, C is the much larger environment. Then the subsystems A and B will separately be found in thermal states, usually product states without internal entanglement (between A and B). This also means that in this limit $S_{vN}(AB) = S_{vN}(A) + S_{vN}(B)$; the entanglement between (AB) and C does hardly depend on the interaction (between A and B), the two effects (interaction, entanglement) are separate. For weak interaction between A and B the two local temperatures tend to be the same.

(5) Multipartite system $(ABCD\ldots)$. The total system may be closed. Often the subsystems will be of the same kind; one may then consider any individual system embedded in the rest, or any pair embedded in the rest, etc. Likewise one may study groupings of different size (coarse-graining), cf. Example 5.7.

5.1.3 *Closed Single System* (A)*: Ergodicity*

Equilibrium means that the system under consideration, (A), evolves toward some particular state and remains in that state for almost all times. But how can this happen?

At the heart of equilibrium thermodynamics are macrostates (thermal states) and—on the micro-level—equilibrium states. Quantum mechanically, these microstates can be taken to be mixtures of pure states. (Such a decomposition is not unique, though.) But such a pure state ensemble would be reminiscent of the classical approach to thermodynamics via ensembles characterized by some probability distribution. The meaning of those ensembles is specified by the quasi-ergodic hypothesis. This hypothesis implies the equivalence between ensemble- and infinite-time average.

Definition 5.3 Ergodicity *Let $p_N(\vec{s})$ be the equilibrium probability distribution for N classical particles (Gibbs ensemble, cf. Eq. (4.164)). $\vec{s}$ is the $6N$-dimensional vector describing position and momentum of all N particles. Then [Diu (1989)]*

$$\overline{F} = \int d^{6N}s \; p_N(\vec{s})F(\vec{s}) = \lim_{\tau\to\infty} \frac{1}{\tau}\int_0^\tau dtF(\vec{s}(t)) \,. \tag{5.10}$$

This equivalence clarifies the operational meaning of the ensemble, given the fact that in any real thermal experiment we have but one ensemble member. Lack of knowledge is interpreted to arise from the fact that a measurement does not refer to a well-defined instant of time (rather, to a "coarse-grained" time).

Proposition 5.1 Long-time average *We define the long-time average of a quantum state $\hat{\rho}(t)$ by (cf., e.g., [Linden (2009)])*

$$< \hat{\rho}(t) >_t \equiv \hat{\rho}_\infty = \lim_{\tau\to\infty} \frac{1}{\tau}\int_0^\tau \hat{\rho}(t)dt \,. \tag{5.11}$$

Now consider a system $\hat{H}$ with energy basis $|E_k>$. The pure state $|\Psi(t)>$ can conveniently be written as a superposition

$$|\Psi(t)> = \sum_k c_k \exp\left(-iE_k t/\hbar\right) , \tag{5.12}$$

implying the density operator

$$\hat{\rho}(t) = \sum_{k,m} c_k c_m^* \exp\left(-i(E_k - E_m)t/\hbar\right)|E_k> < E_m| \,. \tag{5.13}$$

Its long-time average is for non-degenerate states

$$\hat{\rho}_\infty = \sum_k |c_k|^2 |E_k> < E_k| \,. \tag{5.14}$$

Here we have made use of the representation of the delta function

$$\lim_{\tau \to \infty} \frac{1}{\tau} \int_0^\tau \exp\left(i((E_k - E_m)t/\hbar\right)dt = \delta_{k,m} \,. \tag{5.15}$$

The purity (cf. Eq. (3.147)) of this limiting state is

$$P_\infty = \sum_k |c_k|^4 \equiv \frac{1}{d_\infty^{eff}} \,. \tag{5.16}$$

Roughly, d_∞^{eff} tells us, how many pure states contribute appreciably to the mixture [Linden (2009)], cf. Proposition 2.6.

Remark 5.1 Partitions *In classical mechanics complete knowledge of the state of the whole implies complete knowledge of the state of any subsystem. Partitioning as such cannot serve as a route to uncertainty.*

This is in stark contrast to quantum mechanics: Indeed, in quantum thermodynamics ensemble averages and subjective ignorance are not needed. A key ingredient is entanglement between subsystems. With the state of the environment being traced out, there is an objective lack of knowledge about the remaining subsystem. This coarse-graining effect in Hilbert space is very efficient.

5.1.4 *Embedded System $(A)B$: Thermalizing Environment*

We now turn to a bipartite quantum system $(A)C$ with Hilbert space dimension $d(A, B) = d_A d_B$.

Proposition 5.2 *Let the reduced state for subsystem A be $\hat{\rho}(A;t)$, its long-time average $\hat{\rho}_\infty(A)$. Then it holds that*

$$D(\hat{\rho}(A;t), \hat{\rho}_\infty(A)) \leq \sqrt{\frac{d_A^2}{d_\infty^{eff}(A, B)}} \,. \tag{5.17}$$

Here D is the Schmidt–Hilbert distance between the two states, Eq. (3.150). If the right-hand side is small, the subsystem A spends most of its time close to $\hat{\rho}_\infty(A)$.

A proof is given in [Linden (2009)].

Thermalizing environments are the prime target of quantum thermodynamics: Here one is concerned with the question under which conditions an embedding quantum system B is able to impart thermal properties on the embedded system A.

A simple argument is based on Hilbert space statistics (cf. [Gemmer (2009)]): As already noted, any pure state $|\phi>$ in a given d-dimensional Hilbert space can be parametrized as

$$|\phi>=\sum_{j=1}^{d}(\eta_j + i\xi_j)|j> . \tag{5.18}$$

Here, η_j, ξ_j are $2d$ real parameters, and $|j>$ denotes some convenient basis. According to Eq. (3.182) an unbiased ensemble of pure states can then be characterized by the distribution function $w(\eta_1, \xi_1, \dots \eta_d, \xi_d)$. This function allows us to define Hilbert space averages of any function $h(|\phi>) = h(\eta_1, \xi_1, \dots \eta_d, \xi_d)$ as an integral over the $2d$-dimensional parameter space:

$$\overline{h} = \int w(\eta_1, \xi_1, \dots \eta_d, \xi_d) h(\eta_1, \xi_1, \dots \eta_d, \xi_d) d^d\eta d^d\xi . \tag{5.19}$$

Proposition 5.3 Hilbert space average method (HAM) *The function $h(|\phi>)$ can be visualized as a "landscape" over its 2d-dimensional parameter space. We can approximate*

$$\boxed{h(|\phi>) = h(\eta_1, \xi_1, \dots \eta_d, \xi_d) \approx \overline{h} ,} \tag{5.20}$$

provided that landscape of h is "flat" enough.

In particular let us consider some operator $\hat{F}$ in its eigen representation:

$$\hat{F} = \sum_k F_k |f^{(k)}><f^{(k)}| . \tag{5.21}$$

We are free to take these eigenfunctions as the basis for the ensemble of pure states $|\phi>$ and their distribution w,

$$|\phi>=\sum_{j=1}^{d}(\eta_j + i\xi_j)|f^{(j)}> . \tag{5.22}$$

The expectation value $h \equiv <\phi|\hat{F}|\phi>$ of some operator $\hat{F}$ is an example for $h(|\phi>)$; its Hilbert space average is

$$\overline{h} = \overline{<F>} = \sum_j F_j(\overline{\eta_j^2} + \overline{\xi_j^2}) = \frac{1}{d}\mathrm{Tr}\{\hat{F}\} . \tag{5.23}$$

In the last step we have used that by symmetry and normalization $\overline{\eta_j^2} = \overline{\xi_j^2} = 1/(2d)$. Correspondingly one finds for

$$\overline{h^2} = \frac{1}{d(d+1)} \left(\mathrm{Tr}\{\hat{F}^2\} + \mathrm{Tr}\{\hat{F}\}^2 \right) , \qquad (5.24)$$

(for details see [Gemmer (2009)]), which allows us to introduce the Hilbert space variance

$$\Delta^2(h) = \frac{1}{d+1} \left(\mathrm{Tr}\{\hat{F}^2\}/d - \mathrm{Tr}\{\hat{F}\}^2/d^2 \right) . \qquad (5.25)$$

HAM is thus justified for this $h = < F >$, if

$$\Delta(h) \ll \overline{h} . \qquad (5.26)$$

Definition 5.4 Accessible region (AR) *Unless the system is nonintegrable, the Hilbert space accessible from some initially prepared state may severely be constrained by various constants of motion. The system will never be able to leave the respective subspace.*

While the distinction between integrable and nonintegrable models is often considered essential for the resulting statistical behavior, it is of minor concern here. We simply have to make sure that certain conditions of subsystem dimensions and weak inter-subsystem couplings are obeyed. The arguments in terms of HAM then go through.

Remark 5.2 Generalized Gibbs ensemble *Along similar lines of thought the effective equilibrium state for (partially) integrable systems have been suggested [Rigol (2007)] to result from the maximization of the respective entropy subject to the given integrability constraints (i.e., set of conserved quantities $\hat{I}_n$). This simple method based on the maximum entropy approach (MEP) (cf. Proposition 4.1) has recently been applied to ultracold atoms [Cassidy (2011)].*

The HAM approach will now be applied to a closed bipartite system (A, B) under energy exchange; the subsystem A is taken to be a spin.

Example 5.1 Equilibrium via energy exchange *We take system A to be a two-level system with energy-splitting ΔE and B to be a multilevel system with states $|j, k(B)\ >$, where k describes the degeneracies, $k = 1, 2, \ldots g_j$. We single out two energy bands, $j = 0$ and $j + 1 = 1$, and assume weak (resonant) interaction,*

$E_1 - E_0 = \Delta E$. *The total energy of the system* (A, B) *is taken to be* ΔE. *Energy conservation then induces a correlation between allowed states in A and B, that is, not all* $d(A, B)$ *states of the product space are accessible. The accessible region AR can be projected out via the following two projectors,*

$$\hat{P}_0 = |0(A)><0(A)| \otimes \sum_{k=1}^{g_1} |1, k(B)><1, k(B)|,$$
$$\hat{P}_1 = |1(A)><1(A)| \otimes \sum_{m=1}^{g_0} |0, m(B)><0, m(B)|. \quad (5.27)$$

The accessible Hilbert space has dimension $d_{access}(A, B) = g_0 + g_1 = d(B)$. *The respective occupation probabilities* $< P_0 >, < P_1 >$, *averaged according to HAM, Eq. (5.23), are then found to be*

$$\overline{< P_0 >} = \frac{\text{Tr}\{\hat{P}_0\}}{d(B)} = \frac{g_1}{g_0 + g_1},$$
$$\overline{< P_1 >} = \frac{\text{Tr}\{\hat{P}_1\}}{d(B)} = \frac{g_0}{g_0 + g_1}. \quad (5.28)$$

The relative variance, $\Delta(< P_\mu >)/\overline{< P_\mu >}$ *scales with* $1/d_{access}(B)$. *For* $d_{access}(B) \gg 1$ *HAM is thus justified and the equilibrium state of A reflects the degeneracy structure of the embedding subsystem B. Equilibrium via energy exchange leads to the so-called canonical states.*

Definition 5.5 Embedding temperature *Even though the embedding system B is not a bath (i.e., is not in a stationary thermal state) the smaller subsystem A may be said to be in a thermal state with temperature* $T(A)$,

$$\frac{\Delta E}{k_B T(A)} = \ln\left(\frac{\overline{< P_0 >}}{\overline{< P_1 >}}\right) = \ln\left(\frac{g_1}{g_0}\right). \quad (5.29)$$

This effective temperature will be called "embedding temperature", as it derives from the degeneracy structure of the quantum environment B with respect to the transition energy ΔE *selected by A. In general, a different subsystem A (different transition energy) will have a different embedding temperature even within the same B.*

In particular, for $g_0 = g_1$ one finds

$$\overline{< P_0 >} = \overline{< P_1 >} \approx 1/2, \quad (5.30)$$

so that $T(A) = \infty$. For $g_0 = 1 \ll g_1$

$$\overline{< P_0 >} \approx 1 \quad \overline{< P_1 >} \approx 0\,. \tag{5.31}$$

This would imply $T(A) = 0$, apparently giving a simple route to zero temperature.

The scenario is easily generalized to more than two levels for subsystem A and an appropriate embedding. Such models give birth to the concept of (local) temperature as an emergent property: Note that there is neither a temperature for the system as a whole (still being subject to the Schrödinger equation) nor for the subsystem B. A kind of "classical limit" is needed in the sense of $d(B) \gg d(A)$.

Remark 5.3 Embedding temperature and third law *For the weak coupling condition to be valid in the limit $T(A) \to 0$, the interaction energy must be small compared to the energy of subsystem A, which approaches zero. As the thermal relaxation time depends on the interaction strength, this essentially means that the time to reach thermal equilibrium would go to infinity. In this sense the third law is saved: There is no efficient process to reach the absolute zero of temperature.*

Remark 5.4 Negative temperature *Negative embedding temperatures obtain for $g_1 < g_0$. Such a degeneracy structure can be realized by means of a cluster of noninteracting (identical) spins in the high-energy regime. For $g_1 \to 1 \ll g_0$ one would get $T(A) \to -0$ (total inversion).*

It has been claimed [Dunkel (2013)] that negative temperatures were "inconsistent," based on the "micro-canonical" entropy measure $\tilde{S}(E) = k_B \ln \Omega(E)$ with Ω counting the number of eigenstates less or equal E. Obviously, $\tilde{S}(E)$ continues to increase with E and reaches its maximum at the upper energy bound. The associated temperature, $\tilde{T} \equiv \partial\tilde{S}/\partial E$ is always positive. Whether or not this entropy definition is considered acceptable, the argument does not apply here: According to our quantum-thermodynamical considerations the embedding system is typically in a nonthermal state. What counts is its effect on the embedded subsystem.

Proposition 5.4 General equilibrium states *Let the full state of some bipartite system (AB) be described by*

$$|\psi(AB)>=\sum_{K,L}\sum_{k=1}^{g_K(A)}\sum_{l=1}^{g_L(B)}\psi_{kl}^{KL}|K,k(A)>\otimes|L,l(B)>\,. \tag{5.32}$$

Here, $g_K(A)$ is the degeneracy of level K within subsystem A, $g_L(B)$ is the degeneracy of level L within subsystem B.

The probability $w(K, L)$ to find the pair (K,L) in $|\psi(AB)>$ can be calculated from

$$w(K,L)=\sum_{k,l}|\psi_{kl}^{KL}|^2\,, \tag{5.33}$$

from which we obtain the local probabilities

$$w(K(A))=\sum_{L}w(K,L)\quad w(L(B))=\sum_{K}w(K,L)\,. \tag{5.34}$$

The reduced local state for subsystem A is

$$\hat{\rho}(A)=\mathrm{Tr}_B\{|\psi(AB)><\psi(AB)|\}\,, \tag{5.35}$$

with the matrix elements

$$\rho_{Kk,K'k'}=<Kk(A)|\hat{\rho}(A)|K'k'(A)>\,. \tag{5.36}$$

Let us start from an initial state with a given distribution of total energy E, w_E. The distribution is assumed to be non-zero in a finite energy interval ΔE only. Weak coupling requires $E\approx E_K(A)+E_L(B)$, a condition, which selects appropriate state pairs $\{K, L\}$. This selection defines the accessible subspace (AR) for the present problem.

For that AR the HAM most probable value (cf. Proposition 5.3) for the matrix elements $\rho_{Kk,K'k'}$ is

$$\boxed{\begin{aligned}\overline{\rho_{Kk,K'k'}}&=\frac{w(K(A))}{g_K(A)}\delta_{KK'}\delta_{kk'}\,,\\ \frac{w(K(A))}{g_K(A)}&=\textstyle\sum_E\frac{w_E}{g_E}g_L(B)\delta(E_K(A)+E_L(B)-E)\,.\end{aligned}} \tag{5.37}$$

g_E is the total degeneracy of the state E. Based on this result the most probable local state reads [Gemmer (2009)]

$$\hat{\rho}_{equ}(A)=\sum_{K,k}\sum_{K',k'}\overline{\rho_{Kk,K'k'}}|Kk(A)><K'k'(A)|\,. \tag{5.38}$$

One can show that indeed the Hilbert space variance becomes small for $d(B) \gg d(A)$, where the dimensions d characterize the accessible region.

Consider the simple case as of Example 5.1: E is fixed and the accessible state space is characterized by $g_E = g_0(B) + g_1(B)$; the levels of subsystem A are non-degenerate and $K = 0, 1$. The levels of subsystem B are $L = 0, 1$ with degeneracies $g_0(B), g_1(B)$, respectively. Then

$$\overline{\rho_{K,K}} = \frac{1}{g_0(B) + g_1(B)} g_L(B)\delta(E_K(A) + E_L(B) - E), \quad (5.39)$$

$$\overline{\rho_{0,0}} = \frac{1}{g_0(B) + g_1(B)} g_1(B),$$

$$\overline{\rho_{1,1}} = \frac{1}{g_0(B) + g_1(B)} g_0(B).$$

We thus recover the result of Eq. (5.28); the canonical state thus reads

$$\hat{\rho}_{can} = \frac{1}{g_0(B) + g_1(B)} (g_1(B)|0><0| + g_1(B)|1><1|). \quad (5.40)$$

Example 5.2 Equilibrium for exponential degeneracy *We consider a general system (A) embedded in the environment (B) with*

$$g_L(B) = g \exp(\alpha E_L(B)). \quad (5.41)$$

α is some real positive parameter. Then from Eq. (5.37)

$$w(K(A)) = g_K(A) \sum_E \frac{w_E}{g_E} g \exp(\alpha E_L(B))\delta(E_K(A) + E_L(B) - E). \quad (5.42)$$

Replacing $E_L(B) = E_K(A) - E$ we immediately find

$$w(K(A)) = g_K(A) \exp(-\alpha E_K(A)) \sum_E \frac{w_E}{g_E} g \exp(\alpha E). \quad (5.43)$$

Independent of L this is a canonical distribution with temperature $T = 1/(k_B\alpha)$. The last factor (the sum) does not depend on K; it is fixed by normalization.

Example 5.3 Equilibrium for binomial degeneracy *Consider a two-level system (A) embedded in the environment (B) characterized by*

$$g_j = \binom{N}{j} = \frac{N!}{(N-j)!j!}. \quad (5.44)$$

Here, N is an integer > 0; $j = 0, 1, \ldots N$. *For an intuitive realization we could think of an environment consisting of N noninteracting spins with an equidistant spectrum, that is, identical energy splittings* ΔE, *cf. Example 3.10. The index j specifies the "working point," that is, the environmental band (and its neighbor* $j + 1$*) being selected. The environment with fixed "working point" j is a non-thermal state.*

One then finds for system (A),

$$\overline{< P_0 >_j} = \frac{N - j}{N + 1}, \tag{5.45}$$

$$\overline{< P_1 >_j} = \frac{j + 1}{N + 1}. \tag{5.46}$$

Here we have made use of the identity

$$\binom{N}{k} + \binom{N}{k-1} = \binom{N+1}{k}. \tag{5.47}$$

The resulting $\beta(A)$ *depends on j and even changes sign. Indeed,* $\beta(A)$ *approaches minus infinity for* $j \rightarrow N \gg 1$, *cf. Eq. (5.255).*

Proposition 5.5 Equilibrium without energy exchange *In this case all local occupation probabilities w(K(A)) and w(L(b)) are separate constants of motion. We may restrict ourselves to a sharp energy state* $E(B)$ *with* $L = L_0$ *and fixed degeneracy* $g_{L_0}(B)$. *We can then directly apply the previous HAM results for the reduced density operator of the embedded subsystem A:*

$$\overline{\rho_{Kk,K'k'}} = \frac{w(K(A))}{g_K(A)}\delta_{KK'}\delta_{kk'}, \tag{5.48}$$

$$\hat{\rho}_{microcan}(A) = \sum_{K,k} \frac{w(K(A))}{g_K(A)} |Kk(A) > < Kk(A)|. \tag{5.49}$$

This state is diagonal; for a fixed K we have equipartition (occupation probability being $1/g_K(A)$*), as to be expected for a microcanonical state.*

Remark 5.5 Equivalence of ensembles *It is generally expected that the different equilibrium states, in particular those with or without energy exchange (i.e., representing the microcanonical and canonical ensemble, respectively), should become equivalent in the thermodynamic limit. This feature is of minor importance in the context of small thermal systems studied here.*

Proposition 5.6 Embedding temperature versus bath temperature *Consider a more general environment B with discrete energy spectrum $\{E_j, j = 1, 2, \dots\}$ and respective degeneracies g_j. Furthermore, let this system be in a thermal state with partition sum $Z(B)$ and temperature $\beta(B) = 1/k_B T(B)$,*

$$\hat{\rho}_{equ}(B) = \frac{1}{Z(B)} \sum_j \exp(-\beta(B)E_j) \sum_{k=1}^{g_j} |j, k(B) >< j, k(B)| \,. \tag{5.50}$$

Then consider a two-level system A in its excited state $|1(A)>$, which is weakly coupled to B. What will now be the effect of B on A?

Energy conservation implies that the initial pure state $|1(A); j, m(B)>$ with $1 \leq m \leq g_j$ would preferably be coupled to $|0(A); j+1, k(B)>$ with $1 \leq k \leq g_{j+1}$, so that—under HAM—the equilibrium state is expected to be

$$\overline{< P_0 >_j} = \frac{\mathrm{Tr}\{\hat{P}_0\}}{d_j(B)} = \frac{g_{j+1}}{g_j + g_{j+1}} \,, \tag{5.51}$$

$$\overline{< P_1 >_j} = \frac{\mathrm{Tr}\{\hat{P}_1\}}{d_j(B)} = \frac{g_j}{g_j + g_{j+1}} \,. \tag{5.52}$$

But in addition to taking the quantum expectation value and to apply the HAM we have now, in a third step, to perform a thermal averaging over the environmental states j:

$$\hat{\rho}_{equ}(A) = \frac{1}{Z(B)} \sum_j g_j \exp(-\beta(B)E_j) \times \\ \{\overline{< P_0 >_j}|0(A) >< 0(A)| + \overline{< P_1 >_j}|1(A) >< 1(A)|\} \,. \tag{5.53}$$

In general, this equation lacks a simple interpretation. For more transparent results specific models are needed about the environmental spectrum E_j and the degeneracy g_j.

For the following we assume for the environment B an equidistant spectrum,

$$E_j = j\Delta E \,. \tag{5.54}$$

All transitions between adjacent levels are thus resonant with the two-level system A. The following examples concern the degeneracy of B.

Example 5.4 Exponential degeneracy *We assume*

$$g_j = \exp\left(\gamma j \Delta E\right). \tag{5.55}$$

γ is a real positive parameter. Then

$$\overline{< P_0 >} = \frac{1}{1 + \exp\left(-\gamma \Delta E\right)}, \tag{5.56}$$

$$\overline{< P_1 >} = \frac{1}{1 + \exp\left(\gamma \Delta E\right)}, \tag{5.57}$$

independent of j. The resulting temperature is

$$\Delta E / k_B T(A) = \gamma \neq \beta(B).$$

Example 5.5 Binomial degeneracy *We refer to Example 5.3. A thermal environment of (B) obtains after thermal averaging over the bands (working points) j. With the partition sum*

$$Z(B) = \sum_{j=0}^{N} \binom{N}{j} \exp\left(-\beta(B) j \Delta E\right) = \left(1 + \exp\left(-\beta(B)\Delta E\right)\right)^N \tag{5.58}$$

one gets in the limit $N \to \infty$,

$$\beta(A) \approx \beta(B). \tag{5.59}$$

In this large N-limit ("thermodynamic limit") there is no conflict between embedding temperature and the bath temperature. This should hold for positive as well as negative temperatures.

Example 5.6 Subexponential degeneracy *Typically, the degeneracy g_j grows subexponentially, so that the product of degeneracy and occupation probability, $g_j \exp\left(\beta(B) E_j\right)$, exhibits a distinct maximum. In the vicinity of this maximum the degeneracy can often be approximated by an exponential dependence on j with an exponent given by $\beta(B)$, so that, again, $\beta(A) \approx \beta(B)$.*

5.1.5 *Embedded Bipartite System $(AB)C$: Extensivity and Beyond*

Remark 5.6 Embedding versus bath system *In the closed bipartite scenario $(A)B$ the larger subsystem B may impart thermal properties on the smaller system A via entanglement (cf.*

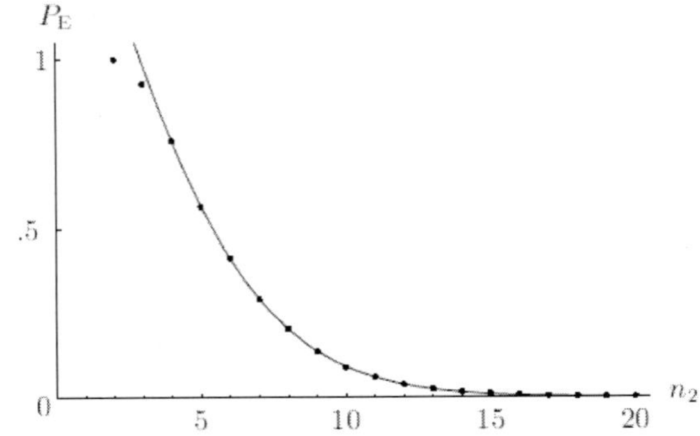

Figure 5.1 Probability P_E for finding entanglement within subsystem AB of dimension $d(A) = 2, d(B) = 3$ embedded in system C of dimension $d(C) = n_2$. The entanglement test is carried out based on the PPT method, cf. Definition 3.93. The solid line is an exponential fit.

Section 5.1.2). The respective von Neumann entropies cannot be additive. This confirms the fact that thermodynamics does not apply to this special internal relation itself—which underlies the local emergence of thermodynamics to begin with. If the embedding system is already in a thermal state, weak coupling may suffice to guarantee thermodynamic relations between embedded and embedding system. This is a kind of "short-cut."

However, to address thermodynamic relations between subsystems within a closed total system we have to consider embedded bipartite (or multipartite) systems, that is, scenarios $(AB)C$ and beyond.

Proposition 5.7 Suppression of entanglement *Consider a composite quantum system $(AB)C$ prepared in a pure state. Then the entanglement within (AB) embedded in a much larger quantum environment C is typically suppressed. A numerical example [Otte (2001)] is shown in Fig. 5.1.*

We confirm that the probability for finding entanglement within (AB), P_E, rapidly approaches zero for $d(C) \gg d(A) \times d(B)$.

Below we turn to a multipartite system consisting of identical subunits, a one-dimensional chain of $n_{tot} \gg 1$ elementary cells with nearest-neighbor interactions. For this model the "operationally

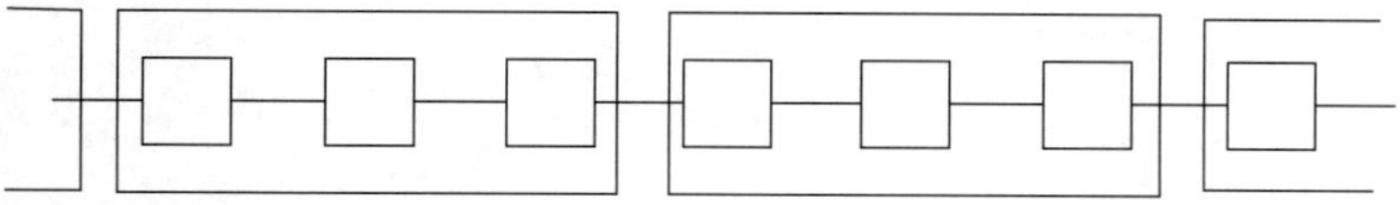

Figure 5.2 Coarse graining of one-dimensional chain. Small squares: elementary units; rectangles: coarse-grained units.

accessible partitioning" (cf. Definition 3.57) can systematically be changed. Coarse graining controls the validity of local thermal parameters such as the temperature. For simplicity we will assume the total system to be prepared in a canonical state.

Weak coupling, Definition 5.2, is not just an annoying requirement: It is a prerequisite for the "local identity" of thermal subsystems. At least for specific scenarios (based on short-range interactions) strong effective coupling might thus be the result of a "wrong" partition.

Example 5.7 Local temperature by partitioning of a one-dimensional chain *The full micro-description of the model is [Hartmann (2004); Hartmann (2005)]*

$$\hat{H} = \sum_{\mu=1}^{n_{tot}} (\hat{h}(\mu) + \hat{v}(\mu, \mu + 1)) . \tag{5.60}$$

This is a chain of n_{tot} elementary cells with nearest neighbor interactions, cf. Fig. 5.2. We now form n_g groups with Δn elementary cells each: $n_{tot} = n_g \Delta n$. The rescaled model can be rewritten as

$$\hat{H} = \hat{H}_0 + \hat{I} , \tag{5.61}$$

$$\hat{H}_0 = \sum_{\nu=1}^{n_g} \hat{H}(\nu) , \tag{5.62}$$

$$\hat{V} = \sum_{\nu=1}^{n_g} \hat{v}(\nu\Delta n, \nu\Delta n + 1) . \tag{5.63}$$

The Hamiltonians $\hat{H}(\nu)$ contain all the interactions within the group ("grain") ν. The respective eigenequation is

$$\hat{H}(\nu)|\psi(\nu)>= E(\nu)|\psi(\nu)> \quad \nu = 1, 2, \ldots n_g . \tag{5.64}$$

Here the additional index for the eigenvalues and eigenvectors is suppressed. The total system (without interactions between the grains) is described by

$$\hat{H}_0|\psi> = E_\psi|\psi>, \tag{5.65}$$

$$|\psi> = \prod_{\nu=1}^{n_g} |\psi(\nu)>, \tag{5.66}$$

$$E_\psi = \sum_{\nu=1}^{n_g} E(\nu). \tag{5.67}$$

Finally, the eigenequation of the full system reads

$$\hat{H}|\Phi>= E_\Phi|\Phi> . \tag{5.68}$$

For the following we assume a finite band with $E_{min} \le E_\Phi \le E_{max}$. *With this full system being prepared in a canonical state* $\hat{\rho}_{can}$, *the respective matrix elements in energy eigenbasis are*

$$<\Phi|\hat{\rho}_{can}|\Phi>= \frac{1}{Z}\exp(-\beta E_\Phi). \tag{5.69}$$

In the simple product-state basis, ψ, *one finds instead*

$$<\psi|\hat{\rho}_{can}|\psi>= \int_{E_{min}}^{E_{max}} dE\; w(E|\psi)\frac{\exp(-\beta E)}{Z}. \tag{5.70}$$

For large n_g *the conditional probability* $w(E|\psi)$ *becomes a Gaussian with variance* δE_ψ

$$(\delta E_\psi)^2 =<\psi|\hat{H}^2|\psi> - <\psi|\hat{H}|\psi>^2 . \tag{5.71}$$

We note that δE_ψ *would be zero, if the product states* ψ *were the energy eigenstates, that is, if the grains would not interact. For local temperatures to exist within each grain we require*

$$\ln <\psi|\hat{\rho}_{can}|\psi> = -\beta_{loc}E_\psi + c. \tag{5.72}$$

This condition can be fulfilled provided [Hartmann (2004)]; [Hartmann (2005)]

$$\boxed{E_\psi + \Delta\epsilon_\psi - E_{min} > \beta(\delta E_\psi)^2 .} \tag{5.73}$$

Here we have defined

$$\Delta\epsilon_\psi =<\psi|\hat{H}|\psi> - <\psi|\hat{H}_0|\psi> . \tag{5.74}$$

The interaction between the grains is a surface effect; it goes down with the number of grain boundaries, that is, for increasing grain size Δn. *However, for very low temperatures,* $\beta \to \infty$, *the condition for local temperatures ceases to be applicable at all: The weak coupling limit can no longer be enforced by a large grain size, cf. Example 5.14. This "no-go" situation may be seen in the context of the third law, cf. Theorem 4.14.*

Remark 5.7 Black box: Encapsulated quantum features *The "grains" define a kind of "black box" for the property of local temperature. While the derivation and characterization of that box requires quantum mechanics, once accepted it may be seen to validate the classical ideas.*

There are more instants like that: The stability of an atom, say, cannot be understood without quantum theory; however, once taken for granted (i.e., the stability as a black box), a classical gas theory can be established. Even superconductivity ("Meissner effect") can be described phenomenologically by a classical two-fluid model (cf. the so-called London theory from 1935).

Proposition 5.8 Extensivity of $U(AB)$ *Let us assume a canonical state for* (AB): $\hat{\rho}(AB) = \hat{\rho}_{equ}(AB)$. *Then, under the condition of weak coupling*

$$\hat{\rho}_{equ}(AB) = \hat{\rho}_{equ}(A) \otimes \hat{\rho}_{equ}(B)\,, \tag{5.75}$$

and the internal energy becomes extensive (additive):

$$U(\mu) = U(\hat{H}(\mu), \hat{\rho}_{equ}(AB)) = \mathrm{Tr}\{\hat{H}(\mu)\hat{\rho}_{equ}(AB)\} \quad \mu = AB,\, A,\, B\,, \tag{5.76}$$

$$U(AB) = U(A) + U(B)\,. \tag{5.77}$$

Proposition 5.9 Non-extensive thermodynamics *For a closed bipartite system (under a reversible transformation) we may formally write:*

$$dU(AB) = dU(A) + dU(B) + d\tilde{U} = 0\,, \tag{5.78}$$

$$dS(AB) = dS(A) + dS(B) + d\tilde{S} = 0\,. \tag{5.79}$$

The terms $d\tilde{U}, d\tilde{S}$ *describe the respective deviations from strict additivity. For the total and the local systems the first law and the Clausius equality read* ($\mu = AB,\, A,\, B$):

$$dU(\mu) = đQ(\mu) + đW(\mu)\,, \tag{5.80}$$

$$dS(\mu) = \frac{đQ(\mu)}{T(\mu)}\,. \tag{5.81}$$

Assuming $\text{đ}W(A) = \text{đ}W(B) = 0$ *the local heat transfer satisfies*

$$\left(\frac{1}{T(B)} - \frac{1}{T(A)}\right) \text{đ}Q(B) = -d\tilde{S} + \frac{d\tilde{U}}{T(A)} . \qquad (5.82)$$

The right-hand side is not zero; this means that, in general, $T(A) \neq T(B) \neq T(AB)$*. The contact equilibrium (zeroth law) does not hold.*

The concept of intensive and extensive variables in thermodynamics depends on the weak coupling limit and on the absence of correlations between the subsystems considered. However, weak coupling is not guaranteed: long-range forces may show up. Coulomb forces, gravitational forces are examples. Such cases are suggested to underly formal thermodynamic entropy models, which are explicitly non-extensive.

Proposition 5.10 Extensivity of $S_{vN}(AB)$ *For a pair of quantum subsystems the joint von Neumann entropy decomposes as*

$$S_{vN}(A, B) = S_{vN}(A) + S_{vN}(B) + \tilde{S} . \qquad (5.83)$$

Here, $\tilde{S} \equiv -I_{vN} \leq 0$ *is the (negative) mutual quantum information. This decomposition is the quantum analogue of the classical Definition (2.19). The mutual quantum information is zero for product states only, that is, only in this case extensivity holds. Note that weak coupling as such would not imply product states. Correlations are a kind of "natural" origin of nonadditivity.*

The Renyi entropies are all extensive for uncorrelated data (see Definition 2.13). The so-called Tsallis entropies, on the other hand, have been designed for "anomalous" thermostatistical systems; they provide explicit though formal models for non-additive entropy situations.

Definition 5.6 Tsallis entropy *The nonextensivity of the Tsallis entropies is quantified by the real parameter q:*

$$S_q^T = \frac{1}{1-q}\left(\sum_j (P_j)^q - 1\right) . \qquad (5.84)$$

Based on this model the entropy of a bipartite system (prepared in a product of mixed states) gets the following non-additive part [Abe (2006)],

$$d\tilde{S} = \frac{1-q}{k_B} S_q^T(A) S_q^T(B) . \qquad (5.85)$$

So-called Tsallis distributions are defined as maximizing the Tsallis entropy. They tend to be heavily tailed. For $q = 1$ the conventional thermodynamic result is obtained and the entropy becomes extensive for uncorrelated subsystems.

It is remarkable that the additivity of entropies could be enforced despite correlations, if one selects Tsallis entropies with specially adopted parameter q [Abe (2006)]. But this is quite artificial and lacks physical justification.

Remark 5.8 Tsallis and MEP *The standard maximum entropy principle (MEP) will not coincide with a Tsallis distribution: The latter appears as asymptotically improbable—unless novel constraints are introduced [Bercher (2008)].*

5.2 First Law: Work and Heat

Definition 5.7 First Law *There are two fundamentally different processes for changing the internal energy of a thermal system: application of mechanical work or heat, cf. Theorem 4.7.*

The distinction between work and heat is one of the major conceptional problems of thermodynamics. These concepts are contextual and relate to an (energy) exchange process, not to a state, and presuppose a partitioning of the total system into subsystems A, B.

Such a partitioning can be symmetrical with respect to those subsystems; then the nature of the exchanged energy will, in general, involve both, work and heat at the same time. Or one of the subsystems, say, B, is taken as an environment. Then limiting conditions can be found under which the energy exchanged must be heat only (heat reservoir) or work only (work reservoir). Without the possibility of such model environments the distinction between heat and work would loose its operational significance.

Let us start with a bipartite quantum system defined by the Hamiltonian,

$$\hat{H}(A, B) = \hat{H}(A) \otimes \hat{1}(B) + \hat{1}(A) \otimes \hat{H}(B) + \hat{V}(A, B)\,. \tag{5.86}$$

This closed system is now subject to a unitary time evolution; in particular, we are interested in the energy transfer between the two subsystems.

Contrary to Section 5.1.4 we assume here that the correlation between A and B can be neglected, at least within some finite evolution time. We further assume that the initial states of the subsystems are pure states. Then the local purities P_ν, $\nu = A, B$ stay close to 1, and the coupled Schrödinger dynamics can be approximated as follows: [Gemmer (2001)]

Proposition 5.11 Factorization approximation (FA) *Let the initial state of a bipartite system be given by the product pure state* $|\Psi_0 >= |\psi_0(A) > \otimes|\psi_0(B) >$. *As long as the purity stays close to 1, the following equations of motion hold,*

$$i\hbar\frac{d}{dt}|\psi(A;t) > = \hat{H}^{eff}(A;t)|\psi(A;t) >, \qquad (5.87)$$

$$i\hbar\frac{d}{dt}|\psi(B;t) > = \hat{H}^{eff}(B;t)|\psi(B;t) >, \qquad (5.88)$$

with the time-dependent effective Hamiltonians

$$\hat{H}^{eff}(A;t) = \hat{H}(A) + < \psi(B;t|\hat{V}(AB)|\psi(B;t) >, \qquad (5.89)$$

$$\hat{H}^{eff}(B;t) = \hat{H}(B) + < \psi(A;t|\hat{V}(AB)|\psi(A;t) > . \qquad (5.90)$$

These equations of motion are (approximate) pure state variants of Eq. (5.129). Unfortunately, due to the interaction the state will, typically, entangle in the course of time, that is, the reduced states will not stay pure, the factorization approximation breaks down.

Models, for which the FA holds for much longer times will somehow have to invoke a quasi-classical limit. There are special models which, indeed, exhibit such a limiting behavior, an important example being the Jaynes–Cummings model [Gemmer (2001)]. Also quantum heat engine cycles may exhibit such a limit, cf. Remark 7.16.

Example 5.8 External force *Let*

$$\hat{V}(AB) = -\hat{x}(A)\hat{F}(B). \qquad (5.91)$$

Then, if the total state remains a product state, the effective Hamiltonian for A simply reads

$$\hat{H}_{eff}(A) = \hat{H}(A) - \hat{x}(A)F(B), \qquad (5.92)$$

$$F(B) = < \hat{F}(B) > .$$

$F(B)$ *acts as an external force, cf. Definition 4.8.*

Remark 5.9 FA for semi-mixed initial states *The FA can easily be generalized for semi-mixed initial states,*

$$\hat{\rho}_0 \to \hat{\rho}_0(A) \otimes |\psi_0(B) > < \psi_0(B)| \,. \tag{5.93}$$

The effective Hamiltonians and the equations of motion are defined correspondingly, for example,

$$\hat{H}^{eff}(B;t) = \hat{H}(B) + \mathrm{Tr}_A\{\hat{V}(AB)(\hat{\rho}(A) \otimes \hat{1}(B))\} \tag{5.94}$$

For the FA to hold it suffices that the purity of subsystem B stays close to 1. (If both subsystems were mixed, we might consider the (approximate) constancy of the local purities as a simple quality test for FA.)

5.2.1 *Embedded System $A(C)$: Mechanical Environment*

Not all environments C can be thermal: There should be alternative scenarios in which an explicit time dependence of the local Hamiltonian for A results, rather than a (local) stationary thermal state.

However, the appearance of an effective time-dependent Hamiltonian might be only half of the story: in a more complete account it could be intimately connected with the emergence of a time-dependent Schrödinger equation as such. This option will be discussed below.

For such a proposition to make sense we tentatively have to assume that (for isoleted systems) the time-independent Schrödinger equation is more fundamental.[a] The reasoning would go something like this: An isolated system should be in a pure state, cf. Remark (3.6). Dynamics makes no sense due to lack of external reference. This implies an energy eigenstate as the appropriate point of departure.

Here we follow the investigation as given in [Briggs (2000)]. It is based on the so-called Born–Oppenheimer approximation, cf. example 3.8.

We again specialize on a bipartite system. It is assumed to be closed, with sharp total energy E. We thus start from the energy

[a]Such an idea has been presented already in [Mott (1931)] for the collision of an α-particle with an atom.

eigenvalue equation, that is, from the time-independent Schrödinger equation.

Each of the two subsystems $\nu = A, C$ is, by itself, taken to be a whole network of N_ν particles. Here we explicitly refer to a quantized form of a classical model rather than an abstract matrix formulation. (Eventually we want to turn to the real-space representation.)

$$\hat{H}(\nu) = \hat{T}_\nu(\hat{\vec{p}}(\nu)) + \hat{V}_\nu(\hat{\vec{r}}(\nu)) . \qquad (5.95)$$

We have used the abbreviation $\hat{\vec{r}}(\nu) = \{\hat{r}_1(\nu), \hat{r}_2(\nu), \ldots \hat{r}_{3N_\nu}(\nu)\}$ and $\hat{\vec{p}}(\nu) = \{\hat{p}_1(\nu), \hat{p}_2(\nu), \ldots \hat{p}_{3N_\nu}(\nu)\}$. The term $\hat{T}_\nu$ represents the kinetic energy, the potential $\hat{V}_\nu(\vec{r}(\nu))$ the particle interactions within each subsystem. The eigenstate problem reads:

$$(\hat{T}_A + \hat{V}_A + \hat{T}_C + \hat{V}_C + \hat{V}(AC))|\Psi(AC)>= E|\Psi(AC)> . \qquad (5.96)$$

Typically, exact solutions are not available.

Proposition 5.12 Born–Oppenheimer approximation (BOA) *For simplicity we assume the particles of each subsystem to have the same mass. Furthermore, the system C is taken to be "large" compared to A, that is, its average energy dominates. Then the wave function for C depends negligibly on A. We thus make the asymmetric Ansatz (real space representation, cf. Definition (3.35)) [Briggs (2000)]*

$$< \vec{r}(A), \vec{r}(C)|\Psi(AC)>= \Psi(\vec{r}(A), \vec{r}(C)) = \phi(\vec{r}(A), \vec{r}(C)) \times \chi(\vec{r}(C)) . \qquad (5.97)$$

This Ansatz does not allow for entanglement. An approximate solution can then be found from the pair of effective eigen equations

$$\boxed{\begin{aligned} &(\hat{H}^{eff}(C) - E)\chi(\vec{r}(C)) = 0 , \\ &\hat{H}^{eff}(C) = \hat{H}(C) + \epsilon(\vec{r}(C)) , \\ &(\hat{H}(A) + \hat{V}(AC; \vec{r}(A), \vec{r}(C)) - \epsilon(\vec{r}(C)))\phi(\vec{r}(A), \vec{r}(C)) \\ &= Q(\vec{r}(A), \vec{r}(C)) . \end{aligned}} \qquad (5.98)$$

$\epsilon(\vec{r}(C))$ are the eigenvalues (index n, say) for subsystem A parametrized by position $\vec{r}(C)$; these so-called Born–Oppenheimer surfaces feed back on the effective eigen equation for subsystem B. It is assumed that there are no excitations of the particles A among various surfaces n. Q is a correction term, which is often set to zero but will turn out to be important below.

Due to the interaction the local energies are not sharp, $\delta E(A) = \delta E(C) > 0$. The eigen equation and thus the eigensolutions for subsystem (A) implicitly depend on the eigensolutions for subsystem (C). Such a correlation can be interpreted to mean that a change of C implies a change of A. It will now be argued that in the classical limit for C this correlation may be mapped onto an effective time dependence for A.

Proposition 5.13 Parametrization of change *For the following the correction Q to the BOA is essential. It can be shown to read [Briggs (2000)]*

$$Q(\vec{r}(A), \vec{r}(C)) \equiv i\hbar \sum_{j=1}^{3N_C} C_j(\vec{r}(C)) \frac{\partial}{\partial r_j(C)} \phi(\vec{r}(A), \vec{r}(C)), \quad (5.99)$$

$$C_j(\vec{r}(C)) = \frac{\hbar}{im(C)} \left(\frac{1}{2} \frac{\partial}{\partial r_j(C)} + \frac{1}{\chi} \frac{\partial \chi}{\partial r_j(C)} \right). \quad (5.100)$$

Based on the time-independent action, $W(\vec{r}(C), E)$, of the classical Hamiltonian $H(C)$, the wave function for subsystem C is now approximated by its semi-classical form,

$$\chi(\vec{r}(C)) \approx a(\vec{r}(C)) \exp(iW/\hbar). \quad (5.101)$$

In this case the operators C_j simplify to

$$C_j(\vec{r}(C)) = \frac{\hbar}{2im(C)} \left(\frac{\partial}{\partial r_j(C)} + \frac{2}{a} \frac{\partial a}{\partial r_j(C)} \right) + \frac{1}{m(C)} \frac{\partial W}{\partial r_j(C)}. \quad (5.102)$$

Keeping only the term of zeroth order in $\hbar$ one finds

$$C_j(\vec{r}(C)) \approx \frac{1}{m(C)} \frac{\partial W}{\partial r_j(C)} = \frac{p_j(C)}{m(C)} \equiv \frac{dr_j(C; t(C))}{dt(C)}. \quad (5.103)$$

In the last step we have used the following parametrization for the change of $r_j(C)$,

$$\Delta r_j(C) \equiv v_j(C)\Delta t(C) = p_j(C)\Delta t(C)/m(C)$$
$$= < \hat{p}_j(C) > \Delta t(C)/m(C). \quad (5.104)$$

Note that the so introduced time increment $\Delta t(C)$ carries no particle index, it functions as a kind of "universal" parameter ("common currency of change"). In this way time $\Delta t(C)$ has entered the scene. The name "time" can be justified by reference to the classical concept *of velocity $v_j(C)$, but also,* post festum, *by its effect as a time derivative in the effective equation of motion (see below). Namely, with $\sum_j C_j \partial/(\partial r_j(C;t(C))) = \partial/\partial t(C)$ the correction term becomes*

$$\boxed{Q(\vec{r}(A), \vec{r}(C)) \approx i\hbar \frac{\partial}{\partial t(C)} \left(\phi(\vec{r}(A); t(C))\right) .} \tag{5.105}$$

The appearance of this term Q indicates that we do not just rename a combination of physical parameters (observables) as a new time variable: Connected is also the generation of dynamics via the time derivative. The index C for $t(C)$ reminds us that these features derive from the correlation with C only, that is, from the outside. The vector $\vec{r}(A)$ is assumed to adjust adiabatically. Likewise,

$\hat{V}(AB; \vec{r}(A), \vec{r}(C; t(C))) \to \hat{V}(AC; \vec{r}(A); t(C))$,

$\epsilon(\vec{r}(C; t(C))) \to \epsilon(t(C))$, *and*

$\phi(\vec{r}(A), \vec{r}(C; t(C))) \to \phi(\vec{r}(A); t(C))$ *that is, also these terms gain an explicit time dependence. Finally, the term $\epsilon(t(C))$ can be removed from the effective equation for subsystem A via the gauge transformation,*

$$\phi = \exp\left(\frac{i}{\hbar}\theta(t(C))\right)\tilde{\phi} , \tag{5.106}$$

$$\theta(t(C)) = \int_{-\infty}^{t} (C)\epsilon(t')dt' . \tag{5.107}$$

The eigen equation for subsystem A then appears as an effective time-dependent Schrödinger equation in real-space representation:

$$\boxed{\begin{aligned} &\left(\hat{H}^{eff}(A; t(C)) - i\hbar\tfrac{\partial}{\partial t(C)}\right) \tilde{\phi}(\vec{r}(A); t(C)) = 0 , \\ &\hat{H}^{eff}(A; t(C)) = \hat{H}(A) + \hat{V}(AC; \vec{r}(A); t(C)) . \end{aligned}} \tag{5.108}$$

The time dependence of $\hat{H}^{eff}(A; t(C))$ and the appearance of the term $\partial/\partial t(C)$ are intimately connected, that is, both appear together—or neither does. The equation (5.108) is not closed: Its solution requires knowledge of the embedding system C. The roles of A and C are not symmetrical.

Definition 5.8 Parameter time in quantum mechanics *Comparing Eq. (5.98) with Eq. (5.108) there is time now, where there has* *been none before. Clearly, the parameter time for A originates from the quasi-classical embedding C; it is a kind of "top-down" effect [Auletta (2008)].*

The explicit time dependence of the effective Hamiltonian means that the subsystem A is nonautonomous. It seems that the effective time derivative is "universal" in the sense that different time parameters (resulting from different embeddings) would nevertheless be at most linear functions of each other (conforming to Mach's principle). For such a frame to be acceptable as a general feature we had to assume that whenever there is time something like a Born–Oppenheimer background should be at work as a time reference.

The "generation" of time via referencing is reminiscent of the quantum measurement scenario: It is not that something pre-existing is merely registered; rather a certain property is generated. Indeed, the generation of time should allow also to measure time, at least in principle.

Definition 5.9 Clock *Special purpose devices such as quantum clocks would have to be based on a similar kind of internal partitioning. Time thus becomes operationally accessible—easy to use for parameterizing the change of any observable of some other system, cf. Section 3.2.7.*

Proposition 5.14 Time-energy uncertainty *C is quantum also, as expressed by the stationary probability distribution* $|\chi(\vec{r}(C))|^2$. *How may such a positional uncertainty be understood? Can the particles of C reside at different positions at the "same time"? Here is where the parametrized change of* $\vec{r}(C)$ *comes in for rescue—based on the quasi-classical limit.*

In fact, for the quantum mechanical uncertainty product, cf. Eq. (3.316),

$$\delta F \delta G \geq \left| \frac{1}{2i} < [\hat{F}, \hat{G}] > \right|, \tag{5.109}$$

one immediately finds with $\hat{F} = \hat{H}(C)$ *and* $\hat{G} = \hat{r}_j(C)$,

$$\delta E(C) \delta r_j(C) \geq \frac{\hbar}{2} \frac{< \hat{p}_j(C) >}{m(C)}. \tag{5.110}$$

Now, reading Eq. (5.104) as a parametrization of the respective quantum variance $\delta r_j(C)$, *one confirms the famous time–energy uncertainty relation [Briggs (2000)]:*

$$\delta E(C)\Delta t(C) \geq \frac{\hbar}{2}. \tag{5.111}$$

With the total energy E being fixed, we may replace $\delta E(C) = \delta E(A)$. This analysis stresses the fact that the time uncertainty is of completely different origin than the quantum mechanical variance for the energy.

The energy uncertainty, though, gives rise to another remarkable identity:

Proposition 5.15 Velocity in Hilbert space *Consider for some system (A) a pure state represented in the form*

$$|\Psi >= \sum_j (\eta_j + i\xi_j)|e^{(j)} > \tag{5.112}$$

The effective Hilbert space velocity can then be defined by

$$v = \sqrt{\sum_j \left(\left(\frac{d\eta_j}{dt}\right)^2 + \left(\frac{d\xi_j}{dt}\right)^2\right)}. \tag{5.113}$$

For unitary state evolution (Schrödinger dynamics) it can be shown that [Gemmer (2009)]

$$\boxed{v = \delta E\,.} \tag{5.114}$$

The Hilbert space velocity is uniquely given by the energy uncertainty. As this uncertainty is constant and given by the initial state preparation, also the velocity is constant.

Remark 5.10 Emergence *Even though the appearance of the effective time is reducible, this very property would not be "visible" in the full description—in fact, the whole (isolated) system is assumed to be timeless, that is, prepared in an eigenstate.*[a]

This context dependence is essential [Briggs (2000)]; [Barbour (2000)]; it is reminiscent of the thermodynamic behavior in embedded quantum subsystems, where the total system need not be thermal at all.

[a]It has been argued that in the full description time is obsolete, cf. Section 2.2.5: The past or future does not contain anything new.

Such a kind of context-dependent property may be called emergent, cf. Section 1.1. It should be noted that emergence is not understood here as an opposing view to reductionism. A useful analogy might be to envision a printed portrait: If looking too close the details represented by those many pixels may suppress the recognition of the person shown.

The "successful" embedding may be hard to find, unless one knows already the phenomenon one is looking for. It constitutes something "new," something unexpected.

5.2.2 *External Driving: Work*

After the effective equation (in the present case: the time-dependent Schrödinger equation for A) has been worked out (cf. Section 5.2.1), one may put aside the details of its origin. The underlying partition is believed to allow for a whole class of similar scenarios. In that sense it would define typicality.

Remark 5.11 Models of time-dependent Hamiltonians *Rather than being concerned with the background system C in full detail (to find the "true" time dependence of $\hat{H}^{eff}(A)$), one may consider classes of such embedding systems producing effective time-dependent effects on A as desired. There can even be scenarios in which $\hat{H}(A)$ is time independent—for some time. Conventional unitary time evolution would result in this case.*

In turn, one might argue that at the end of the day, any nonautonomous behavior should be the result of restricting attention on part of the total system only, that is, the result of a partitioning (A, C). It remains to be seen whether any such time dependence can be traced back to a scenario as suggested here.

Definition 5.10 Work *Subject to the conditions as specified in Section 5.2.1 the subsystem (C) may become a quasi-classical driving for subsystem (A). The Heisenberg equation of motion for the respective Hamiltonian $\hat{H}_H(A;t)$ reads (cf. Eq. (3.277))*

$$\frac{d}{dt}\hat{H}_H(A;t) = \hat{S}^\dagger(A;t)\frac{\partial \hat{H}_H(A;t)}{\partial t}\hat{S}(A;t)\,, \tag{5.115}$$

and the change of the expectation value for the energy from time $t = 0$ to t for given initial state $|\psi(A; 0)>$

$$\boxed{\Delta E_A = \int_0^t <\psi(A;t')|\frac{\partial \hat{H}_H(A;t')}{\partial t'}|\psi(A;t')> dt' \equiv \Delta W_A\,.} \tag{5.116}$$

Here we have used that $|\psi(A;t')> = \hat{S}(A;t')|\psi(A;0)>$; the result is easily generalized for states given by a density operator. A system (A) described by an effective time-dependent Hamiltonian $\hat{H}(A;t)$ can only exchange work ΔW_A with its environment (C). For a closed system this work would be zero, that is, the energy would be a constant of motion.

Remark 5.12 Work is not an operator *Under the very special condition,*

$$\left[\frac{\partial \hat{H}_H(A;t')}{\partial t'}, \hat{S}(A;t')\right] = 0\,, \tag{5.117}$$

and using $|\psi(A;t)> = \hat{S}(A;t)|\psi(A;0)>$ the work could be rewritten as

$$\begin{aligned}\Delta W_A &= <\psi(A;0)|(\hat{H}(A;t) - \hat{H}(A;0))|\psi(A;0)> \\ &\equiv <\psi(A;0)|\Delta \hat{W}_A|\psi(A;0)>\,,\end{aligned} \tag{5.118}$$

that is, as the expectation value of a "work operator," $\Delta \hat{W}_A$. In general, this is not the case (cf. [Talkner (2007)]). In fact, as should be clear already from classical thermodynamics, work has to refer to a process (trajectory C) and thus cannot be a state property. The same holds for the concept of heat.

Definition 5.11 Classical limit *The general classical limit may conveniently be obtained by writing the integrand in Eq. (5.116) as $(\partial H(A;t')/\partial t')_{expl}$. Given a control $\lambda(t)$ and thus a trajectory $C = \{\vec{p} = \vec{p}(\lambda), \vec{r} = \vec{r}(\lambda)\}$ the work ΔW_A imparted on a classical Hamilton system $H(A;\vec{p},\vec{r};t)$ can be inferred from [Jarzynski (1997]*

$$\Delta W_A = \int_0^t dt' \frac{d\lambda}{dt'}\frac{\partial H(A;\lambda)}{\partial \lambda}\,. \tag{5.119}$$

Example 5.9 Driven quantum oscillator *A standard, though somewhat special example for work obtains for the linear harmonic* *oscillator,* $\hat{H}(A;t) = \hat{H}_0 + \hat{V}^{ex}(t)$*, in which case the external driving can be modeled as*

$$\hat{V}^{ex}(t) = -F_C(t)\hat{x}\,, \tag{5.120}$$

where $\hat{x}$ *is the displacement and* $F_C(t) = -\partial\hat{V}^{ex}/\partial\hat{x}$ *is the external force.*

With the notation $< x(t') > \equiv < \psi(A;t')|\hat{x}|\psi(A;t') >$ *one thus obtains from Eq. (5.116)*

$$\begin{aligned}\Delta W_A &= -\int_0^t dt' \left(< x(t') > \frac{dF_C}{dt'} \right) = \int_0^t dt' \left(\frac{d < x(t') >}{dt'} F_C(t') \right) \\ &= \int_C F_C d < x > \,. \end{aligned} \tag{5.121}$$

In the second step we have applied partial integration. Eventually, the integration is taken along the trajectory C *of the average position,* $< x(t') >$*, just as expected classically. The associated power is* $\dot{W}_A = F_C d < x(t' > /dt'$*.*

Example 5.10 Parametric quantum oscillator *The classical parametric oscillator is known to describe the amplification of a (damped) swing, a children's favorite. Its description differs from that of the driven oscillator insofar as the effective force* $F_C(t)$ *depends on the momentary elongation x(t), in fact, for the amplification to become operative the time dependence of the force has to be correlated with x(t). The quantum version is:*

$$\hat{H}(t) = \frac{1}{2m_0}\hat{p}^2 + \frac{1}{2}m_0\omega^2(1 + f(t))\hat{x}^2 = \hat{H}_0 + \hat{V}^{ex}(t)\,, \tag{5.122}$$

$$\hat{V}^{ex}(t) = -f(t)\omega^2\hat{x}^2\,. \tag{5.123}$$

From the point of view of work, though, the two model types, driven and parametric oscillator, are equivalent.

Remark 5.13 Work in thermal systems *Up to now we have restricted ourselves to pure states,* $S_{VN}(A) = 0$*. We can imagine that a further environmental system B would impart thermal properties on A—without suppressing the time-dependent driving via subsystem C. This would complete the quantum-thermodynamical approach to general embedded quantum systems.*

5.2.3 *Combination $(A)BC$: Thermal and Mechanical Environment*

We consider subsystem A under the influence of both types of embeddings, B being the thermal and C the mechanical environment. For simplicity we assume that there is no interaction between B and C. What is their combined effect on A?

Detailed numerical studies do not exist yet. However, it seems plausible that the mechanical environment still produces an effective time-dependent potential (driving) for subsystem A, while the thermal environment B should still give rise to a local thermal state, conditioned by the momentary parameters controlled by C. For this to be the case one will have to assume a quasi-adiabatic scenario in the sense that the driving should be slow compared to the thermal relaxation times. In this sense the two embeddings are taken to be "modular," like devices making up a computer chip.

Example 5.11 Nano-electromechanical systems (NEMS) *A pertinent class of experiments is based on specially designed cantilevers or doubly clamped beams, subject, for example, to piezoelectric actuation (the time-dependent control) [Matheny (2013)]. As the geometric dimensions shrink, the mechanical and thermal degrees of freedom become less and less separable; this leads to increased thermomechanical noise. The excited vibrational mode couples to other modes and to phonons (the proper "bath environment"). With the damping being proportional to velocity, the low-frequency modes tend to survive.*

Such environmental modules underly the first law of thermodynamics: The transfer of heat and work. In fact, typical quantum-thermodynamic processes are based on respectively generalized embeddings producing explicitly time-dependent boundary conditions. Just as in the case of classical thermodynamics, we will not be able to come up with a complete (i.e., autonomous) model—including the respective quantum environments. Such a full description will only be possible under simplifying assumptions such as those considered in the following.

5.2.4 *Local Effective Measurement Basis (LEMBAS)*

Rather than considering a large environment split into a thermal (B) and a (mechanical) function (C) we can try to study the energy exchange between A and a second subsystem B. Both parts need not be large and can thus be treated on equal footing. The energy exchange will, in general, involve both work and heat.

The bipartite quantum system (AB) is defined by the Hamiltonian,

$$\hat{H}(A, B) = \hat{H}(A) \otimes \hat{1}(B) + \hat{1}(A) \otimes \hat{H}(B) + \hat{V}(A, B). \quad (5.124)$$

This closed system is now subject to a unitary time evolution (parameter time being given); in particular, we are interested in the energy transfer between the two subsystems.

A general state for Eq. (5.124) can be written as

$$\hat{\rho}(A, B) = \hat{\rho}(A) \otimes \hat{\rho}(B) + \hat{C}(A, B). \quad (5.125)$$

The $\hat{\rho}(\mu)$ are the respective reduced density operators. The correlation term $\hat{C}(A, B)$ describes the deviation from the the simple product form. For the following we consider subsystem A as the system of interest and subsystem B as its environment. However, the environment need not be large. So, the roles of A and B can be interchanged. The LEMBAS approach [Weimer (2008)] addresses two interrelated questions:

- How could we define a local energy for A in the presence of non-negligible interactions with B?
- What could be the meaning of work and heat in relation to a unitary energy exchange between A and B?

Note that the second problem cannot be discussed without a clear concept of local energy: One has to take a local perspective, the total energy stays constant, by definition. In a first step we therefore try to introduce a local perspective into the dynamic evolution. Based on the effective Hamiltonian for A,

$$\hat{H}^{eff}(A; t) = \hat{H}(A) + \Delta\hat{H}(A; t), \quad (5.126)$$

$$\Delta\hat{H}(A; t) = \mathrm{Tr}_B\{(\hat{1}(A) \otimes \hat{\rho}(B; t))\hat{V}(A, B)\}, \quad (5.127)$$

and the effective "dissipator"

$$\hat{\mathcal{L}}_A^{eff}(t) = -\frac{i}{\hbar}\mathrm{Tr}_B\{[\hat{V}(A, B), \hat{C}(A, B)]\}, \quad (5.128)$$

the effective equation of motion of subsystem A can be written as

$$\boxed{\dot{\hat{\rho}}(A;t) = -\frac{i}{\hbar}[\hat{H}^{eff}(A;t), \hat{\rho}(A;t)] + \hat{\mathcal{L}}_A^{eff}(t).} \tag{5.129}$$

If we could neglect the dissipator term, the FA was exact and the above equation would reduce to the respective FA case for semi-mixed states, Remark 5.9. Anyway, these effective schemes are not closed; they still refer to the total dynamics. A corresponding equation can be written down for subsystem B. The name "dissipator" is motivated by the fact that the local entropy changes right through this term:

$$\begin{aligned}\dot{S}_{vN}(A) &= -\mathrm{Tr}_A\{\frac{d}{dt}(\hat{\rho}(A))\ln\hat{\rho}(A) + \hat{\rho}(A)\frac{d}{dt}\ln\hat{\rho}(A)\} \\ &= -\mathrm{Tr}_A\{\hat{\mathcal{L}}_A^{eff}(t)\ln\hat{\rho}(A)\}.\end{aligned} \tag{5.130}$$

In going from the first to the following line we have observed that the second term gives zero.

In a second step we introduce an operational aspect into the model, a quantum measurement. For this purpose we take the well-defined energy eigen basis of the free subsystem A as the "effective measurement basis," even when the subsystem is coupled to its environment B. This motivates us to split the effective Hamiltonian, Eq. (5.126), into

$$\hat{H}_0^{eff}(A;t) = \hat{H}(A) + \Delta\hat{H}_0(A;t), \tag{5.131}$$

$$\hat{H}_1^{eff}(A;t) = \Delta\hat{H}_1(A;t). \tag{5.132}$$

Here, $\Delta\hat{H}_0(A;t)$, is the part that commutes with $\hat{H}(A)$, while $\Delta\hat{H}_1(A;t)$ is the rest that does not. The term $\hat{H}_0^{eff}(A;t)$ thus acts as the effective local "observable," its eigenvalues (and associated eigenvectors) as the possible measurement results.

In a final step we consider the energy balance based on the local equation of motion.

$$\dot{U} = \frac{d}{dt}\mathrm{Tr}_A\{\hat{H}_0^{eff}(t)\hat{\rho}(A;t)\} = \mathrm{Tr}_A\{\dot{\hat{H}}_0^{eff}(t)\hat{\rho}(A;t) + \hat{H}_0^{eff}(t)\dot{\hat{\rho}}(A;t)\}. \tag{5.133}$$

This balance can now be split into a term $\dot{W}(A)$, by which the local entropy $S_{vN}(A)$ remains constant, and a term $\dot{Q}(A)$, for which the entropy changes:

$$\dot{W}(A) = \mathtt{Tr}_A\{\frac{d}{dt}(\hat{H}_0^{eff})\hat{\rho}(A;t) - \frac{i}{\hbar}[\hat{H}_0^{eff}(A;t), \hat{H}_1^{eff}(A;t)]\hat{\rho}(A;t)\}, \tag{5.134}$$

$$\dot{Q}(A) = \mathtt{Tr}_A\{\hat{H}_0^{eff}(A;t)\hat{\mathcal{L}}_A^{eff}(t)\}. \tag{5.135}$$

$\dot{W}(A)$ describes the exchange of work, $\dot{Q}(A)$ the exchange of heat (within an otherwise closed system). The total system entropy is constant.

Remark 5.14 Relational aspects of thermodynamics *The LEMBAS approach is explicitly relational: Work and heat exchanged between A and B are defined with respect to one of these subsystems. This means that their values will, in general, change, as we change the reference. This feature is reminiscent of relational quantum mechanics [Rovelli (1996b)].*

Remark 5.15 Additional embedding *The subsystem A might, in addition, be subject to external dissipation, as expressed by the Lindblad equation Eq. (3.368). In this case we would have to replace*

$$\hat{\mathcal{L}}_A^{eff}(t) \rightarrow \hat{\mathcal{L}}_A^{eff}(t) + \hat{\mathcal{L}}_{inc}(\hat{\rho}(A)). \tag{5.136}$$

Definition 5.12 Local temperature *Assuming the Clausius equality to apply also for the quasi-static (effective) evolution, we get*

$$T^{eff}(A) = \frac{đQ(A)}{dS(A)}, \tag{5.137}$$

and thus from Eqs. (5.130, 5.135, 5.136), and observing $S(A) = k_B S_{vN}(A)$,

$$T^{eff}(A) = -\frac{\mathtt{Tr}_A\{\hat{H}_0^{eff}(A)(\hat{\mathcal{L}}_A^{eff}(t) + \hat{\mathcal{L}}_{inc}(\hat{\rho}(A)))\}}{\mathtt{Tr}_A\{(\hat{\mathcal{L}}_A^{eff}(t) + \hat{\mathcal{L}}_{inc}(\hat{\rho}(A)))\ln\hat{\rho}(A)\}k_B}. \tag{5.138}$$

This result includes the case, where we have external embedding only, leading to the canonical state

$$\ln(\hat{\rho}(A)) = -\beta\hat{H}(A) - \ln Z. \tag{5.139}$$

Then we immediately confirm

$$T^{eff}(A) = \frac{1}{k_B\beta} = T. \tag{5.140}$$

Here we have used $\mathtt{Tr}_A\{\hat{\mathcal{L}}_{inc}(\hat{\rho}(A)\ln Z\} = \ln Z\,\mathtt{Tr}_A\{\hat{\mathcal{L}}_{inc}(\hat{\rho}(A)\} = 0$.

In general, that is, in the presence of an additional (strong) coupling to some system B as a part of the total environment, the two temperatures may deviate from each other (cf. Remark (5.19)).

Example 5.12 Fixed eigenfunctions of A *If the eigenfunctions of* $\hat{H}^{eff}(A;t)$ *do not change with time, that is, under the condition*

$$\hat{H}_1^{eff}(A;t) = \Delta\hat{H}_1(A;t) = 0 \tag{5.141}$$

it follows with $[\hat{H}^{eff}(A;t), \hat{\rho}(A;t)] = 0$ *that the momentary state remains incoherent (with respect to the measurement basis), and the above result, Eqs. (5.134, 5.135), simplifies to*

$$\boxed{\begin{aligned} \dot{W}(A) &= \mathrm{Tr}_A\left\{\frac{d}{dt}(\hat{H}(A;t))\hat{\rho}(A;t)\right\}, \\ \dot{Q}(A) &= \mathrm{Tr}_A\left\{\hat{H}(A;t)\frac{d}{dt}(\hat{\rho}(A;t))\right\}. \end{aligned}} \tag{5.142}$$

This is what one would have expected classically and what is often used also in the quantum domain.

However, fixed eigenfunctions are not the general case.

Example 5.13 Rotating eigenfunctions of A *Consider a two-level system (spin-1/2) described by*

$$\hat{H}(A) = \frac{\Delta E}{2}\hat{\sigma}_3, \tag{5.143}$$

which, due to the interaction with some subsystem B is subject to a time-independent driving force

$$\Delta\hat{H}(A) = g\hat{\sigma}_1 = \Delta\hat{H}_1(A). \tag{5.144}$$

According to the LEMBAS approach one finds

$$\dot{W}(A) = -i\mathrm{Tr}_A\{[\hat{H}(A), \Delta\hat{H}_1(A)]\hat{\rho}(A;t)\}. \tag{5.145}$$

which, in general, is non-zero. Indeed, if originally prepared in the ground state, the system A would oscillate between ground and excited state.

On the other hand, according to Eq. (5.142) the work would always have to be zero.

5.3 Second Law

Definition 5.13 Second Law *In a closed thermal system processes that decrease entropy do not occur. Kelvin–Planck formulation: It is impossible that a machine, operating in a cycle, produces no effect other than the conversion of heat from a single bath into an equal amount of work.*

Remark 5.16 Maximum entropy principle *It has been argued that for a bipartite (closed) quantum system the smaller part A will exhibit thermal properties, if some weak conditions are fulfilled. When the global state of A and B is pure, the local entropies* $S_{vN}(A) = S_{vN}(B)$ *reflect entanglement. But is the entropy* $S_{vN}(A)$ *necessarily extremal, and if so, under what constraints?*

One may say that the MEP, indeed, follows for "typical" quantum embeddings, insofar as the resulting equilibrium states for A approach the Boltzmann form. However, stable equilibrium states could be envisioned that would strongly deviate from this form. Such an unusual behavior of the embedded subsystem A would reflect unusual energy spectra of its environment B. It is interesting to note that within the context of quantum thermodynamics the MEP (cf. Proposition 4.1) looses some of its "fundamental glamor" [Gemmer (2009)].

This underlines the fact that the Jaynes' principle (cf. Proposition 2.4) is not directly rooted in physics: It cannot replace detailed physical analysis.

Proposition 5.16 Quantum confirmation of MEP *We distinguish the following scenarios:*

- *Microcanonical constraint: With the occupation probabilities* $w(K(A))$ *of all local energy states* $E_K(A)$ *being constants of motion, little space remains for relaxation behavior. Besides the loss of coherence (disappearance of off-diagonal matrix elements) equipartition (within each degenerate energy band) shows up; this indeed maximizes the entropy under the given constraints.*
- *Canonical constraint: As has been shown, it is the effective temperature, which is fixed by the degeneracy structure of*

the embedding quantum system. The result is the canonical distribution function. This distribution can be reinterpreted in the sense of MEP as the function maximizing the von Neumann entropy subject to the average internal energy as a macro-constraint, cf. Theorem 4.2.

Are there ways to break the second law? And what would this mean? Such problems cannot be tackled from within thermodynamics; one rather needs a generalized scheme like statistical physics or quantum thermodynamics.

Obviously, breaking the second law has to be distinguished from situations in which the second law simply does not apply (like in simple mechanical devices). Such cases have to be put aside here. We deliberately restrict ourselves to scenarios in which non-thermal behavior has to be "created" by external manipulation: Non-equilibrium states are not constrained by the second law (maximum entropy), but their preparation (and stabilization) requires extra costs in the form of physical resources. Reduced entropy means additional information.

We recall that additional information has, indeed, been suggested as a possible route to circumvent restrictions imposed by the second law (cf. Maxwell's demon). However, to transform this potential resource into an operational advantage, the retrieval of information has to be included as a physical process. This turns out to be so costly, that the net effect does no longer outperform the standard thermodynamic procedure, cf. Remark 4.23.

Weak coupling between system and environment is one of the principle conditions for the validity of thermodynamics. Quite often the tentative violation of this condition can be removed by a change of partitioning, that is, by a change of the definition of the system properly, cf. Example 5.7.

Nevertheless, it is tempting to relax this condition to study the consequences. A violation of at least some of the basic thermodynamic relations cannot be excluded. Indeed, a violation of the second law seems to be imminent in cases of strong coupling. This would also threaten Landauer's erasure principle (cf. Theorem 4.18), a rather disturbing perspective.

Again, one can show that the violation of the second law under the condition of strong coupling is not "really useful": If the hidden costs are included, the standard constraints reappear. In this sense the laws of thermodynamics (or, at least, its consequences) turn out to be astonishingly robust.

5.3.1 *Strong Coupling and Entanglement*

Remark 5.17 Hidden costs *Neither additional information nor strong coupling appear to allow a successful attack on the second law. Nevertheless, it has to remain open whether the second law could be considered "unconditional save"—like quantum cryptography being unconditional secure: It can become quite challenging to make sure that all costs have properly been included.*

Some pertinent examples will be discussed below. They show that oversimplified treatments are prone to significant error.

For the following we specialize on a bipartite system AB, where A is a quantum oscillator, and B a larger system operating as a heat bath.

Example 5.14 Cost of entanglement *Entanglement may exist without interaction; it can therefore survive spatial separation between the entangled subsystems without limits. The generation and exploitation of entanglement is a different story, though. In fact, if the interaction is neglected, a cyclic machine could be designed which would extract work from a single bath (thus violating the second law in the Kelvin–Planck formulation). This is shown in the following example, a bipartite quantum system (AB).*

The environment system (B) *is taken to consist of N noninteracting oscillators,*

$$\hat{H}(B) = \sum_{j=1}^{N} \left(\frac{\hat{p}_j^2}{2m_j} + \frac{m_j \omega_j^2}{2} \hat{x}_j^2 \right), \tag{5.146}$$

prepared in a thermal state with some temperature $T(B) = 1/(k_B \beta(B))$. *(We will mainly be interested in the small temperature limit.) Let us add now an additional oscillator* (A) *with frequency* ω_A

and mass M,

$$\hat{H}(A) = \frac{\hat{p}^2}{2M} + \frac{M\omega_A^2}{2}\hat{q}^2 . \tag{5.147}$$

When in contact with system (B),

$$\hat{H}(AB) = \hat{H}(A) + \hat{H}(B) + \hat{V}(AB) , \tag{5.148}$$

where the interaction Hamiltonian is given by (C_j are the coupling constants)

$$\hat{V}(AB) = -\hat{q}^2 \sum_{j=1}^{N} m_j\omega_j^2\hat{x}_jC_j + \frac{\hat{q}^2}{2}\sum_{j=1}^{N} m_j\omega_j^2\hat{x}_jC_j^2 \tag{5.149}$$

the resulting (N + 1) oscillator system is expected to approach the equilibrium state

$$\hat{\rho}_{equ}(AB) = \frac{1}{Z(AB)}\exp\{-\beta\hat{H}(AB)\} \tag{5.150}$$

with the same temperature, $T = T(B)$. Based on this $\hat{\rho}_{equ}(AB)$ the total energy expectation value can be rewritten as [Kim (2006)]

$$< H(AB) > - < H(A) + H(B) >=< V(AB) >\equiv V_T . \tag{5.151}$$

But because of the interaction and for small enough temperatures the additional oscillator (A) and the oscillator system (B) are entangled, the reduced state of (A)

$$\hat{\rho}(A) = \mathrm{Tr}_B\{\hat{\rho}_{equ}(AB)\} , \tag{5.152}$$

is non-Gibbsian, a local temperature $T(A)$ does not exist. Locally this means a mixed state, even for the global temperature, T, going to zero. As a consequence,

$$\Delta E_A \equiv< H(A) >_{T\to 0} - \frac{\hbar\omega_A}{2} \geq 0 , \tag{5.153}$$

where $\hbar\omega_A/2$ is the ground state energy of the uncoupled oscillator.

Can that surplus energy ΔE_A be extracted and used? It seems as if that was possible by immersing (A) into the bath (B) and then letting (A) give away its surplus energy "after having been detached from the bath." Such a procedure would violate the second law [Ford (2006)].

However, in so doing the interaction energy, V_T, should be taken into account, irrespective of its size: Indeed, for $T \to 0$ one shows that V_T is positive and larger than ΔE_A, that is, the generation *of entanglement is not for free. For each cycle the energy needed to produce entanglement thus overcompensates for the postulated energy gain [Ford (2006); Kim (2006)]; the second law is saved.*

Example 5.15 Exchange of heat under strong coupling *We return to the previous Example (5.14). The mean energy of the coupled oscillator (A) is given by*

$$\tilde{U}(A) = \mathrm{Tr}\{\hat{H}(A)\hat{\rho}(A)\} = \frac{< p^2 >}{2M} + \frac{M\omega_A^2}{2} < q^2 > . \tag{5.154}$$

The tilde should remind us that this "internal energy" is based on a non-equilibrium state, that is, presents a generalization. The first law is taken to hold in the conventional form

$$d\tilde{U}(A) = đQ(A) + đW(A), \tag{5.155}$$

$$đQ(A) = \mathrm{Tr}\{d\hat{\rho}(A)\hat{H}(A)\}, \tag{5.156}$$

$$đW(A) = \mathrm{Tr}\{\hat{\rho}(A)d\hat{H}(A)\}. \tag{5.157}$$

Now suppose that the coupled oscillator (A) undergoes the quasi-static mass variation (cf. spectral deformation, Example 4.1)

$$M_0 \rightarrow M_1 = M_0 + \Delta M \tag{5.158}$$

with $\Delta M \geq 0$. We are interested in the associated heat exchanged with (B) (cf. [Hilt (2011); Kim (2010); Allahverdyan (2005a)]),

$$\Delta Q(A) = \int_{M_0}^{M_1} \frac{1}{2M}\frac{\partial}{\partial M} < p^2 > + \frac{M\omega_A^2}{2}\frac{\partial}{\partial M} < q^2 > dM . \tag{5.159}$$

$\Delta Q(A)$ is non-negative. With

$$\Delta S_{vN}(A) = S_{vN}(A; M_1)) - S_{vN}(A; M_0) \tag{5.160}$$

and identifying $S_{vN}(A) = \tilde{S}(A)/k_B$ one finds for very low temperatures (cf. Eq. (4.137))

$$\Delta W_{diss}(A) = T\Delta\tilde{S}(A) - \Delta Q(A) < 0 . \tag{5.161}$$

The Clausius inequality for system (A) is violated—and with it the Landauer principle. However, as indicated, the internal energy, $\tilde{U}(A)$, as well as the entropy $\tilde{S}(A)$ involve generalizations in that the underlying state is no longer a thermal state. The temperature T is the global temperature, not the temperature of (A). And contrary to the LEMBAS approach the interaction energy in a strict sense, V_T (see Eq. (5.151)), has not been considered. Violations of fundamental relations (like the

Clausius inequality) based on such modified input terms should thus not be surprising.

Can this deficiency be cured? There are, at least, two ways out: Considering a substitute process, cf. Definition 4.27, or using an effective description.

Remark 5.18 Substitute process *One rather intuitive way is to essentially stick to the above description, but include the built-up of interaction as an adiabatic pre-process [Hilt (2011)], similar to what has been suggested in the previous Example (5.14).*

Let the coupling $C_j = C_j(\eta)$ be quasi-statically switched on by varying η, starting from $C_j(\eta = 0) = 0$. In the uncoupled case, $\eta = 0$, both subsystems are in a local thermal state with the same temperature $T(A) = T(B)$. Again, we are interested in the heat exchanged with (B), but now for the coupling process, in which we increase η up to its final value η_0 for fixed M,

$$\Delta U(A) = U(A; \eta_0) - U(A; 0), \tag{5.162}$$

$$\Delta Q(A) = \Delta U(A) - \Delta W(A) = \Delta U(A) + F(A; 0) - F(A; \eta_0). \tag{5.163}$$

Here, $F(A; \eta_0)$ is the free energy for the coupled oscillator. One finds $\Delta Q(A) < 0$ and $\Delta S(A) > 0$. Adding both processes together, that is, $(\eta : 0 \to \eta_0)$ and then $(M : M_0 \to M_1)$, the Clausius inequality is no longer violated.

Remark 5.19 Effective thermodynamics *An alternative way to save the Clausius inequality is to introduce effective thermodynamic parameters [Kim (2010)]. One first observes that the reduced density operator of the coupled oscillator is not only nonthermal, but even nondiagonal. After a state-dependent renormalization of the product of frequency and mass of the oscillator, the diagonalized state becomes thermal and can be assigned a new, effective temperature $\beta_{eff}(A)$. The von Neumann entropy does not change under this transformation. One is free to choose the effective frequency, ω_{eff}, such that $U_{eff}(A) = U(A)$. Then the Clausius inequality holds in the standard form*

$$T_{eff}(A)\Delta S - \Delta Q_{eff}(A) \geq 0. \tag{5.164}$$

This is what one would expect for an uncoupled oscillator (subject to a change of mass, say), but here with the effective thermal parameters as given.

5.3.2 *Relaxation Dynamics*

The characterization of equilibrium states is incomplete without knowing the way equilibrium is reached, if we start outside equilibrium. In particular, one would have to demonstrate the thermalization to be independent of the initial state (of the thermalizing subsystem) and to occur with a finite relaxation time. Detailed dynamical models (including numerical simulations) are thus indispensable.

However, pertinent models should be "typical": We are not interested in special models; these might happen to prevent thermal behavior completely—or, to the contrary, support an overoptimistic view not justified "in general." This brings us to the expectation that there should be an ensemble of models, from which we pick one member randomly as an input for a concrete calculation. Such a procedure would render the actual choice of model parameters "unbiased."

Definition 5.14 Bipartite matrix models *Let the subsystem Hamiltonians be given in eigen representation*

$$\hat{H}(A) = \sum_K \sum_{k=1}^{g_K} E_K |K, k(A) > < K, k(A)| \,, \qquad (5.165)$$

$$\hat{H}(B) = \sum_L \sum_{l=1}^{g_L} E_L |L, l(B) > < L, l(B)| \,. \qquad (5.166)$$

The g_K and g_L specify the respective degeneracies. The product states

$$|K, k(A) > \otimes |L, l(B) > \equiv |K, k, L, l > \qquad (5.167)$$

can then be taken as the basis also for the matrix representation of the subsystem interaction $\hat{V}(AB)$:

$$V(K, k, L, l; K', k', L', l') =< K, k, L, l|\hat{V}(AB)||K', k', L', l' > \,. \qquad (5.168)$$

But how to model the form of this matrix? To avoid any bias we will proceed as follows: Unless stated otherwise we will assume the entries of this matrix to be taken from the Gaussian unitary ensemble, GUE, Eq. (3.104), with variances $\sigma_d = \sqrt{2}\Delta E(A)\alpha$, $\sigma_{od} = \Delta E(A)\alpha$, for the diagonal elements and off-diagonal elements, respectively. $\Delta E(A)$ is a characteristic energy scale of subsystem A, $\alpha \ll 1$ guaranties *small variance of the coupling V around zero, that is, weak coupling, cf. Eq. (5.8).*

The accessible Hilbert space is constrained by the conservation of the total energy; the Schrödinger dynamics of the total system (AB) can now be considered under various interaction types and initial states.

Example 5.16 Embedded two-level system: Microcanonical conditions *For simplicity we consider subsystem A as a non-degenerate two-level system. The energy-splitting of the two-level system is* $\Delta E(A)$*. Under microcanonical coupling no energy can be exchanged between A and environment B. To enforce this constraint, all blocks within* $V(K, kL, l; K', k', L', l')$ *are set to zero which would connect different energy levels within A. As the initial state we assume a product state with sharp energy of subsystem B,* $E_{L_0}(B)$*, degeneracy* $g_{L_0} \gg 1$*, and a coherent state for A.*

We expect the coherence to disappear and the local entropy S(A) to approach its maximum value subject to the constraint of constant local energy distribution, that is, constant diagonal matrix elements, $\rho_{KK}(A)$*.*

This is exactly what can be seen in the numerical simulation of Fig. 5.3 [Borowski (2003)]. The final attractor state of subsystem (A) is independent of the initial state. The remaining temporal fluctuations are due to finite-size effects; these would disappear for $g_{L_0} \rightarrow \infty$*.*

Example 5.17 Embedded two-level system: Canonical conditions *The same model, but now under canonical constraints is considered, cf. Example 5.1. As the initial state we assume a product state: The initial state of (A) is arbitrary, the initial state of (B) is* $E_{L_1}(B)$ *(sharp); because of the energy exchange with (A) two more levels of (B) have to be included,* $E_{L_0} = E_{L_1} - \Delta E(A)$*,* $E_{L_2} = E_{L_1} + \Delta E(A)$*, with corresponding degeneracies* $g_{L_0}, g_{L_1}, g_{L_2}$*. A numerical simulation is shown in Fig. 5.4*

The final attractor state of subsystem (A) is independent of its initial state. The remaining temporal fluctuations are due to finite-size effects; these would disappear for $g_{L_0} \rightarrow \infty$*.*

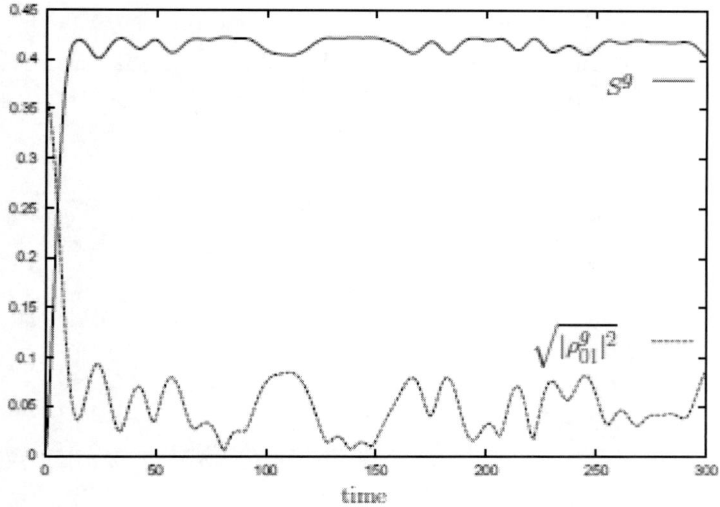

Figure 5.3 Relaxation of two-level system under microcanonical constraints. Time in units $\hbar/\Delta E(A)$, degeneracy of environmental level $g_{L_0}(B) = 50$. Solid line: von Neumann entropy S^g of (A) in units of k_B. Broken line: Absolute value of the off-diagonal matrix element of the reduced density matrix of (A).

We see that, indeed, the expected canonical occupation probabilities built up in subsystem A, independent of its initial state. The fluctuations (variance of the occupation probabilities) can be shown to decrease proportional to $1/\sqrt{g_{L_1}(B)}$. Such models can easily be generalized to larger state spaces for A [Borowski (2003)].

The above models have been designed to capture typical relaxation behavior. But what might happen under more special constraints? We look at the influence of initial correlations, inversion in the environment, structural constraints, and strong coupling.

Example 5.18 Influence of inversion *We turn to a "hybrid" description, where only part of the Hamiltonian is based on a matrix model. The total system Hamiltonian is [Waldherr (2010b)]*

$$\hat{H}(AB) = \hat{H}(A) + \hat{H}(B) + \hat{V}(AB) + \hat{V}(BB). \qquad (5.169)$$

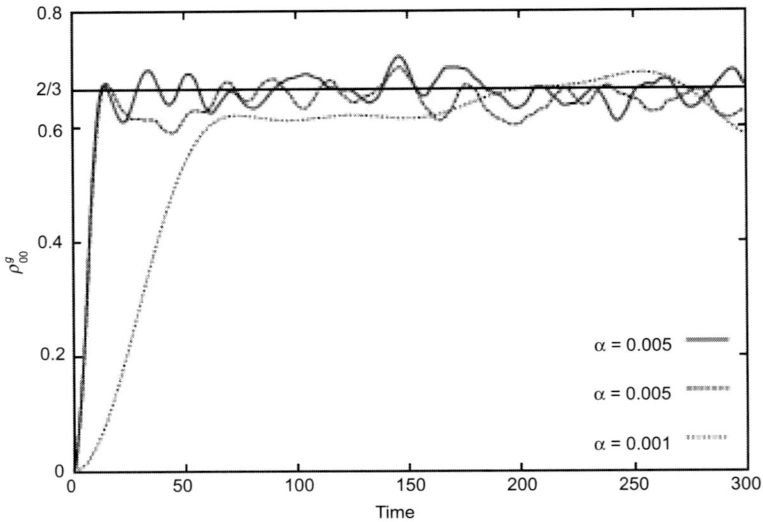

Figure 5.4 Relaxation of two-level system under canonical constraints. Time in units $\hbar/\Delta E(A)$, the pertinent environmental degeneracies are $g_{L_0} = 50$, $g_{L_1} = 100$, $g_{L_2} = 200$. Shown is the ground-state matrix element of reduced density matrix of (A), ρ^g_{00}, for various coupling constants α.

Neglecting the zero-point energy the Hamiltonian and spectrum for the harmonic oscillator (A) can be written as

$$\hat{H}(A) = \hbar\omega(A)\hat{b}^\dagger\hat{b}, \tag{5.170}$$

$$E_n(A) = \hbar\omega(A)n \quad n = 0, 1, \ldots. \tag{5.171}$$

with the eigenstates $|n>$*, see Eq. (3.6). These states are non-degenerate,* $g_n(A) = 1$*. For the environment (B) consisting of N(B) spins one has,*

$$\hat{H}(B) = \sum_{\nu}^{N(B)} \frac{\hbar\omega(B)}{2}\hat{\sigma}_3(\nu) + \frac{\hbar\omega(B)N(B)}{2}, \tag{5.172}$$

$$E_s(B) = s\hbar\omega(B) \quad s = 0, 1, 2 \ldots N(B), \tag{5.173}$$

$$g_s(B) = \binom{N(B)}{s}. \tag{5.174}$$

$g_s(B)$ *is the respective degeneracy, cf. Example 3.10. The eigenbasis of* $\hat{H}(B)$ *thus reads* $|s, i>$, $i = 1, 2, \ldots g_s(B)$*. The interaction between*

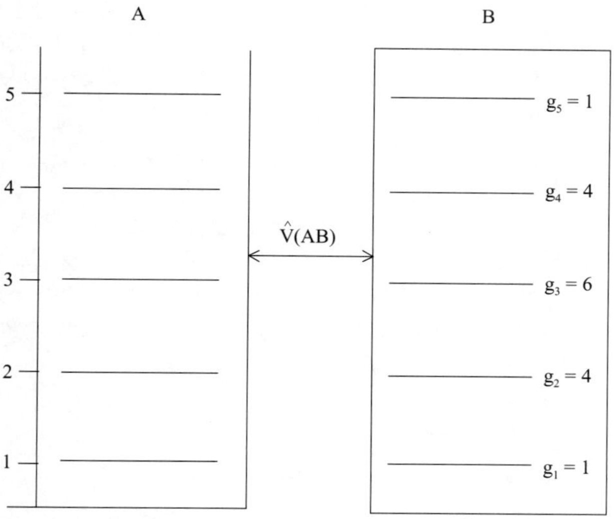

Figure 5.5 Oscillator (A) coupled to spin environment (B), here of N(B) = 4 spins. Note the non-monotonous behavior of the degeneracy.

oscillator and spin network is a local Jaynes–Cummings coupling to spin $\nu = 1$ *with* $\omega(A) = \omega(B) = \omega$:

$$\hat{V}(AB) = \frac{\hbar g_0}{2}(\hat{\sigma}^-(1)\hat{b}^\dagger(A) + \hat{\sigma}^+(1)\hat{b}(A)) \tag{5.175}$$

(cf. Eq. (3.121) and Fig. 5.5).

For the interspin interaction we use a random matrix model with respect to the eigenbasis $|s, i>$*—under the additional constraint*

$$[\hat{H}(B), \hat{V}(BB)] = 0\,. \tag{5.176}$$

The interaction then simplifies to the block-diagonal form

$$\hat{V}(BB) = \kappa\hbar\sum_s\sum_{i,j}(V^{(s)})_{ij}|s, i> < s, j|\,. \tag{5.177}$$

The parameter κ *(as well as* g_0*) has to be chosen small enough to guarantee weak coupling. As* $\hat{H}(A)$ *commutes with* $\hat{V}(BB)$ *and the Jaynes–Cummings coupling with* $\hat{H}(A) + \hat{H}(B)$ *the total Hamiltonian in the product-state basis* $|n; s, j>$ *decomposes into non-interacting blocks indexed by* $K = n + s = n' + s'$:

$$\hat{H}(AB) = \sum_{nsj;n's'j'}(H_K)_{nsj;n's'j'}\delta_{n+k,K}\delta_{n'+k',K}|n; sj> < n'; s'j'|\,. \tag{5.178}$$

The entries of the blocks will be chosen from the Gaussian unitary ensemble with the variances $\sigma_d = \sqrt{2}\hbar\omega$, $\sigma_{od} = \hbar\omega$, for the diagonal elements and off-diagonal elements, respectively, cf. Definition 5.14. The reduced density operator for the oscillator is

$$\hat{\rho}(A) = \sum_{n,n'} \sum_{s,j} < n; sj|\hat{\rho}|n'; s, j > |n> < n'| . \tag{5.179}$$

If the initial state lies within a subspace K, it will remain so for all times. For this fixed $K = n + s = n' + s'$ the matrix elements in Eq. (5.179) with $s = s'$ imply that $n = n'$; the oscillator state is diagonal in the Fock basis. According to Eq. (5.37) the asymptotic equilibrium state should be given by

$$w(E_n(A)) = g_n(A) \sum_E \frac{g_s(B)w(E)}{g(E)} \delta(E_s(B) + E_n(A) - E) . \tag{5.180}$$

Here the δ-function selects s for given n and E.

We specialize this model to a sharp total energy $E \equiv K\hbar\omega$; the sum over E disappears. In the weak coupling limit the interaction energies are neglected. Then, observing $K = n + s$, we find for the total degeneracy:

$$\begin{aligned} g(E) \rightarrow g(K\hbar\omega) &= \sum_{s=0}^{K} \sum_{n=0}^{\infty} \delta_{n+s,K} g_s(B) g_n(A) \\ &= \sum_{s=0}^{K} g_s(B) = \sum_{s=0}^{K} \binom{N(B)}{s} . \end{aligned} \tag{5.181}$$

The second sum in the first line has broken down to a single term $g_{K-s}(A) = 1$, for each s. Eq. (5.180) thus reduces to

$$w(E_n(A)) = \binom{N(B)}{K-n} \frac{1}{g(K\hbar\omega)} . \tag{5.182}$$

For the following we interpret the oscillator (A) as representing a photon mode. The initial oscillator state is the vacuum state, n(A) = 0. The initial spin-network state is homogeneous, that is, each spin is in the same local state. As usual, the final photon distribution is governed by the binomial degeneracy of the spin system. Numerical results are depicted for N(B) = 12 and two different cases for K: The case K = 6 = s (all spins at infinite temperature) is shown in Fig. 5.6.

The case K = 12 = s (the maximum energy for N(B) = 12) implies complete occupation inversion in the spin system ("negative

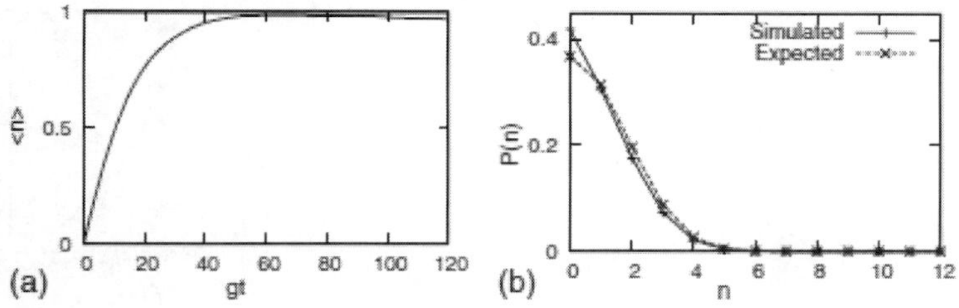

Figure 5.6 Non-lasing relaxation for N(B) = 12 and K = 6 = s. Time in units g_0^{-1}. Coupling parameters are $g_0 = 0.0001$, $\kappa = 0.35g_0$. (a) Average photon number n; (b) photon number distribution $P(n)$ in equilibrium.

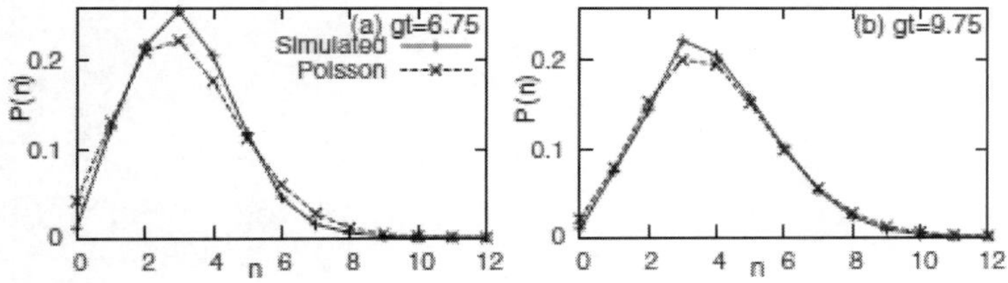

Figure 5.7 Lasing relaxation for N(B) = 12 and K = 12 = s. Coupling parameters are $g_0 = 0.0001$, $\kappa = 0.35g_0$. Transient photon distribution P(n) for (a) time $g_0t = 6.75$; (b) time $g_0t = 9.75$.

temperature"). This leads to a transient lasing transition, as shown in Fig. 5.7. It is interesting to note that the photon distribution goes through a Poisson-type statistics (non-equilibrium), before it eventually settles in the equilibrium specified above. Poisson statistics is typical for laser fields; however, the field state is incoherent.

The approach to the asymptotic state is shown in Fig. 5.8.

This example shows that lasing is a phenomenon that may even show up in the relaxation behavior within a closed bipartite quantum system. Because of the finite energy content of the environment B such a lasing process (amplification) must be transient, though.

For a comment on the thermodynamic significance of negative temperatures, see Remark 7.10.

Example 5.19 Influence of structural constraints *The thermal behavior of an embedded system depends on the degeneracy structure of the quantum environment. This structure is easy to predict based on*

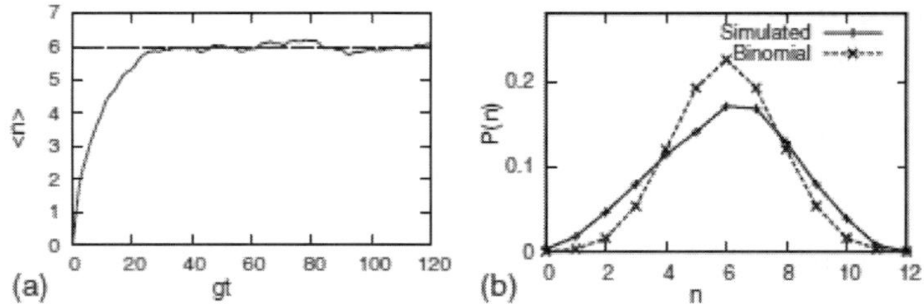

Figure 5.8 Lasing relaxation for N(B) = 12 and K = 12 = s. Time in units g_0^{-1}. Coupling parameters are $g_0 = 0.0001$, $\kappa = 0.35 g_0$. (a) Average photon number n; (b) photon number distribution $P(n)$ in equilibrium.

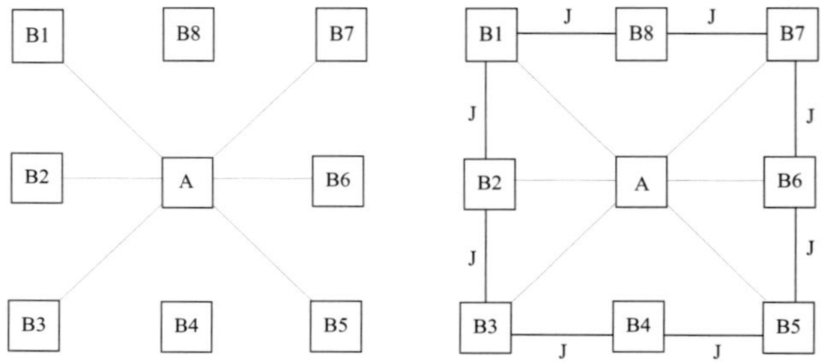

Figure 5.9 Single spin (A) in its N-spin environment (B): Spin-star (J = 0; left) versus ring-star ($J \neq 0$; right)

random matrix models for the coupling between the subsystems A and B. However, the situation may change drastically, once the accessible region becomes subject to additional conservation laws, that is, for interaction models with structural constraints.

As a well-known example we compare the so-called spin-star with the ring-star configuration, cf. Fig. 5.9. In any case the system A is a single spin, which is in contact with a ring of N(B) other spins, the embedding system B.

The Hamiltonian model is [Schmidt (2005)]

$$\hat{H}(AB) = \hat{H}(A) + \hat{H}(B) + \hat{V}(BB) + \hat{V}(AB), \qquad (5.183)$$

$$\hat{H}(A) = \frac{\delta(A)}{2}\hat{\sigma}_3(A), \qquad (5.184)$$

$$\hat{H}(B) = \frac{\delta(B)}{2} \sum_{\nu=1}^{N} \hat{\sigma}_3(\nu; B) + \frac{\delta(B)N}{2}, \tag{5.185}$$

$$\hat{V}(AB) = \alpha \sum_{\nu=1}^{N} \sum_{i,j=1}^{3} \gamma_{ij}(A, \nu)\hat{\sigma}_i(A) \otimes \hat{\sigma}_j(\nu). \tag{5.186}$$

Weak coupling requires $\alpha \ll \delta(A; B) = \hbar\omega(A; B)$. The interaction within the environment B is taken to be

$$\hat{V}(BB) = \frac{J}{2} \sum_{\nu=1}^{N} (\hat{\sigma}_1(\nu) \otimes \hat{\sigma}_1(\nu+1) + \hat{\sigma}_2(\nu) \otimes \hat{\sigma}_2(\nu+1)) . \tag{5.187}$$

For $J = 0$ we have the "spin-star configuration"; the environment has the binomial degeneracy, Eq. (5.44)

$$g_k = \binom{N}{k} \quad E_k = k\delta(B) \quad k = 0, 1, 2 \ldots N. \tag{5.188}$$

Here k is the band index; the interaction between the spins is severely constrained.

For $J \neq 0$ the embedding spins form an interacting ring (nearest neighbor coupling, "ring-star configuration"). Because of the intra-subsystem interaction the resonance condition, $\delta(A) = \delta(B)$, has slightly to be adapted. For the initial state

$$|\Psi(AB) >= |1(A) > \otimes |k, m(B) > \tag{5.189}$$

and for $N = 14$, $k = 2$ the pertinent degeneracies are $g_2 = 91$, $g_3 = 364$ and thus the expected equilibrium value for the inversion is (see Eqs. 5.51 and 5.52)

$$\overline{< \sigma_3(A) >} = \overline{< P_1 >} - \overline{< P_0 >} = \frac{g_2 - g_3}{g_2 + g_3} = -3/5 \approx -0.6 . \tag{5.190}$$

For large enough inter-system coupling J within the environment this average value is confirmed, the distribution becomes narrower as N is increased. For reduced coupling within the environment the effective degeneracies start to deviate significantly from the simple binomial formula. The inversion parameter shifts to higher values [Schmidt (2005)], see Fig. 5.10.

Structural constraints are, in fact, quite common: Any functional design will require a special arrangement of subsystems in real space.

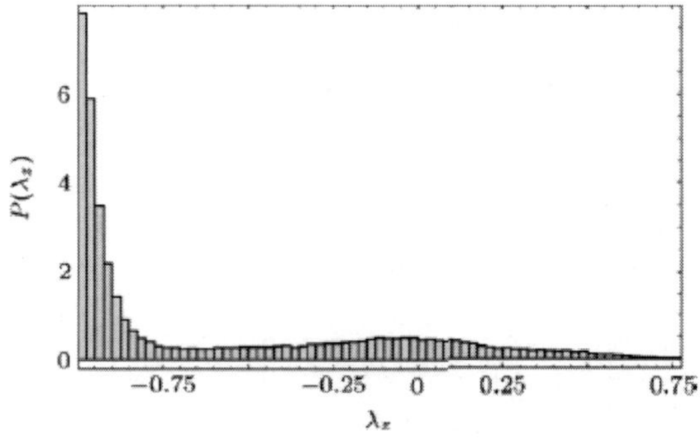

Figure 5.10 Distribution of inversion of central spin A. Parameters: N = 14; k =2, $\delta(A) \approx \delta(B); \alpha = \delta/5000,\ J = 0$.

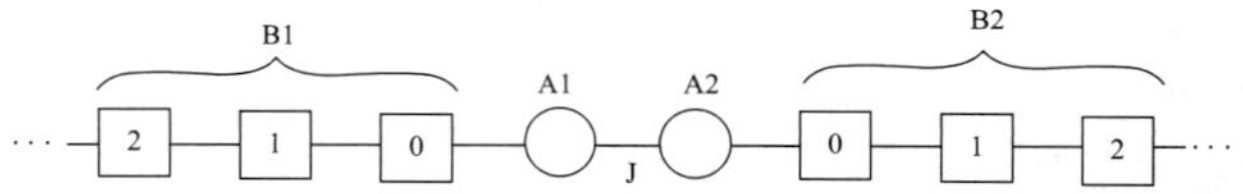

Figure 5.11 Dimer (A1, A2) selectively coupled to two chains, B1, B2, respectively.

In the last two examples we will encounter dynamical deviations from the second law.

Example 5.20 Influence of strong coupling *Another model, now aimed at the influence of strong system–environment coupling, is shown in Fig. 5.11. Here each spin A1, A2 of an interacting two-spin system ("dimer") is selectively coupled to its own environment, a semi-infinite chain of oscillators, B1, B2. These oscillators have varying local energies ϵ; next neighbors are coupled (parameter t).*

The detailed model reads [Prior (2010)]

$$\hat{H}(A1, A2, B1, B2) = \hat{H}(A1) + \hat{H}(A2) + \hat{H}(B1) + \hat{H}(B2) + \\ + \hat{V}(A1, A2) + \hat{V}(A1, B1) + \hat{V}(A2, B2), \quad (5.191)$$

$$\hat{H}(Aj) = \frac{\delta(Aj)}{2}\hat{\sigma}_3(Aj), \tag{5.192}$$

$$\hat{V}(A1, A2) = 2J\left(\hat{\sigma}_1(A1) \otimes \hat{\sigma}_1(A2) + \hat{\sigma}_2(A1) \otimes \hat{\sigma}_2(A2)\right), \tag{5.193}$$

$$\hat{H}(Bj) = \sum_{\nu=0}^{\infty} \epsilon_\nu(Bj)\hat{b}_\nu^\dagger(Bj)\hat{b}_\nu(Bj) + \\ + \sum_{\nu=0}^{\infty}(t\hat{b}_{\nu+1}^\dagger(Bj)\hat{b}_\nu(Bj) + c.c.), \tag{5.194}$$

$$\hat{V}(Aj, Bj) = \sqrt{(\eta/(16\pi))}(1 + \hat{\sigma}_3(Aj))(\hat{b}_0(Bj) + \hat{b}_0^\dagger(Bj)). \tag{5.195}$$

The oscillator chains $j = 1, 2$ operate as baths with identical spectral functions. Chain j is locally coupled to spin Aj at the respective chain boundary at site n=0. The coupling strength between system Aj and environment Bj is characterized by

$$\eta = 8\lambda\gamma \int \frac{\omega}{\omega^2 + \gamma^2} d\omega \tag{5.196}$$

γ specifies the dynamical response time of each bath. What is the relaxation dynamics?

A numerical example [Prior (2010)] is shown in Fig. 5.12 as a function of the AjBj-coupling strength λ. The chains are initially at zero temperature. We see how the excitation probability of site A1 decays from its initial value 1. As the coupling is increased the oscillatory behavior (coherence) gets more and more suppressed, while the asymptotic value first decreases and then increases again.

This is believed to be typical for non-perturbative and non-Markovian effects implied by strong coupling. The model has been motivated by the quest for quantum coherence in photosynthesis, cf. Section 9.2.2.

Remark 5.20 Strong coupling effects *Violations of the weak coupling condition typically lead to apparent "thermodynamic anomalies." Their origin can be traced back to the fact that the subsystem under consideration is no longer in a thermal equilibrium state; examples are the non-existence of local temperatures, cf. Example 5.7, or negative specific heats [Campisi (2010)].*

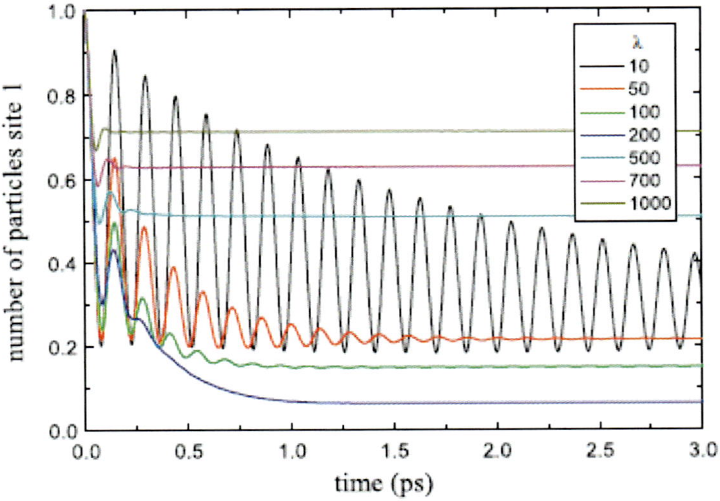

Figure 5.12 Relaxation dynamics of an embedded two-spin system (A1,A2), see text. Parameters are $J = 100$, $\gamma = 53$, $\delta(A1) - \delta(A2) = 100$, all in wave-number units cm^{-1}.

While strong coupling may not always be avoidable, its treatment in terms of thermodynamics must be considered with great care.

Remark 5.21 Substitute process *If the detailed time-dependent behavior is not of interest, one may replace the actual relaxation dynamics in the above model by its weak coupling variant, supplemented by an additional final step, in which the coupling is quasi-statically increased from weak to strong. The asymptotic state is expected to be the same.*

Example 5.21 Influence of initial correlations *The emergence of thermal behavior is usually demonstrated based on a bipartite quantum system: It is taken for granted that initially the two subsystems are weakly correlated; often one assumes a simple product state with zero correlations.*

Such a choice is not just a matter of convenience. It reflects an asymmetry imposed on us—eventually—by cosmological constraints. Thermodynamics would look quite differently from a more general point of view. In fact, strong initial correlations between the parts of an otherwise closed quantum system could induce highly anomalous relaxation behavior [Partovi (2008)].

We start from the quantum relative entropy, Eq. (4.47).

$$\frac{1}{\beta} S_{vN}(\hat{\rho}_1 \| \hat{\rho}_{can}) = \mathcal{F}(\hat{\rho}_1) - F(\hat{\rho}_{can}) . \tag{5.197}$$

Here, $\hat{\rho}_1$ is a general state, $\hat{\rho}_{can}$ a canonical state. This equation can be rewritten as

$$S_{vN}(\hat{\rho}_1 \| \hat{\rho}_{can}) = \beta \Delta U - \Delta S_{vN} \geq 0 . \tag{5.198}$$

The non-negativity results from Klein's inequality, Eq. (3.167). This equation is now applied to a closed bipartite system AB:

$$\beta(\mu)\Delta U(\mu) \geq \Delta S_{vN}(\mu) \quad \mu = A, B , \tag{5.199}$$

subject to (weak coupling)

$$\Delta U(A) + \Delta U(B) \approx 0 . \tag{5.200}$$

From Eq. (5.199) we thus conclude

$$\boxed{(\beta(A) - \beta(B))\Delta U(A) \geq \Delta S_{vN}(A) + \Delta S_{vN}(B) .} \tag{5.201}$$

We distinguish two cases:

Case i.) Uncorrelated initial state: *In this case, $S_{vN}(AB;initial) = S_{vN}(A;initial) + S_{vN}(B;initial)$. We assume local equilibrium states. For the final state we have $S_{vN}(AB;final) \leq S_{vN}(A;final) + S_{vN}(B;final)$. The evolution between initial and final state is unitary, so that the total entropy does not change. As a result,*

$$\begin{aligned} &\Delta S_{vN}(A) + \Delta S_{vN}(B) = \\ &S_{vN}(A;final) + S_{vN}(B;final) - S_{vN}(AB;initial) \geq 0 . \end{aligned} \tag{5.202}$$

Combining this result with Eq. (5.201) and interpreting the exchanged energy as heat, heat must flow from the (originally) hotter to the colder subsystem. This is in accord with the zeroth law, Eq. (4.112).

Case ii.) Highly correlated initial state: *Let us consider the initial pure state*

$$|\Psi(AB)> = \frac{1}{\mathcal{N}} \sum_j \exp(-\alpha\epsilon_j/2)|j(A)> \otimes |j(B)> , \tag{5.203}$$

$$\epsilon_j = \gamma^{-1}(A)E_j(A) = \gamma^{-1}(B)E_j(B) . \tag{5.204}$$

Here, $\mathcal{N}$ is the normalization, $\gamma(\mu)$ are scale factors. The two subsystems have essentially the same spectra $E_j(\mu)$, cf. Eq. (4.7). If we trace out subsystem B, we obtain a local thermal state for A with temperature $\beta(A) = \gamma^{-1}(A)\alpha$. The corresponding result is obtained for subsystem B. Equation (5.201) still applies. But now $S_{vN}(A) = S_{vN}(B)$ at all times and thus $\Delta S_{vN}(A) = \Delta S_{vN}(B)$. From Eq. (5.199) and $\Delta U(A) = \Delta U(B)$ we conclude that $\Delta S_{vN}(\mu) \leq 0$. We can no longer guarantee that energy (heat) flows from the hotter to the colder subsystem; there can be anomalous behavior [Partovi (2008)].

5.3.3 *Arrow of Time: Loschmidt Echo*

Remark 5.22 Arrow of time and the second law *Back in 1928 A. Eddington introduced the notion "arrow of time" to describe the obvious directionality of events in the macro-domain of physics. According to [Jarzynski (2011)] this observation can be cast into a "game": Is a movie being shown in "correct" or "reversed" order? A reliable test can be based on the second law, Eq. (4.140). If $\Delta W > \Delta F$ the events proceed in the correct order, if $\Delta W < \Delta F$ we see the reversed order. Unfortunately, this test does not* explain *the origin of this asymmetry.*

Remark 5.23 Causal arrow *The temporal asymmetry is reminiscent of the asymmetry between cause and effect [Allahverdyan (2008b)]. The causal arrow refers to a dynamical situation, in which one degree of freedom C (the cause) influences another degree, E (the effect), without back-reaction. Causality appears to be an "emergent phenomenon," c.f. Remark 5.10. A typical example is the so-called slaving principle [Haken (1983)].*

Definition 5.15 Loschmidt echo *The irreversibility of thermodynamics is in stark contrast with the underlying time-symmetric microscopic dynamics. This contradiction is expressed by the so-called Loschmidt paradox: Consider the evolution of a classical gas, say, for some time τ. In principle, one could reverse all the velocities to recover the initial state after another waiting time τ. (To be precise, one would at this point have to reverse the velocities again to end up with the same mechanical state, not only the same particle positions.)*

Note that the reversal of velocities does not require to know the velocities, that is, to measure. Rather than changing the state (inverting the velocities) one can change the Hamiltonian, which generates the underlying dynamics. This is a special form of time-dependent control.

Definition 5.16 Time reversal symmetry *The time-dependent Schrödinger equation, Eq. (3.227),*

$$i\hbar\frac{d}{dt}|\Psi(t)>= \hat{H}|\Psi(t)>, \tag{5.205}$$

is invariant under the joint transformation $t \rightarrow -t$ and taking the complex conjugate of the equation. It thus holds that with $|\Psi(t)>$ also $<\Psi(-t)|$ is solution.

Definition 5.17 Time-reversal operation *In the Schrödinger equation the transformation $t \rightarrow -t$ can be substituted by the transformation $\hat{H} \rightarrow -\hat{H}$. Assuming that such a sign change of the Hamiltonian (at time τ) can be implemented, but at finite precision only, we write*

$$\hat{H}(t=\tau) \rightarrow -\hat{H}' \equiv -\hat{H} - \epsilon\hat{V}. \tag{5.206}$$

Here, $-\epsilon\hat{V}$ specifies the unavoidable perturbation (imprecision).

Based on these two Hamiltonians we introduce the respective time-evolution operators [Waldherr (2010b)]

$$\hat{U}(\tau) = \exp(-i\hat{H}\tau/\hbar), \tag{5.207}$$

$$\hat{U}'(\tau) = \exp(-i\hat{H}'\tau/\hbar). \tag{5.208}$$

For $\hat{H}$ we take now a weakly interacting N-spin system. (Weak interaction will be required to allow for arbitrary partitions into system A and environment B in the sense of quantum thermodynamics, see below.) Let the initial total state be $\hat{\rho}_\alpha(AB)$. Specific unitary transformations of interest are

$$\hat{\rho}_\beta = \hat{U}(\tau)\hat{\rho}_\alpha\hat{U}^\dagger(\tau), \tag{5.209}$$

$$\hat{\rho}'_\beta = \hat{U}'(\tau)\hat{\rho}_\alpha\hat{U}'^\dagger(\tau), \tag{5.210}$$

$$\hat{\rho}'_\alpha = \hat{U}'(-\tau)\hat{U}(\tau)\hat{\rho}_\alpha\hat{U}^\dagger(\tau)\hat{U}'^\dagger(-\tau). \tag{5.211}$$

These evolutionary steps allow to specify two different tasks, each characterized by a fidelity measure F as defined in Eq. (3.154):

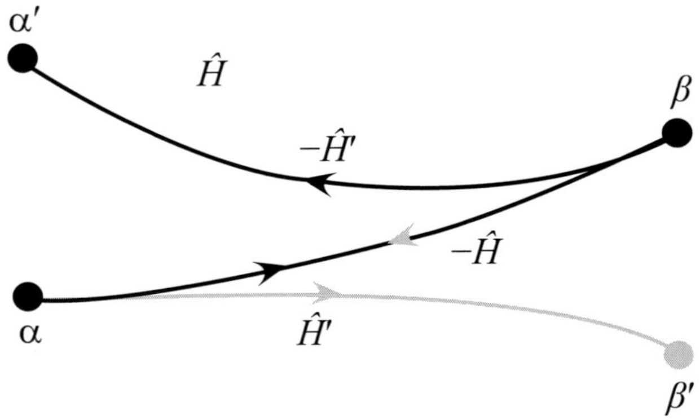

Figure 5.13 Loschmidt echo versus forward stability.

(i) Forward stability: The task to set up the generating Hamiltonian in such a way that the system evolves from state α into the target state β in given time τ. The success probability is $F(\hat{\rho}_\beta, \hat{\rho}_{\beta'})$.

(ii) Loschmidt echo: The task to return to the initial state α after time τ from that state β, by manipulating the Hamiltonian for the way back. The success probability is $F(\hat{\rho}_\alpha, \hat{\rho}_{\alpha'})$.

Because the fidelity is invariant under unitary transformation and thus [Waldherr (2010b)]

$$F(\hat{\rho}_\alpha, \hat{\rho}_{\alpha'}) = F(\hat{U}'\hat{\rho}_\alpha \hat{U}'^\dagger, \hat{U}'\hat{\rho}_{\alpha'}\hat{U}'^\dagger) = F(\hat{\rho}_\beta, \hat{\rho}_{\beta'}), \tag{5.212}$$

the success probabilities for the two tasks are always the same, independent of the perturbation and the time τ:

$$\kappa \equiv F(\hat{\rho}_\beta, \hat{\rho}_{\beta'}) - F(\hat{\rho}_\alpha, \hat{\rho}_{\alpha'}) = 0. \tag{5.213}$$

There is not yet an arrow of time, see Fig. 5.13.

This situation changes once we change perspective and restrict attention to a part A out of the total system: The total system of N spins is split according to N = N(A) + N(B), and we consider the reduced density operator

$$\hat{\rho}_\nu(A) = \mathrm{Tr}_B\{\hat{\rho}_\nu(A, B)\} \tag{5.214}$$

for the total states ν : $\alpha, \alpha', \beta, \beta'$ as introduced before. The local "asymmetry measure" now reads

$$\kappa(A) \equiv F(\hat{\rho}_\beta(A), \hat{\rho}_{\beta'}(A)) - F(\hat{\rho}_\alpha(A), \hat{\rho}_{\alpha'}(A)). \tag{5.215}$$

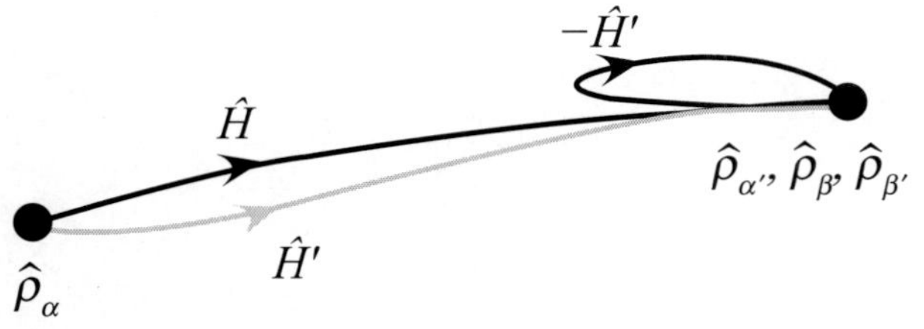

Figure 5.14 Loschmidt echo versus forward stability in reduced Hilbert space (A) for $N(A) \ll N$ and $\tau > \tau - echo$.

One easily convinces oneself that this $\kappa(A)$ will, in general, no longer be zero: We first note that

$$F(\hat{\rho}_\alpha(A), \hat{\rho}_{\alpha'}(A)) \to 0 \quad \text{for} \quad \tau > \tau_{echo}\,. \tag{5.216}$$

By definition, for $\tau \ll \tau_{echo}$ the fidelity stays close to 1, that is, the echo works. In the limit of no perturbation, $\epsilon = 0$, the echo time would approach infinity. If subsystem A is small enough compared to B, we expect the initial state of the total system to relax into an equilibrium state $\hat{\rho}_{equ}(A)$ with respect to B, independent of the details of the (weak) interaction. So, if the perturbation does not change the structure of the Hamiltonian, $\hat{\rho}_{\alpha'}(A) \to \hat{\rho}_{equ}(A)$ and, in particular

$$F(\hat{\rho}_\beta(A), \hat{\rho}_{\beta'}(A)) \to F(\hat{\rho}_{equ}(A), \hat{\rho}_{equ}(A)) \to 1 \quad \text{for} \quad \tau > \tau_{relax}\,. \tag{5.217}$$

The qualitative behavior changes from Fig. 5.13 to Fig. 5.14 .

As a concrete example we use the spin network as introduced in Example 5.18. The respective partition is shown in Fig. 5.15.

Results are shown for N(A) + N(B) = 13 in Fig. 5.16 for various N(A). With increasing N(A) the asymmetry $\kappa(A)$ changes from approximately 1 to zero. The decrease for small N(A) is due to the fact that in this limit the probability for the subsystem A to stay close to its initial state increases, that is, $F(\hat{\rho}_\alpha(A), \hat{\rho}_{\alpha'}(A)) \approx 1/d(A)$, where $d(A)$ is the Hilbert space dimension of subsystem A.

It is interesting to note that according to this analysis the arrow of time is entirely operational (i.e., agent-based), not "fundamental": With unlimited resources time reversal would always be possible.

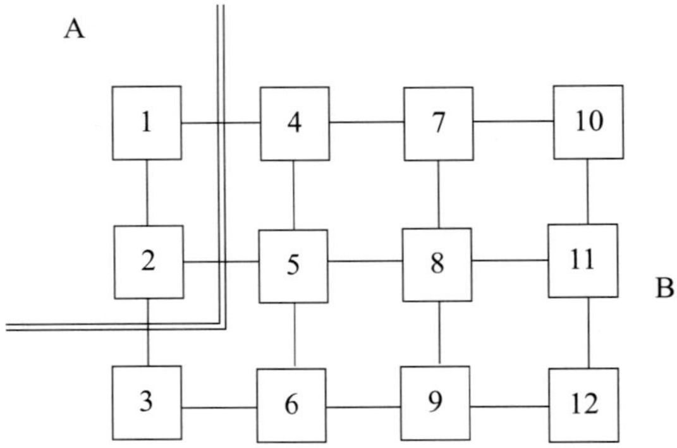

Figure 5.15 Partition of weakly interacting spin network.

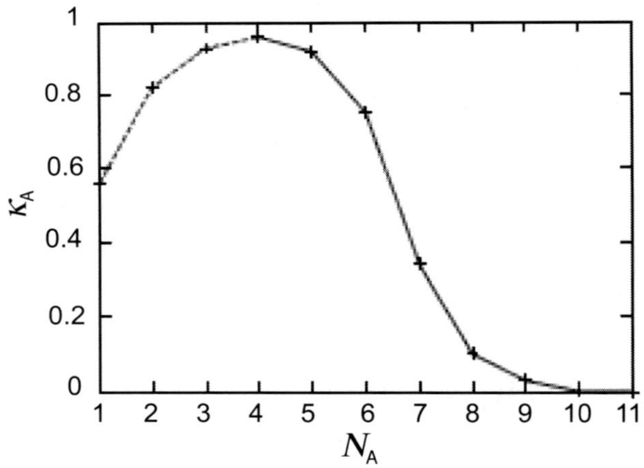

Figure 5.16 Asymmetry measure $\kappa(A)$.

5.4 Third Law: Unattainability of Zero Temperature

Definition 5.18 Physical tractability *In the following we wish to discuss processes P, which are "tractable" (i.e., efficient) under the condition C. Condition C typically refers to limited resources in terms of energy, time, (memory-)space, etc. Tractability is an operational criterion for a process to be carried out (implemented) "in practice."*

Remark 5.24 Algorithmic tractability *Tractability is also an important feature of information processing tasks, here often termed computability, cf. Section 2.3.3.*

Both tractabilities are related: The implementation of an intractable task ("hard problem") would require a physically intractable process. The common characteristic feature for intractability is the need of exponential resources [Nielsen (2000)].

The third law deals with the intractability of the physical process of "cooling" in the approach to zero temperature. Cooling means extracting energy from a quantum system initially prepared outside its ground state, that is, the system is loaded with some amount of energy (acts like a "battery"). Fundamental limitations should reflect themselves operationally, that is, in a testable way.

Corollary 5.1 Third Law *(Cf. Theorem 4.14) "There is no efficient process for reaching the asymptotic limit $T = 0$." Formulated as a "no-go theorem": It is impossible to reach the limit $T = 0$.*

Example 5.22 Quantum fingerprints of the third law *What does it mean to lack an efficient process? What are the allowed resources? The answer to these questions can become quite challenging, actual costs may easily be overlooked.*

Hard problems associated with the limit $T \to 0$ appear in various forms:

Local temperatures cease to exist: The embedding becomes destructive, cf. Example 5.7.

Temperature measurements become unreliable, cf. Example 7.5.1.

Machine limitations (Otto cycle): Cooling rate goes to zero, cf. Section 7.5.1.

Dynamical cooling: Work expenditure tends to "explode," cf. Eq. (5.247).

Algorithmic cooling: number of fresh auxiliary spins increases exponentially, cf. Section 6.5.

In the following we will analyze two cooling processes based on Ref. [Allahverdyan (2011b)]: The Nernst set-up and dynamical cooling. In the Nernst case limitations are somewhat hidden behind ad hoc assumptions. In the latter case fundamental constraints eventually derive from the limited availability of work. Also operation time is a precious commodity; we will return to the so-called dynamical version of the third law in Section 7.5.1 and in Section 8.3.1.

5.4.1 *Nernst Set-Up*

We turn to thermal processes with respect to cooling. Refrigerator cycles (and their relation to the third law) are studied in Section 7.5.1.

Proposition 5.17 Two-step cooling: Nernst set-up *Consider a thermal system A described by $U(S, V)$ subject to an isothermal process (temperature $T_0 > 0$) followed by an adiabatic step, see Fig. 5.17: $\hat{\rho}_{can}(t_0) \to \hat{\rho}_{can}(t_1) \to \hat{\rho}_{can}(t_2)$. Then the work ΔW needed for cooling is given by the relative entropy of the final state $\hat{\rho}_{can}(t_2)$ with respect to the initial state $\hat{\rho}_{can}(t_0)$:*

$$\boxed{\Delta W = k_B T_0 S_{vN}(\hat{\rho}_{can}(t_2) \| \hat{\rho}_{can}(t_0))\,.} \tag{5.218}$$

The relative entropy has been defined in Eq. (3.166).

Proof. Based on the free energy [Allahverdyan (2011b)]

$$F(T, V) = U(S(T, V), V) - T S(U(T, V), V)\,, \tag{5.219}$$

we consider the two-step process (also cf. Example 4.8):

i. Isothermal step ($t_0 \to t_1$): $V_0 \to V_1$ for $T_1 = T_0$

$$S(T_0, V_0) > S(T_0, V_1) \quad \text{(required for cooling)}\,, \tag{5.220}$$

$$\Delta W_{01} = F(T_0, V_1) - F(T_0, V_0)\,. \tag{5.221}$$

Here, ΔW_{01} is the respective work.

ii. Adiabatic step ($t_1 \to t_2$): $V_1 \to V_2 = V_0$ for $S(T_2, V_2) = S(T_1, V_1)$,

$$S(T_2, V_0) = S(T_0, V_1)\,, \tag{5.222}$$

$$\Delta W_{12} = U(T_2, V_0) - U(T_0, V_1)\,. \tag{5.223}$$

This second step makes the external driving cyclical, $V_0 \to V_1 \to V_0$. Of course, this does not mean that the target system is also back in its initial state. The total work needed for cooling is thus given by the

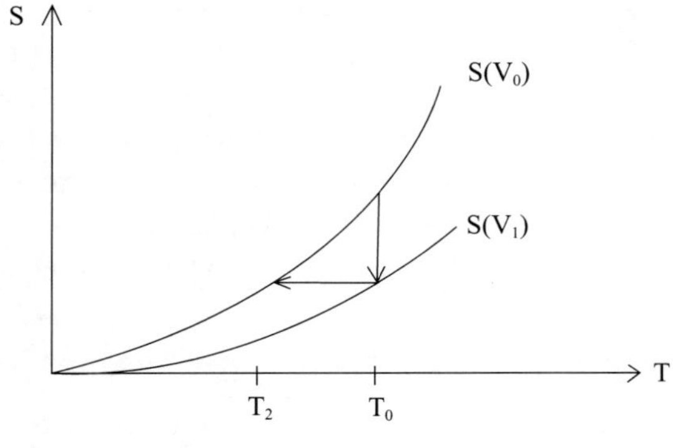

Figure 5.17 Nernst set-up. Process steps in entropy/temperature-space.

relative entropy of the final state $\hat{\rho}_{can}(t_2)$ with respect to the initial state $\hat{\rho}_{can}(t_0)$:

$$\begin{aligned}\Delta W_{02} &= \Delta W_{01} + \Delta W_{12} \\ &= U(T_2, V_0) - T_0 S(T_2, V_0) - (U(T_0, V_0) - T_0 S(T_0, V_0)) \\ &= \mathcal{F}(\hat{\rho}_{can}(t_2)) - F(\hat{\rho}_{can}(t_0)) . \end{aligned} \tag{5.224}$$

The relation of the generalized free energy $\mathcal{F}$ with the relative entropy has been introduced in Eq. (4.48). It thus follows that

$$\Delta W_{02} = k_B T_0 S_{vN}(\hat{\rho}_{can}(t_2) \| \hat{\rho}_{can}(t_0)) \geq 0 . \tag{5.225}$$

□

Alternatively, the work can be written as (cf. Eq. (4.158))

$$\Delta W_{02} = F(T_2, V_0) - F(T_0, V_0) + (T_2 - T_0) S(T_2, V_0) . \tag{5.226}$$

Remark 5.25 Iteration *If this two-step process is followed by a second one leading from temperature T_2 to $T_4 < T_2$ we find, using $S(T_4, V_0) < S(T_2, V_0)$,*

$$\begin{aligned}\Delta W_{04} &\geq F(T_4, V_0) - F(T_0, V_0) + (T_4 - T_0) S((T_4, V_0) \\ &= k_B T_0 S_{vN}(\hat{\rho}_{can}(t_4) \| \hat{\rho}_{can}(t_0)) \geq 0 . \end{aligned} \tag{5.227}$$

This means that a sequence of such cooling processes will require a total work bounded from below by the relative entropy of the final state relative to the initial state.

The above lower bound for the work input does not appear to define any severe constraint. This is misleading. The Nernst set-up is based on equilibrium thermodynamic processes; it must be criticized on the grounds that with each subsequent isothermal step a further bath of decreased temperature would have to be provided. This unrealistic requirement can be avoided by *dynamical cooling*; such a cooling process, however, has to allow for non-equilibrium.

5.4.2 *Dynamical Cooling of Partitioned Systems*

Proposition 5.18 Passivity as state property *Let a quantum system A be prepared in a canonical state, $\hat{\rho}_{can}(t_0)$ with temperature $T = 1/(k_B\beta)$. (Contact with a heat bath may have been used to prepare the state; for the following process, however, the system is decoupled from the bath.) The system is then subject to a unitary transformation $\hat{\rho}_{can}(t_0) \rightarrow \hat{\rho}(t_1)$ implying the energy change*

$$\Delta E = \mathrm{Tr}\{(\hat{\rho}(t_1) - \hat{\rho}_{can}(t_0))\hat{H}\}\,. \tag{5.228}$$

Then the mean energy cannot decrease (cf. [Thirring (2002)]):

$$\boxed{\Delta E = k_B T\, S_{vN}(\hat{\rho}(t_1)\|\hat{\rho}_{can}(t_0)) \geq 0\,.} \tag{5.229}$$

The increase of energy has to be fed in as work from the outside, $\Delta E = \Delta W$. In general, $\hat{\rho}(t_1)$ is no longer a canonical state.

Proof. According to Eq. (4.36) the initial canonical state implies

$$\hat{H} = -\frac{1}{\beta}(\ln \hat{\rho}_{can}(t_0) + \ln Z_{can})\,, \tag{5.230}$$

so that

$$\beta\Delta E = \mathrm{Tr}\{(\hat{\rho}_{can}(t_0)\ln \hat{\rho}_{can}(t_0) - \hat{\rho}(t_1)\ln \hat{\rho}_{can}(t_0)\}\,. \tag{5.231}$$

Due to unitarity the entropy is a constant of motion. The first operator product under the trace can thus be replaced by $\hat{\rho}(t_1)\ln \hat{\rho}(t_1)$ and the proposition follows. □

Definition 5.19 Dynamical cooling *Dynamical cooling is based on unitary transformations. (As, in general, we move out of equilibrium, these transformations are non-thermal processes.) From the property of passivity it follows that no "cooling" is possible for any closed system originally prepared in a thermal equilibrium state. There are two* *possible ways out: Consider a bipartite system (only one subsystem being cooled), or start from non-equilibrium initial states. We will discuss both these options. These scenarios should be distinguished from those aimed at "maximal work from a single bath," Proposition 4.5, where the system remains in contact with the bath.*

For the preparation of the dynamical cooling set-up based on (AB)-scenarios we first show:

Proposition 5.19 Decomposition of joint relative entropy *For a bipartite system AB we assume $\hat{\rho}(t_0) = \hat{\sigma}_A(t_0) \otimes \hat{\sigma}_B(t_0)$. Then the joint relative entropy of $\hat{\rho}(t_1)$ relative to $\hat{\rho}(t_0)$ can be decomposed into the respective local ones according to*

$$\begin{aligned} S_{vN}(\hat{\rho}(t_1) \| \hat{\rho}(t_0)) &= S_{vN}(\hat{\sigma}_A(t_1) \| \hat{\sigma}_A(t_0)) + S_{vN}(\hat{\sigma}_B(t_1) \| \hat{\sigma}_B(t_0)) \\ &\quad + I_{vN}(\hat{\rho}(t_1) : \hat{\sigma}_A(t_1), \hat{\sigma}_B(t_1)) . \end{aligned} \tag{5.232}$$

The strict additivity is thus spoiled by the mutual quantum information of the state $\hat{\rho}(t_1)$.

Here the mutual quantum information is given by (cf. Def. (2.19))

$$I_{vN}(\hat{\rho} : \hat{\sigma}_A, \hat{\sigma}_B) = \mathrm{Tr}\{\hat{\rho} \ln \hat{\rho}\} - \mathrm{Tr}_A\{\hat{\sigma}_A \ln \hat{\sigma}_A\} - \mathrm{Tr}_B\{\hat{\sigma}_B \ln \hat{\sigma}_B\} \geq 0 . \tag{5.233}$$

Proof. Plugging in the definitions of relative entropy and mutual information we get for the right-hand side X of Eq. (5.232):

$$X = \mathrm{Tr}\{\hat{\rho}(t_1) \ln \hat{\rho}(t_1)\} - \mathrm{Tr}_A\{\hat{\sigma}_A(t_1) \ln \hat{\sigma}_A(t_0)\} - \mathrm{Tr}_B\{\hat{\sigma}_B(t_1) \ln \hat{\sigma}_B(t_0)\} . \tag{5.234}$$

As $\hat{\rho}(t_0)$ is uncorrelated,

$$\ln \hat{\rho}(t_0) = \ln \hat{\sigma}_A(t_0) \otimes \hat{1}_B + \hat{1}_A \otimes \ln \hat{\sigma}_B(t_0) , \tag{5.235}$$

and thus

$$\begin{aligned} \mathrm{Tr}\{\hat{\rho}(t_1) \ln \hat{\rho}(t_0)\} &= \mathrm{Tr}\{\hat{\rho}(t_1) \ln \hat{\sigma}_A(t_0)\} + \mathrm{Tr}\{\hat{\rho}(t_1) \ln \hat{\sigma}_B(t_0)\} \\ &= \mathrm{Tr}_A\{\hat{\sigma}_A(t_1) \ln \hat{\sigma}_A(t_0)\} + \mathrm{Tr}_B\{\hat{\sigma}_B(t_1) \ln \hat{\sigma}_B(t_0)\} . \end{aligned} \tag{5.236}$$

In the last step we have used that $\mathrm{Tr}_A\{\hat{\rho}(t_1)\} = \hat{\sigma}_B(t_1)$ and $\mathrm{Tr}_B\{\hat{\rho}(t_1)\} = \hat{\sigma}_A(t_1)$. With the help of Eq. (5.236) the term X takes the form of the joint relative entropy, as claimed. □

Proposition 5.20 Dynamical cooling within (AB) *We consider a bipartite quantum system consisting of a subsystem A and a partner system B with* $\hat{H} = \hat{H}(A) + \hat{H}(B)$*, originally (i.e., at and before time* t_0*) prepared in a canonical state,* $\hat{\rho}_{can}(t_0)$ *with temperature* $T_0 = 1/(k_B\beta_0)$*. This state is uncorrelated, the subsystems are uncoupled. Then, for some finite time, the system is perturbed,*

$$\hat{H}'(t) = \hat{H}(A) + \hat{H}(B) + \hat{V}(t) \quad \text{for } t_0 \leq t \leq t_1\,, \tag{5.237}$$

where $\hat{V}$ *acts on both subsystem A and B. This leads to a unitary transformation* $\hat{\rho}_{can}(t_0) \rightarrow \hat{\rho}(t_1)$*. The new state is, in general, no longer canonical. The work* ΔW *supplied by the perturbation from the outside is bound from below by the relative entropy of subsystem A:*

$$\boxed{\Delta W \geq k_B T_0 S_{vN}(\hat{\sigma}_A(t_1)\|\hat{\sigma}_A(t_0)) \geq 0\,.} \tag{5.238}$$

The passivity condition has become an inequality; the lower bound is identical with the result found for the Nernst set-up, Eq. (5.218).

Proof. The conditions for proposition 5.19 apply. We thus obtain from Eq. (5.229) with the help of Eq. (5.232)

$$\begin{aligned}\Delta W &= k_B T_0 S_{vN}(\hat{\rho}(t_1)\|\hat{\rho}_{can}(t_0))\\ &= k_B T_0\,(S_{vN}(\hat{\sigma}_A(t_1)\|\hat{\sigma}_A(t_0)) + S_{vN}(\hat{\sigma}_B(t_1)\|\hat{\sigma}_B(t_0)) +\\ &\quad + I_{vN}(\hat{\rho}(t_1); \hat{\sigma}_A(t_1), \hat{\sigma}_B(t_1)) \geq k_B T_0 S_{vN}(\hat{\sigma}_A(t_1)\|\hat{\sigma}_A(t_0))\,.\end{aligned} \tag{5.239}$$

In the last step we have used the positivity of the mutual information and of the relative entropy. The work input is now estimated by means of a lower bound exclusively expressed by the local relative entropy of subsystem (A). □

For the subsystem A cooling is, indeed, possible though severely constrained. We consider an example.

Proposition 5.21 Concrete system-reservoir model *Under the conditions of proposition 5.20 we restrict ourselves to a two-level system A and a M-level system B. If dynamical cooling was to reach zero temperature for subsystem A, the respective ground state probability* $p_0(t_1)$ *would have to approach 1. However, this cannot happen, that is,* $p_0(t_1) < 1$*.*

Proof. At t_0 both systems are in a diagonal state (density matrix) with the matrix elements ($p_0 < 1, p_1$) and ($\pi_0, \pi_1, \ldots \pi_\alpha \cdots$), respectively [Allahverdyan (2011b)]. Then the state $\hat{\rho}$ of the total system is a simple tensor product of these two states with the following set of diagonal matrix elements:

$$\{\hat{\rho}\}_{t_0} = \{p_0\pi_0,\, p_1\pi_0,\, p_0\pi_1,\, p_1\pi_1,\, p_0\pi_2, \ldots\}\,. \tag{5.240}$$

Let us now apply the unitary transformation $\hat{U}(t_1)$ specified by the matrix elements $< j, \alpha|\hat{U}|i, \gamma >$. Then the diagonal matrix elements transform as

$$(p_j\pi_\alpha)_{t_1} = \sum_{i,\gamma} p_i\pi_\gamma| < j, \alpha|\hat{U}(t_1)|i, \gamma > |^2 \leq 1\,. \tag{5.241}$$

We thus obtain for the marginal probability

$$p_0(t_1) = \sum_{i,\gamma} p_i\pi_\gamma \sum_\alpha | < 0, \alpha|\hat{U}|i, \gamma > |^2\,. \tag{5.242}$$

Because

$$\sum_{j,\alpha} | < j, \alpha|\hat{U}|i, \gamma > |^2 = 1 \tag{5.243}$$

for any $\{i, \gamma\}$, we must have, in general

$$\sum_\alpha | < j, \alpha|\hat{U}|i, \gamma > |^2 < 1 \tag{5.244}$$

for at least some of the $\{i, \gamma\}$. With

$$\sum_{i,\gamma} p_i\pi_\gamma = 1\,, \tag{5.245}$$

it follows that

$$p_0(t_1) < 1 \;\rightarrow\; T_A(t_1) > 0\,. \tag{5.246}$$

□

Dynamical cooling does not allow to reach zero temperature.

Remark 5.26 Relation to finite resources *But how does this limitation come about? One can show that as the above target temperature would be forced to zero, the required work ΔW had to approach infinity.*

A simple, qualitative argument should suffice here (for more details see [Allahverdyan (2011b)]). We reduce B (like A) to a two-level system. The unitary quantum SWAP operation, Example 3.22, is known to cyclically permutate the state between A and B. Assume both subsystems to be in a diagonal state, with subsystem R in its ground state. By means of such a SWAP it would thus be easy to get system A into its ground state, irrespective of its original state.

However, as we require both subsystems to be originally at the same temperature, this transformation does not help, unless the energy splitting of B approaches infinity: In this case its original ground-state occupation would go to 1, despite finite temperature. Unfortunately, though, the energy needed to drive this process from the outside would then also tend to infinity. Indeed, one finds

$$\Delta W \approx 1/T_A(t_1) \rightarrow \infty\,. \tag{5.247}$$

It is in this sense that there is no effective process for reaching zero temperature. The third law of thermodynamics essentially follows from the finiteness of resources.

5.4.3 *Dynamical Cooling of Unpartitioned Systems: Ensemble Effects*

We return to the second option for dynamical cooling: the use of non-equilibrium within an unpartitioned system. Here the notion of "cooling" is somewhat ill-defined; there is no initial temperature. For the following we generalize the concept of passivity, Proposition 5.18, to systems composed of identical subsystems (homogeneous ensembles).

We first note that given any non-equilibrium state $\hat{\rho}(t_0)$, we can construct the corresponding passive state which maximizes the extractable work as follows:

Proposition 5.22 Maximum extractable work *For the state* $\hat{\rho}(t_0) = \sum_\nu \rho_{\nu\nu}|e^{(\nu)} >< e^{(\nu)}|$ *and the Hamiltonian* $\hat{H} = \sum_\nu E_\nu|e^{(\nu)} >< e^{(\nu)}|$ *there is the passive state,* $\hat{\rho}(t_1)$ *given by [Alicki (2012)]*

$$\hat{\rho}(t_1) = \sum_{\nu=1}^{d} r_\nu|e^{(\nu)} > < e^{(\nu)}| \tag{5.248}$$

with the eigenvalues r_ν *being those of* $\hat{\rho}(t_0)$ *arranged in non-increasing order with the energy, that is,*

$$r_{\nu+1} \leq r_\nu \quad for \quad E_{\nu+1} \leq E_\nu \,. \tag{5.249}$$

Indeed, the energy

$$E(t_1) = \sum_\nu r_\nu E_\nu = E_{min} \tag{5.250}$$

is then minimal under all permutations of the occupation probabilities $\rho_{\nu\nu}$. *Permutations define a set of unitary transformations, cf. Example 3.2.*

Remark 5.27 *For a pure initial state the above construction is not needed, the minimum energy is the ground state of the system, which is passive by definition. This is in accordance with the fact that "full knowledge" (zero entropy) allows maximum control; we can always find a unitary transformation extracting the original energy completely. With respect to the task of energy retrieval this is the ideal case.*

In general, however, the scheme based on Proposition 5.22 has its limitations: the (tensor-) product of two (identical) passive states is not necessarily passive. This leaves room for more efficient energy retrieval.

To get a first idea, consider the following case:

Example 5.23 Two three-level systems *Let the energies E of the individual three-level system be arranged in increasing order,* $i = 1, 2, 3$. *The respective occupation probabilities* r_i *are non-increasing. Based on the energy spectrum of the pair,* $E_{ij} = E_i + E_j$, *and its occupation probabilities* $r_{ij} = r_i \times r_j$ *one may easily encounter a situation in which* $r_{13} > r_{22}$ *while* $E_{13} > E_{22}$. *The total system is not passive, while the individual system states are.*

Such a result can be excluded if the individual (as well as the total) state is canonical:

Definition 5.20 Complete passivity *Consider a system composed of N identical sub-units prepared in identical local states. There are no mutual interactions. A state is called completely passive, if for any N*

$$\hat{\rho} = \bigotimes_{\mu=1}^{N} \hat{\rho}(\mu) \tag{5.251}$$

is passive (see Proposition 5.18). This only holds if $\hat{\rho}$ is a canonical state [Alicki (2012)].

As two-level systems can always be considered to be in canonical state, we have to look, at least, to three-level systems to find violations of complete passivity, cf. Example 5.23. In turn, the optimization of energy retrieval (for ensembles) should aim at complete passivity. Unfortunately the pertinent unitary transformation is hard to construct. We briefly sketch a method.

Definition 5.21 Effective passive state *Let the von Neumann entropy of the initial state $\hat{\rho}(t_0)$ be given by $S_{vN}(t_0)$. The canonical state $\hat{\rho}_{can}(T_{eff})$ with the same entropy is specified by some temperature T_{eff}. This state is passive for any N and can, in principle, be reached from the initial state by means of a unitary transformation.*

Proposition 5.23 Entangling unitary transformation *A unitary $\hat{U}$ able to transform the initial state into the effective passive state $\hat{\rho}_{can}(T_{eff})$ would produce the maximum possible cooling effect. Methods to achieve this can be based on the Proposition 5.22, but have to go beyond the product form, that is, require to entangle the members of the ensemble [Alicki (2012)]. In this way, more energy can be extracted per subsystem than would be possible from individual "batteries." This surplus increases with the size of the ensemble and and saturates for $N \to \infty$ at the optimum value defined by $\hat{\rho}_{can}(T_{eff})$.*

Remark 5.28 A perfect power source? *This proposal has been welcomed as the "perfect power source," supposedly outperforming any classical batteries [Yirka (2012)]. This expectation is far from reality.*

The theoretical analysis only concerns the retrieval of energy, not the design of the devices as such. A molecule, for example, does not yet make a useful battery. The energy retrieval is partial only, even under ideal conditions: The final energy is finite, not zero. And the conditions for the entangling control to be beneficial are quite demanding: There must be an ensemble of identically prepared quantum systems; the ensemble must be isolated (shielded against baths); their initial state must be non-equilibrium; the unitary transformation sensitively depends on the initial state, which thus must be known exactly; the specification and implementation of the unitary transformation will be difficult. All these requirements are very "costly" and most likely will overcompensate possible benefits.

5.5 Observational Quantum Thermodynamics

Despite occasional reference to information, quantum thermodynamic settings do not include the observer: No means whatsoever are provided to transfer information to the outside world. In that sense the scenarios are similar to those studied in abstract quantum theory, that is, based on isolated systems, Section 3.3. It thus appears quite natural to add a section on observational quantum thermodynamics, just as we added a section on observational quantum mechanics, Section 3.4.

In the following we will be concerned with a thermal system under permanent ("stroboscopic") supervision [Jahnke (2010)] and the measurement of the intensive thermodynamic variables, temperature T and pressure p. We close with some remarks on measurement-induced work fluctuations.

Remark 5.29 Fluctuation dissipation *Interactions with the environment are the fundamental source of noise in both classical and quantum systems [Nielsen (2000)]. These interactions give rise to so-called fluctuation-dissipation theorems [Callen (1951)]. They can be derived as relations between fluctuations in a system at thermal equilibrium and the (linear) response of the system state to weak external perturbations (non-conservative forces).*

Quantum measurements produce fluctuating results, their action is dissipative (irreversible). This is a very intuitive and powerful metaphor for the intimate connectedness between fluctuation and dissipation. In fact, open systems, that is, systems coupled to certain environments, or often said to be "continuously measured" by those environments [Pike (1995); Kurchan (2005)]. Operational details are typically missing, though.

Remark 5.30 Quantum versus thermal fluctuations *While in any stationary quantum state there have to be many unsharp properties (observables), there are no (temporal) fluctuations unless measured (in the form of a time-indexed protocol).*

In a full measurement context one has to be very clear about the process to retrieve information: the thermal variables are only indirectly related to proper quantum observables. Detailed models are needed to clarify how and to what extent the observational process works.

According to quantum thermodynamics the separation between quantum and thermal properties becomes blurred. Nevertheless, we may want to distinguish two basic scenarios:

(i) Quantum fluctuations: "Ensemble interpretation" of uncertainty relations for closed Hamilton systems. The lower bound is the "standard quantum limit, Remark 3.19.

(ii) Thermal fluctuations: "Ensemble interpretation" of the mixed state generated by entanglement between system and environment.

In each case the ensemble properties have to be verified via measurements. As already remarked in Section 4.2.7 the so-called thermal fluctuations are thus traced back to measurement-induced quantum fluctuations.

5.5.1 *Periodic Measurements*

We will be concerned with the following bipartite system,

$$\hat{H}(AB) = \frac{\delta(A)}{2}\hat{\sigma}_3(A) + \hat{H}_0(B) + \hat{V}(BB) + \lambda\hat{V}(AB). \qquad (5.252)$$

The system A is a single spin with energy-splitting $\delta(A)$, the environment B consists of $N \gg 1$ spins. Its product states,

$$|\psi(B)> = |m(1), m(2), \ldots, m(N)> \quad m(\mu) = \mp 1, \qquad (5.253)$$

are eigenfunctions to $\hat{H}_0(B)$ with eigenvalues $E_k(B) = \delta(B)k$, where k denotes the number of spins in state $m = 1$. For given band-index k all the states, $|n_k>$, $n_k = 1, 2, \ldots g_k$, have the same energy. Now focus on some index k_0, the "working point" in energy space. For $N > k_0 \gg 1$ the binomial degeneracies (cf. Example 5.3) can be approximated by

$$g_k \approx g_0 \exp\left(\beta(k_0)E_k\right), \qquad (5.254)$$

$$\beta(k_0) \approx \frac{1}{\delta(B)} \ln\left(\frac{N}{k_0} - 1\right). \qquad (5.255)$$

Due to the weak interaction $\hat{V}(BB)$, each degenerate energy level k becomes a band of width $\Delta\epsilon_k \ll \delta(B)$.

The interaction between subsystem A and B is scaled by the strength factor λ and given by

$$\hat{V}(AB) = \hat{\sigma}_1(A) \otimes \hat{I}(AB), \tag{5.256}$$

$$\hat{I}(AB) = \sum_k \sum_{n_k, m_{k+1}} C_{k+1,k}(n_k, m_{k+1})|n_k > < m_{k+1}| + c.c.\,. \tag{5.257}$$

Here c.c. means complex conjugate term to be added. Only transitions between next neighbor bands are allowed. The coupling parameters $C_{k+1,k}(n_k, m_{k+1})$ form a set of hermitian matrices $C_{i,j}(a, b)$. The respective entries are taken from a Gaussian distribution normalized to $\overline{|C_{i,j}(a, b)|^2} = \sqrt{(g_i g_j)}$.

The total state will be written as

$$\hat{\rho}(AB) = \hat{\rho}(A) \otimes \hat{\rho}(B) - \hat{C}(AB), \tag{5.258}$$

$$\hat{\rho}(A) = \mathrm{Tr}_B\{\hat{\rho}(AB)\} \quad \hat{\rho}(B) = \mathrm{Tr}_A\{\hat{\rho}(AB)\}. \tag{5.259}$$

Here $\hat{C}(AB)$ specifies the correlation between A and B. Let the initial state at time t_0 have zero correlation; in particular, we start from the special product state

$$\hat{\rho}(AB; t_0) = \hat{\rho}(A; t_0) \otimes \hat{\rho}(B; k_0), \tag{5.260}$$

$$\hat{\rho}(B; k_0) \equiv \frac{1}{g_{k_0}} \sum_{n_{k_0}=1}^{g_{k_0}} |n_{k_0} > < n_{k_0}|. \tag{5.261}$$

Resonance will be assumed, that is, $\delta(B) = \delta(A) \equiv \delta$. Then the subsystem B acts as a thermalizing environment, correlation $\hat{C}(AB)$ builds up, and A relaxes to a state with temperature

$$\frac{1}{k_B T(A)} = \beta(A) = \beta(k_0), \tag{5.262}$$

independent of the initial state of A. Up to this point this is just a typical scenario for quantum thermodynamics, cf. Section 5.3.2.

Now we intend to retrieve information about A. This could be done in two different ways: Direct measurements on A or indirect measurements via B exploiting the correlation between A and B.

Measuring the energy of subsystem A directly would mean to find it in the ground- or excited state with probability $P_m(A)$ given by the thermal distribution. Immediately after measurement the subsystem would be found in the respective energy eigenstate $|m(A)>$, $m \mp 1$. This momentary state and the associated information gain would not last for long, though: With the embedding still present system A would again relax to its equilibrium state.

We switch now to the indirect measurement scenario. We restrict ourselves to the measurement of the band index, k, that is, to an incomplete measurement. The projection by $\hat{P}_{k_1}(B)$ at time $t_1 = t_0 + \Delta t$ also influences subsystem A (co-jump, see Eq. (3.340)):

$$\hat{\rho}'(A; t_1) = \frac{\mathrm{Tr}_B\{\hat{P}_{k_1}(B)\hat{\rho}(AB; t_1\}}{\mathrm{Tr}_B\{\hat{P}_{k_1}(B)\hat{\rho}(B; t_1)\}} \tag{5.263}$$

$$= \hat{\rho}(A; t_1) + \frac{\mathrm{Tr}_B\{\hat{P}_{k_1}(B)\hat{C}(AB; t_1\}}{\mathrm{Tr}_B\{\hat{P}_{k_1}(B)\hat{\rho}(B; t_1)\}} . \tag{5.264}$$

After measurement we simplify the total state as the product state

$$\hat{\rho}(AB; t_1) \approx \hat{\rho}'(A; t_1) \otimes \hat{\rho}(B; k_1) . \tag{5.265}$$

This is an approximation as, due to the incomplete measurement, some correlations between A and B could still have survived. The whole process is now iterated: unitary evolution steps of duration Δt are interrupted by instantaneous measurement projections as described above. The result is a stochastic trajectory for A as shown in Fig. 5.18.

These trajectories correspond to the classical statistical idea, that the thermal system A is always in some well-defined state, but "fluctuates" between those such that the time-average is identical with the ensemble average. We now investigate the ensemble average over such trajectories after a certain number of measurements (i.e., after the memory about the initial state has been lost). The result represents an attractor state, which can be expressed analytically [Jahnke (2010)]:

$$\overline{\rho}_{00}(\Delta t) = \frac{\exp(-\beta\delta/2)\sin^2 \delta\Delta t + \exp(\beta\delta/2)\delta^2(\Delta t)^2}{2\cosh \beta\delta/2(\sin^2 \delta\Delta t + \delta^2(\Delta t)^2)} . \tag{5.266}$$

The corresponding behavior is shown in Fig. 5.19.

This attractor state can alternatively be calculated as a time-average over a single trajectory; it has two remarkable bounds:

$$\lim_{\Delta t \to 0} \overline{\rho}_{00}(\Delta t) = 1/2 . \tag{5.267}$$

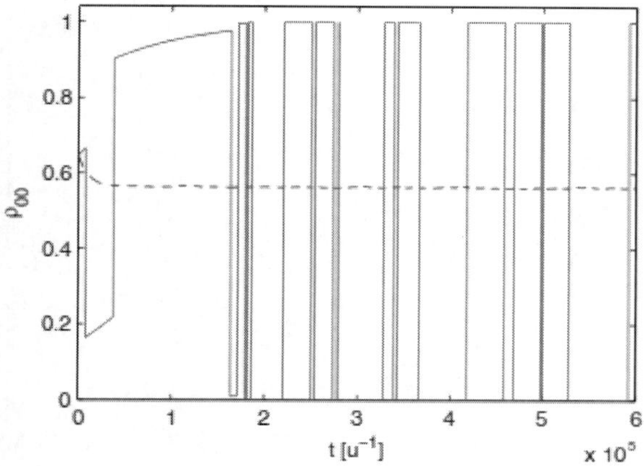

Figure 5.18 Trajectory for subsystem A. $\Delta t = 2u^{-1}$ with u being an arbitrary energy unit. Thermodynamic temperature $\beta = 0.75u^{-1}$, $\delta = 0.8u$, $\lambda = 4 \cdot 10^{-3}u$. Broken line: Ensemble average. Initial state: Unperturbed attractor.

This lower bound means that for very rapid repetitions of measurements the system eventually heats up to $T(A) \to \infty$. Only apparently is this in conflict with the so-called Zeno effect (cf. Proposition 3.25), that is, the freezing-in of the original state under "continuous" measurement: Indeed, for $\Delta t \to 0$ it would take infinitely long to reach the attractor state.

An upper bound for $\overline{\rho}_{00}$ is approached for

$$\Delta t = \frac{n\pi}{\delta} \qquad n = 1, 2, \ldots . \tag{5.268}$$

This bound corresponds to $k_B T(A) = 1/\beta(k_0)$; it is the same temperature subsystem A would obtain in the absence of any measurements. At these waiting times one finds a strict correlation between measuring the band index k to have gone up by 1 (compared to its previous value) or down by 1 and the state of A: For "up" A is in the ground state, for "down" A is in the excited state. This is the underlying "measurement logic."

For $\Delta t \to 0$, on the other hand, the correlation goes to zero: System B has no information about A. The measurements of the environment B lead to an additional perturbation of A.

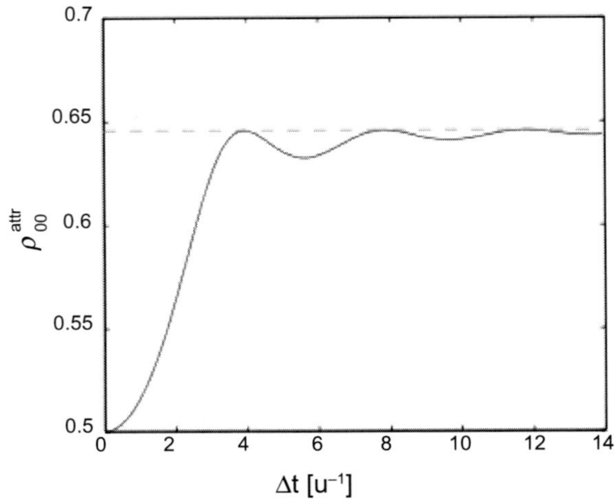

Figure 5.19 Effective attractor for subsystem A as a function of time between measurements. Broken line: Unperturbed attractor. Parameters same as above.

Remark 5.31 Peaceful coexistence *Observational quantum-thermodynamics provides a remarkable link between the abstract thermal properties, which are stationary, and the measured thermal properties, which are fluctuating. The long-time average and the ensemble average of the latter agree with the former result, provided that the measurement is run under optimal conditions, that is, there is a clear and definite measurement logic.*

This establishes a "peaceful coexistence" between the quantum and the classical descriptions—despite their different points of departure. Similar connections can be established in the context of thermodynamic machines and of fluctuations.

5.5.2 *Measurement-Induced Temperature Fluctuations*

Remark 5.32 The classical picture *A system at non-zero temperature does not stay in its equilibrium state but randomly samples all possible states, with probabilities given by the Boltzmann distribution. Thermodynamic variables, such as pressure, temperature, or entropy, likewise would undergo thermal fluctuations.*

At first sight, this picture is in conflict with quantum thermodynamics. The temperature T is a fixed parameter in the canonical distribution, as is the pressure p in the T–p ensemble. By definition, these cannot fluctuate[a]. Nevertheless (experimental) estimates of those parameters do suffer from fundamental uncertainties; these limitations cannot be avoided and will be discussed below.

Let us consider a thermalized quantum system specified by $U(S, V)$; in the quantum-thermodynamic context we may envision this system A to be embedded in a larger quantum environment, C. Under these conditions measurements on the subsystem A can be repeated—after it has rethermalized. The measurement apparatus as such will not explicitly be described.

The temperature is defined as the partial derivative, Eq. (4.98),

$$T(S, V) = \left(\frac{\partial U(S, V)}{\partial S}\right)_V . \tag{5.269}$$

Inverting this function, we get $S = S(T, V)$, by which we can rewrite the internal energy as a function of temperature: $U(S(T, V), V) = U(T, V)$. For the following the mechanical control, V, is kept constant. We can thus write for the response in U to the perturbation $T \to T + \Delta T$,

$$\Delta U = \left(\frac{\partial U(T, V)}{\partial T}\right)_V \Delta T \equiv C_V \Delta T . \tag{5.270}$$

C_V is the heat capacity at constant volume, a thermodynamic response function. ΔT is thus related to ΔU:

$$\boxed{\Delta T = \frac{\Delta U}{C_V} .} \tag{5.271}$$

Now U is the thermal average of the operator $\hat{H}$. Quantum measurements of the energy are random and will thus result in fluctuations also of ΔT. Writing $U = f(T)$ we determine T from U [Jahnke (2011a)],

$$T = f^{-1}(U) . \tag{5.272}$$

An estimate of the temperature can thus be obtained from the single(!) energy measurement ("single shot"), resulting in E_m:

$$\boxed{T_{est} = f^{-1}(E_m) .} \tag{5.273}$$

[a] Here we do not consider generalizations in terms of "superstatistics," where distributions of temperature are allowed.

Obviously, this estimated temperature will fluctuate, as the measurement results fluctuate. We assume a Gaussian distribution peaked at energy U

$$w(E_m) \approx \frac{1}{\sqrt{2\pi}\sigma} \exp\left(-(E_m - U)^2/2\sigma_U^2\right). \tag{5.274}$$

For the respective variance we exploit Eq. (4.42),

$$\mathrm{var}(E_m) = \sigma_U^2 = -\frac{\partial U}{\partial \beta} = -\left(\frac{\partial U}{\partial T}\right)\frac{dT}{d\beta} = C_V(T)k_B T^2. \tag{5.275}$$

We thus expect that

$$\overline{T_{est}} = \int_{-\infty}^{\infty} w(E_m) T_{est}(E_m) dE_m \tag{5.276}$$

approaches the true temperature T, in which case the estimation method would be called "unbiased." This is not necessarily the case, though. One finds instead (with $C_V' \equiv dC_V/dT$ etc.),

$$\overline{T_{est}} = T - k_B T^2 \frac{C_V'}{2(C_V(T))^2} + \dots . \tag{5.277}$$

The fluctuation of the measured temperature is given by [Jahnke (2011a)]

$$\boxed{\mathrm{var}(T_{est}) = \Delta_0^2 + T^4 k_B^2 \left(\frac{7(C_V')^2}{2(C_V)^4} - \frac{C_V''}{(C_V)^3}\right) + \dots ,} \tag{5.278}$$

$$\Delta_0^2 = \frac{k_B T^2}{C_V} \equiv \mathrm{var}(T_{est})_{cl} . \tag{5.279}$$

To lowest order we recover the standard classical result, $\mathrm{var}(T_{est})_{cl}$, cf. [Landau (1980)]. Within quantum thermodynamics this result gets a new and entirely transparent operational meaning. Thermal states are stationary, they do not fluctuate; nevertheless, measurements do. Deviations from the standard result are conveniently described by the ratio

$$X(T) \equiv \frac{\mathrm{var}(T_{est})}{\mathrm{var}(T_{est})_{cl}} . \tag{5.280}$$

Proposition 5.24 Energy–temperature uncertainty product *We infer from Eq. (5.278) and Eq. (5.275)*

$$\mathrm{var}(E_m)\mathrm{var}(T_{est}) \approx k_B^2 T^4 . \tag{5.281}$$

The equation is an operational statement on ensembles of estimates T_{est} each inferred from a single energy measurement E_m. This thermodynamic uncertainty relation is reminiscent of fundamental quantum variants, cf. Eq. (3.326); its significance has been overestimated, though (cf. [Lavanda (1991)]). In fact, it merely specifies a correlation rather than any incompatibility in a strict sense. For a given system with fixed C_V, and fixed temperature T there is no choice of the individual uncertainties. Both are strictly related, knowing one we can immediately calculate the other. Recall that T itself cannot fluctuate.

Example 5.24 N-spin model *The energy spectrum and binomial degeneracy are ($k = 0, 1 \ldots N$)*

$$E_k = k\epsilon \quad g_k = \binom{N}{k} . \tag{5.282}$$

The average energy is thus given by

$$U(T) = < H > = \frac{1}{Z} \sum_{k=0}^{N} \binom{N}{k} \exp(-\epsilon k/k_B T)\epsilon k = \frac{N\epsilon}{1 + \exp(\epsilon/k_B T)} . \tag{5.283}$$

We assume $0 \geq T < \infty$ and thus $U(T) < N\epsilon/2$. Inverting this function leads to the temperature estimate T_{est} from a single energy measurement $U \to E_m$:

$$k_B T_{est} = \frac{\epsilon}{\ln\left(\frac{N\epsilon}{E_m} - 1\right)} . \tag{5.284}$$

Under the above conditions this estimate only works for $m \leq m_c \equiv N/2 - 1$; accidental energy measurements m above this limit have to be substituted by the fixed cut-off m_c. A numerical calculation [Jahnke (2011a)] of $\overline{T_{est}}$ and X according to Eq. (5.280) are shown in Figs. 5.20 and 5.21, respectively.

Deviations from the classical behavior dominate in the limit $T \to 0$. This may be seen as a fingerprint of the third law, cf. Theorem 4.14. In addition, there is an upper temperature limit beyond which deviations show up again; this limit increases with system size.

5.5.3 *Measurement-Induced Pressure Fluctuations*

Again we start from a thermalized subsystem A. Pressure, like temperature, is an intensive thermodynamic variable, defined as a

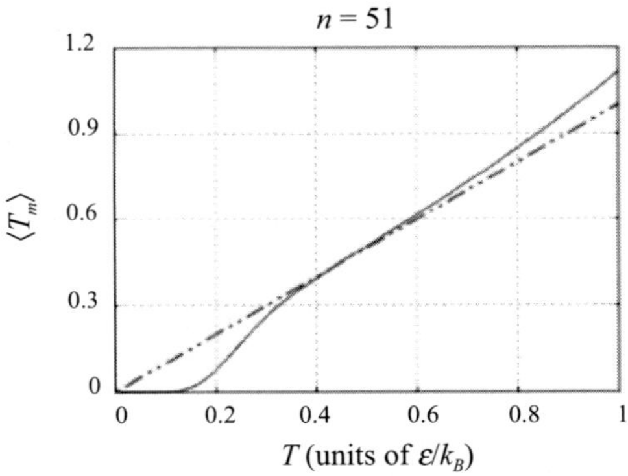

Figure 5.20 Spin model (N = 51): Average estimated temperature, as a function of thermodynamic temperature T.

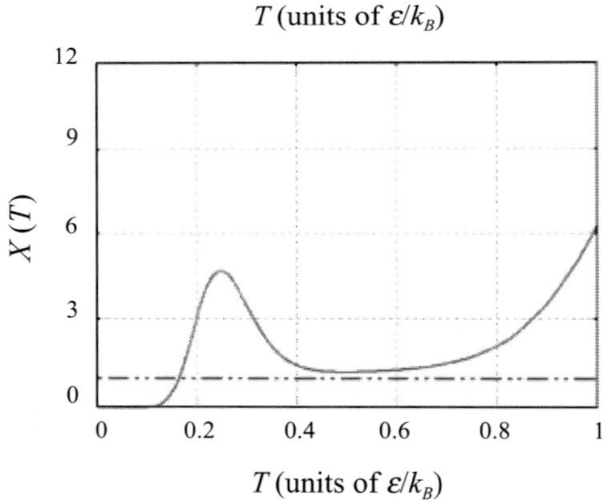

Figure 5.21 Spin model (N = 51): Variance of estimated temperature, as a function of thermodynamic temperature T.

partial derivative of the state function $U(S, V)$.

$$p(S, V) = -\left(\frac{\partial U(S, V)}{\partial V}\right)_S . \tag{5.285}$$

This partial derivative can be interpreted as the response p to an external perturbation, $V \rightarrow V + \delta V$, subject to the constraint of

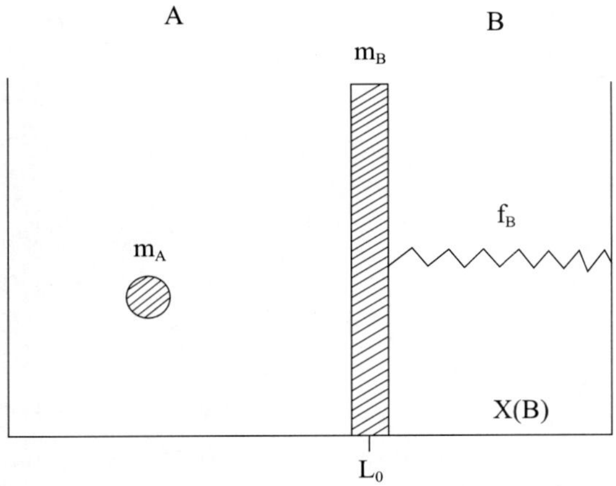

Figure 5.22 Particle (A) in a box with movable wall (oscillator B). f_B is the spring constant.

constant entropy:

$$\boxed{\Delta p = -\left(\frac{\partial p(S,V)}{\partial V}\right)_S \Delta V \equiv \frac{1}{V\kappa_S}\Delta V\,.} \tag{5.286}$$

$\kappa_S > 0$ is the so-called isentropic compressibility, the (negative) relative change of volume per change of pressure. But what is ΔV? In the following this perturbation will be related to measurement fluctuations. For this purpose the mechanical parameter V has to be interpreted as the average over some quantum mechanical operator.

Example 5.25 Particle in a box with movable wall (oscillator) *In the following we consider the bipartite system (AB) as depicted in Fig. 5.22. System A is supposed to represent the system under consideration. It is a particle with mass m_A in a (one-dimensional) potential box with a fixed wall at position zero and a movable wall at position*

$$L_A = L_0 + < x(B) > \,. \tag{5.287}$$

The operator $\hat{x}(B)$ describes the position of particle B (= the movable wall) with mass m_B within a harmonic potential centered at position L_0. Subsystem B acts as a mechanical environment.

Without inter-subsystem interaction the energy eigenvalues of the particle in a box (A) with $L_A = L_0$ is given by

$$E^A_{n_A} = \frac{\hbar^2\pi^2(n_A)^2}{2m_A L_A^2} \,. \tag{5.288}$$

We assume thermal equilibrium at some temperature $T(A)$,

$$U(A) = \sum_{n_A} \frac{\hbar^2\pi^2 n_A^2}{2m_A L_A^2} w(n_A) \,. \tag{5.289}$$

With inter-subsystem interaction the force (energy per length)

$$F^A = -\left(\frac{\partial U(A)}{\partial L_A}\right)_{S(A)} = \sum_{n_A} \frac{\pi^2\hbar^2 n_A^2}{m_A L_A^3} w(n_A) > 0 \,. \tag{5.290}$$

will shift the equilibrium position of particle B such that $\Delta L_A =< x(B) >$; system A transfers the work

$$\Delta W(A) = -F^A < x(B) > < 0 \tag{5.291}$$

to subsystem B. In a quasi-classical approach[a] *we assume that this force F^A acts as a parameter on the contact-oscillator B according to*

$$\begin{aligned} \hat{H}_{eff}(B) &= \frac{\hat{p}^2(B)}{2m_B} + \frac{1}{2} m_B \omega_B^2 \hat{x}^2(B) - F^A \hat{x}(B) \\ &= \frac{\hat{p}^2(B)}{2m_B} + \frac{1}{2} m_B \omega_B^2 \left(\hat{x}(B) - \frac{F^A}{m_B\omega_B^2}\right)^2 - \\ &\quad - \frac{(F^A)^2}{2m_B\omega_B^2} \,. \end{aligned} \tag{5.292}$$

We observe that the "restoring force" is given by

$$F^B = < \partial\hat{H}(B)/\partial\hat{x}(B) >= -f_B < \hat{x}(B) > < 0 \,, \tag{5.293}$$

$$f_B \equiv m_B \omega_B^2 \,. \tag{5.294}$$

f_B is the spring constant. The respective energy spectrum is

$$E^B_{n_B} = \hbar\omega_B\left(n_B + \frac{1}{2}\right) - \frac{(F^A)^2}{2m_B\omega_B^2} \,. \tag{5.295}$$

Due to F^A the average oscillator coordinate is no longer zero but reads

$$x(B) \equiv< x(B) >= \frac{F^A}{f_B} \,, \tag{5.296}$$

that is, $F^A = -F^B$.

---

[a]It has been shown in [Borowski (2003b)] that entanglement can indeed be neglected.

Just as we did in Eq. (5.273) for the temperature as a function of U*, we now apply the relationship between force and shift, Eq. (5.296), to infer the estimated force from a positional measurement of the harmonic oscillator ("single-shot experiment" [Borowski (2003b)]):*

$$\boxed{F^A_{est} = f_B x(B)_m .} \tag{5.297}$$

The index m specifies the actual measurement result. Averaging over those results, $x(B)_m$*, we recover Eq. (5.296).*

For given n_B *the positional variance of the oscillator is—independent of this shift:*

$$\mathrm{var}(x(B)_m) = \frac{\hbar\omega_B}{f_B}(n_B + \frac{1}{2}) . \tag{5.298}$$

Based on the variance relation, Eq. (2.18), the measurement-induced force fluctuations can be written as

$$\mathrm{var}(F^A_{est}) = f_B^2 \mathrm{var}(x(B)_m) = f_B \hbar\omega_B (n_B + \frac{1}{2}) . \tag{5.299}$$

For the thermal average we replace the oscillator energy for fixed n_B *by* $U(B)$ *according to Eq. (4.45):*

$$\mathrm{var}(F^A_{est}) = f_B U(B) = \begin{cases} f_B \hbar\omega_B/2 & \text{if } T(B) \to 0 \\ f_B k_B T & \text{if } T(B) \gg \hbar\omega_B . \end{cases} \tag{5.300}$$

In equilibrium we expect $T(A) = T(B) = T$*. While the intensive thermodynamic variable* F^A_{est} *is not a direct observable (quantum mechanically it is not an operator), its value can be inferred from a conventional quantum measurement on the position of some reacting external particle, the oscillator B. As the individual outcomes for such a measurement are random, also* F^A_{est} *has to fluctuate. Again, that is, similar to the temperature case, the fluctuations have found an intuitive (operational) interpretation.*

In the limit $T \to 0$, $L_0 \to \infty$ the variance $\mathrm{var}(x(B)$ is fixed and finite, while F^A and the average $< x(B) >$ go to zero, that is, such single-shot pressure measurements can no longer be carried out in a reliable way.

The present model bears some similarities with the Brownian motion. The mechanical oscillator (B) of our model and the Brownian particle are "environments" but at the same time act as measurement devices (their position being accessible from the outside). But as is typical for quantum thermodynamics the system (A) does not have to be macroscopic at all. We expect that all standard thermal fluctuation relations can be provided with a similar operational meaning.

Remark 5.33 Classical limit *Finally, writing the negative work, Eq. (5.291), in the standard form, see Definition 4.23,*

$$\Delta W(A) = -(-p\Delta V)\,, \tag{5.301}$$

that is, $F^A_{test} \to -p_{test}$, $< x(B) >= \Delta L_A \to \Delta V$ *and*

$$f_B = \frac{\partial F^A}{\partial < x(B) >} \to -\left(\frac{\partial p}{\partial V}\right)_S\,, \tag{5.302}$$

we find in the high-temperature limit

$$\boxed{\mathrm{var}(F^A_{est}) \to \mathrm{var}(p_{est})_{cl} = -k_B T \left(\frac{\partial p}{\partial V}\right)_S = \frac{k_B T}{V \kappa_S}\,.} \tag{5.303}$$

This is in agreement with the classical statement according to [Landau (1980)]; κ_S *is the isentropic compressibility. The result has the same formal structure as Eq. (5.279).*

Well-known results from classical thermostatistics are thus confirmed and reinterpreted by appropriate quantum mechanical single-shot experiments.

Remark 5.34 Length (volume) fluctuations *According to Eq. (5.299) force (pressure) and length (volume) fluctuations are directly related. In the high-temperature limit we find*

$$\mathrm{var}(x(B)_m) = \frac{1}{f_B^2}\mathrm{var}(F^A_{est}) = \frac{1}{f_B} k_B T\,, \tag{5.304}$$

which implies the uncertainty product $\mathrm{var}(F^A_{est})\ \mathrm{var}(x(B)_m) = k_B^2 T^2$. *Again, this product merely indicates the correlation between the respective pair of variables, here displacement and force.*

5.5.4 *Measurement-Induced Work Fluctuations*

Definition 5.22 Work measurement *Work is energy, but not an observable, cf. Remark 5.12. It relates to a thermodynamic process rather than to a state. Single-time measurements do not suffice: one rather has to settle for differences of energy between two times.*

For this purpose we consider an adiabatic process imposed on a system initially in a canonical equilibrium state. As the Hamiltonian depends explicitly on time, energy conservation does not apply. The change of energy between two instants of time, $\{t, 0\}$, must be supplied by the external control and has to count as work,

$$W_{mn}(t, 0) \equiv E_m(t) - E_n(0) \,. \tag{5.305}$$

This notion of work is thus based on a correlated two-time measurement (of the system energy). $W_{mn}(t, 0)$ is a "single-shot" result. Such measurements underlie, for example, the Jarzynski relation, see Section 6.3. Obviously, no new statistical theory is needed for that. The respective "process average" will be specified by $<< \cdots >>$; it is to be taken over the various event pairs (n,m) and given by

$$\begin{aligned} << W(t,0) >> &\equiv \sum_{mn} W_{mn}(t,0) K_{mn}(t) P_n(0) \\ = \sum_m E_m(t) \sum_n K_{mn}(t) P_n(0) &- \sum_n E_n(0) P_n(0) \sum_m K_{mn}(t) \,. \end{aligned} \tag{5.306}$$

Because of the sum rule, Eq. (6.14), and the definition of $P_m(t)$ according to Eq. (6.12) we immediately confirm that

$$\begin{aligned} << W(t,0) >> &= \sum_m E_m(t) P_m(t) - \sum_n E_n(0) P_n(0) \\ &= < H(t) > - < H(0) > \,. \end{aligned} \tag{5.307}$$

The process average (two-time average) decomposes into two time-local expectation values, as expected. With the final state still being an equilibrium state the process average of work can be identified as the change of internal energy.

Remark 5.35 Quantum fluctuations of work *It is clear that—given the initial state of the system and given the imposed dynamics—individual work values will, in general, fluctuate, as each single energy measurement ("single shot") fluctuates. This is a true quantum effect. Fluctuations would be absent only if both, initial and final state, coincide with some energy eigenstate—specified already before measurement. Such situations might be inforced by proper design of the scenario.*

5.5.5 *Quantum Games*

Games as rule-based algorithms have been introduced in Definition 2.28; in the following single-player games are considered as tools to increase or modify control within a physical context.

Remark 5.36 Measurement scenarios as quantum games *Specific measurement scenarios can be considered "games." Erasure has been introduced as such a game, see Definition 4.35. In this case, the averaged work needed to perform the task can be derived based on a thermodynamic substitute process, a refrigeration scheme.*

Maxwell's demon, cf. Theorem 4.19, is another example. Here we are concerned with single-shot work extractions each based on a "closed-loop" feedback control. Fluctuations of the second measurement can be avoided if the initial energy measurement is followed by an adiabatic evolution (in the quantum sense, Definition 6.2).

A third example is the maximal work extraction from a single bath, cf. Proposition 4.5. For given initial state this work can be calculated. From the point of view of measurements this result must be considered as the respective average. Based on two-time energy measurements as defined above also single-shot values could easily be obtained for a given strategy. Again, no new "single-shot statistical mechanics" is needed for that. This may change only as we move toward different and artificial scenarios and by that away from standard thermal physics. We close with two pertinent examples.

Example 5.26 Quantum work deficit *The amount of extractable work from a bipartite system (AB) in state $\hat{\rho}(AB)$ is expected to depend on whether the players are able to operate only locally on A or B or else have access to the system state as a whole. Though the problem of extractable work appears to be close to a thermal task, see Section 4.3.7, it is not. In fact, the initial state does not need to be thermal, it can be any state.*

The problem is formalized with the help of two different strategies: Strategy (i) is defined to be based on global operations (closed operations CO on (AB)) and strategy (ii) on local operations (closed local operations CLO available to A,B separately, supplemented by classical communication CC between A,B). For example, let (AB) *consist of $N_A + N_B = N$ qubits (Hilbert space dimension $d = 2^N$). Then the retrievable Shannon information under strategy (i) is in units of bits given by*

$$S_{CO}(\hat{\rho}(AB)) = N - S_I(\hat{\rho}(AB)), \qquad (5.308)$$

which is simply the difference between maximum and initial entropy. On the other hand, for strategy (ii) that is, CLO plus CC, one finds [Oppenheim (2002)]

$$S_{CLOCC}(\hat{\rho}(AB)) \leq N - \max\{S_I(\hat{\rho}(A)), S_I(\hat{\rho}(B)\}. \qquad (5.309)$$

The so-called quantum work deficit is defined as

$$\Delta S_I(AB) = S_{CO}(\hat{\rho}(AB)) - S_{CLOCC}(\hat{\rho}(AB)) \geq 0. \qquad (5.310)$$

$\Delta S_I(AB)$ counts the "missing information" that can be attributed to correlations between A and B; it is a kind of non-locality measure.

Example 5.27 Maximal work extraction under risk *For the following we restrict access to the work according to Eq. (5.305) by assuming that the result of both measurements on system A remains unknown to the experimenter. What is known to him are the initial and final states each specified in terms of an inhomogeneous ensemble (representing his subjective ignorance). Both these mixed states, $\hat{\rho}_{ini}(A)$, $\hat{\rho}_{fin}(A)$, have to be diagonal in the respective energy basis, but need not be thermal. Of course, "in reality" there is initially one energy eigenstate and one final energy eigenstate only. The measured difference would be the actual single-shot physical work value, W.*

Based on this incomplete knowledge, what is the maximal work the experimenter can get by choosing various strategies? According to [Dahlsten (2011); Egloff (2012)]) a possible answer can be found subject to the following additional constraints:

(i) *The initial state and the initial spectrum of system (A) are given.*
(ii) *The final state and the final spectrum are given. These constraints present a kind of pre- and post-selection, respectively.*
(iii) *To extract work the "agent" can follow a strategy of his choice. But any such strategy must observe the above constraints. And it may exclusively consist of the following partial processes: For fixed energy levels, change the occupation probabilities of selected levels via coupling to the bath of given temperature; for fixed occupation probabilities, change the spectrum thus exchanging work with some work reservoir.*

(iv) *"Predicted work": Before the experimenter starts to carry out his strategy, he is forced to specify what he thinks he can achieve, that is, to predict a work value,* W_p*. The physical work is then accepted as work proper only, if its actual value is larger than the predicted* W_p*. The amount of* W *above the predicted value is discarded. If the actual value* W *is less than* W_p*, no work in the present sense obtains, which thus counts as a "failure." The idea apparently is that work below expectation was useless. Just as if the predicted work was the minimal work needed to perform a given task.*

Obviously, W_p is fixed, it does not fluctuate, contrary to W. For a given strategy and a given realization there will be a certain probability of failure. Not surprisingly, the maximal extractable work value W_p depends on the risk (high or low failure probability ϵ) one is ready to accept. It turns out [Egloff (2012); Dahlsten (2011)]) that these upper bounds can be derived based on the so-called "smooth entropies", which, in turn, can be expressed via the Renyi entropies S_∞, $S_{1/2}$ (cf. Eq. (3.155, cf. also Example 4.5).

The above game is rather artificial, and so is the formalism. No doubt, the "predicted work" is alien to thermodynamics and even alien to its statistical foundation in terms of fluctuations. The use of non-equilibrium states further underlines the deviation from thermal physics. Indeed, the latter feature is generic for the subjective ignorance interpretation: incomplete knowledge does, in general, not map on ensembles equivalent to thermal states. The handling of such generalized states may require novel entropy measures but these will hardly have anything to do with a thermal entropy. Conventional thermostatistics becomes useless. But—contrary to what is claimed in [Egloff (2012)]—this is not because we are interested in single-shot experiments (cf. Section 5.5.2); rather it is because the rules of the game are in conflict with the laws of thermal physics.

Chapter 6

Finite Processes

Jonas Burgert: Untitled, 2005, based on photography by Peter Sander [Reprinted with permission from Schreiber and Wipplinger (2011)].

Quantum Thermodynamic Processes: Energy and Information Flow at the Nanoscale
Günter Mahler

ISBN 978-981-4463-73-7 (Hardcover), 978-981-4463-74-4 (eBook)
www.panstanford.com

Einstein's notion of reality (cf. Section 3.2.1) combined with Feynman's quest for machinery (cf. Remark 2.3) brings us back to the delicate relationship between physics and information: To what extent is a physical process "real?" Clearly, we always need a mathematical description—but does this description define what is "really" going on?

There are good reasons to doubt that: The simulation based on such a mathematical description may become computationally hard—with little relevance for its actual physical realization, see, for example [Crespi (2012)]. On the other hand, there will be control limits to cope with; at the end of the day such operational constraints may destroy scalability. This is "agent-based" physics, cf. Section 1.5.

In the following we will focus on non-autonomous systems (i.e., explicitly time-dependent Hamiltonians). In general, their evolution is governed by a delicate combination of eigendynamics and imposed control dynamics. Under adiabaticity conditions the latter can be made to dominate: A finite process (with definite starting and end point) shows up as a path in the respective control space.

6.1 Unitary Dynamics of Non-Autonomous Systems

A non-autonomous Hamilton system is specified by $\hat{H}(\lambda(t))$, where $\lambda(t)$ defines the external time-dependent control function. In the instantaneous diagonal basis the eigenequation for this Hamiltonian reads,

$$\hat{H}(t)|e^{(n)}(t)>= E_n(t)|e^{(n)}(t)> . \tag{6.1}$$

We then consider the time evolution of the density operator, $\hat{\rho}$, within this basis:

$$\rho_{nn'}(t) = < e^{(n)}(t)|\hat{\rho}(t)|e^{(n')}(t)>, \tag{6.2}$$

$$\frac{d}{dt}\rho_{nn'} = < e^{(n)}(t)|\frac{d}{dt}\hat{\rho}(t)|e^{(n')}(t)> + \tag{6.3}$$

$$< \dot{e}^{(n)}(t)|\hat{\rho}(t)|e^{(n')}(t)> + < e^{(n)}(t)|\hat{\rho}(t)|\dot{e}^{(n')}(t)> . \tag{6.4}$$

For the first term on the RHS we apply the Liouville equation, Eq. (3.270),

$$\frac{d}{dt}\hat{\rho}(t) = \frac{1}{i\hbar}(\hat{H}(t)\hat{\rho}(t) - \hat{\rho}(t)\hat{H}(t))\,. \tag{6.5}$$

With the help of the completeness relation, $\hat{1} = \sum_j |e^{(j)}(t) > < e^{(j)}(t)|$, cf. Eq. (3.14), we obtain:

$$\begin{aligned}\frac{d}{dt}\rho_{nn'} &= -\omega_{nn'}\rho_{nn'} + \sum_k \rho_{nk} < e^{(k)}(t)|\dot{e}^{(n')}(t) > \\ &\quad + \sum_j < \dot{e}^{(n)}(t)|e^{(j)}(t) > \rho_{jn'} \\ &= -\omega_{nn'}\rho_{nn'} + \sum_{kj} S_{nn';kj}\ \rho_{jk}\,, \end{aligned} \tag{6.6}$$

$$S_{nn';kj} \equiv < e^{(k)}(t)|\dot{e}^{(n')}(t) > \delta_{nj} + < \dot{e}^{(n)}(t)|e^{(j)}(t) > \delta_{n'k}\,, \tag{6.7}$$

$$\hbar\omega_{nn'} = E_n - E_{n'}\,. \tag{6.8}$$

Proposition 6.1 *For the matrix elements the following property applies:*

$$\begin{aligned}\sum_k S_{nn';kk} &= < e^{(n)}(t)|\dot{e}^{(n')}(t) > + < \dot{e}^{(n)}(t)|e^{(n')}(t) > \\ &= \frac{d}{dt} < e^{(n)}(t)|e^{(n')}(t) >= 0\,,\end{aligned} \tag{6.9}$$

for any n, n'.

Proposition 6.2 *The totally mixed state (d = dimension of Hilbert space),*

$$\rho_{jj'} = \frac{1}{d}\delta_{jj'}\,, \tag{6.10}$$

solves the Liouville equation Eq. (6.6) at all times. The lhs is obviously zero, $d\rho_{jj'}/dt = 0$, *as is the RHS:*

$$\frac{1}{d}\sum_j S_{nn';jj} = 0\,. \tag{6.11}$$

Here we have made use of Eq. (6.9).

Definition 6.1 Transition matrix *Let* K_{fi} *be the probability for the system to be in state* $|e^{(f)}(t) >$ *at time t given that it started at state*

$|e^{(i)}(0)>$ at time t = 0. Then, the (unconditional) probability for state f is [Mukamel (2003)]:

$$\rho_{ff}(t) = \sum_i K_{fi}(t)\ \rho_{ii}(0)\,. \tag{6.12}$$

Alternatively this equation can be written as $P_f(t) = \sum_i K_{fi}(t) P_i(0)$, cf. Eq. (3.158).

Proposition 6.3 Sum rules *The transition matrix satisfies two important sum rules: First, summing over all final states gives*

$$1 = \sum_f \rho_{ff}(t) = \sum_{fi} K_{fi}(t)\ \rho_{ii}(0)\,. \tag{6.13}$$

Then the pure state $\rho_{ii} = \delta_{ii_0}$ leads to

$$\boxed{s(i_0) = \sum_f K_{fi_0}(t) = 1 \quad \text{for any } i_0\,.} \tag{6.14}$$

This sum rule does not require unitary evolution. Now consider the initial state $\rho_{ii}(0) = 1/d$, for $i = 1, 2 \ldots d$. In this case

$$\rho_{ff} = \frac{1}{d} \sum_i K_{fi}(t)\,. \tag{6.15}$$

But this initial state is the totally mixed state known to be a solution of the Liouville equation for all times, that is, we must have $\rho_{ff} = 1/d$, for any f. It thus follows the sum rule

$$\boxed{s(f) = \sum_i K_{fi}(t) = 1 \quad \text{for any } f\,.} \tag{6.16}$$

This second sum rule requires unitarity.

Proposition 6.4 Representation of $K_{fi}(t)$ *Let us assume for the unitary evolution*

$$|e^{(j)}(t)> = \hat{U}(t)|e^{(j)}(0)> \tag{6.17}$$

for any j, that is, a simple non-crossing relation for all time-dependent eigenstates. Then the conditional probability can be written as

$$\begin{aligned} K_{fi}(t) &= |<e^{(f)}(t)|e^{(i)}(t)>|^2 \\ &= |<e^{(f)}(t)|\hat{U}(t)|e^{(i)}(0)>|^2 = |U_{fi}(t)|^2\,. \end{aligned} \tag{6.18}$$

Both sum rules follow immediately:

$$\sum_i < e^{(f)}(t)|\hat{U}(t)|e^{(i)}(0) >< e^{(i)}(0)|\hat{U}^\dagger(t)|e^{(f)}(t) > =$$
$$< e^{(f)}(t)|\hat{U}\hat{U}^\dagger|e^{(f)}(t) > = 1,$$
$$\sum_f < e^{(i)}(0)|\hat{U}^\dagger(t)|e^{(f)}(t) >< e^{(f)}(t)|\hat{U}(t)|e^{(i)}(0) > =$$
$$< e^{(i)}(0)|\hat{U}^\dagger\hat{U}|e^{(i)}(0) > = 1. \quad (6.19)$$

6.2 Adiabaticity

6.2.1 *Definition and Conditions*

A concept of central interest is the adiabaticity, which decomposes into the following aspects:

(1) Adiabatic theorem (for closed systems)
(2) Adiabatic approximation: Conditions for state occupation levels remaining (almost) constant
(3) Non-adiabaticity measures: Quantification of deviations from adiabatic behavior
(4) Adiabaticity for open systems

Definition 6.2 Adiabatic dynamics *"A closed quantum system is said to undergo adiabatic dynamics if its Hilbert space can be decomposed into decoupled Schrödinger eigenspaces with distinct, time-continuous, and non-crossing instantaneous eigenvalues of $\hat{H}(t)$" [Sarandy (2004)]. In the adiabatic limit we thus expect*

$$\boxed{K_{fi}(t) = \delta_{fi}\,.} \quad (6.20)$$

This means that if at a given time a quantum system (with non-degenerate spectrum) is found to be in a specific instantaneous eigenstate of a slowly varying Hamiltonian it will remain so at any later time. There is an independent evolution of the instantaneous eigenvectors, the corresponding eigenvalue will evolve continuously; eigenenergies do not cross.

Under what conditions does this happen? To simplify matters we consider the time-dependent Schrödinger equation rather than the

Liouville equation,

$$i\hbar\frac{d}{dt}|\Psi(t)>=\hat{H}(t)|\Psi(t)> . \qquad (6.21)$$

Let us diagonalize the instantaneous Hamiltonian by means of a (time-dependent) unitary transformation:

$$\hat{H}_d(t) \equiv \hat{U}^{-1}(t)\hat{H}(t)\hat{U}(t), \qquad (6.22)$$

$$|\Psi>_d \equiv \hat{U}^{-1}|\Psi> . \qquad (6.23)$$

Multiplying Eq. (6.21) from left by $\hat{U}^{-1}$, we obtain

$$\hat{H}_d(t)|\Psi>_d= i\hbar\hat{U}^{-1}\frac{d}{dt}|\Psi>\approx i\hbar\frac{d}{dt}|\Psi>_d . \qquad (6.24)$$

In the last step we have neglected the term $-\dot{\hat{U}}^{-1}|\Psi>$ as "slowly varying." Then, as $\hat{H}_d(t)$ is diagonal by construction, the system would, indeed, evolve separately in each eigenenergy level, that is, "adiabatically." However, the notion of "slow" has still to be qualified.

Proposition 6.5 Adiabatic condition *We assume the spectrum of $\hat{H}(t)$ to be discrete and non-degenerate. Then the instantaneous eigenbasis of the Hamiltonian is given by*

$$\hat{H}(t)|e^{(n)}(t)> = E_n(t)|e^{(n)}(t)>, \qquad (6.25)$$

$$g_{nk}(t) \equiv E_n(t) - E_k(t) . \qquad (6.26)$$

The $g_{nk}(t)$ specify the various energy gaps. Let Δt be the total evolution time under consideration. Then the condition for adiabatic evolution reads

$$\boxed{\max_{0\leq t\leq\Delta t}\left|\frac{<e^{(k)}(t)|\dot{\hat{H}}|e^{(n)}(t)>}{g_{nk}(t)}\right| \ll \min_{0\leq t\leq\Delta t}|g_{nk}(t)/\hbar|} \qquad (6.27)$$

for all $n \neq k$. Note that the inequality has dimension of 1/time, that is, of a frequency; it thus compares two frequencies. Also note the dot inside the matrix element as a short-hand notation for time derivative.

Proof. We expand $|\Psi(t)>$ in terms of the instantaneous eigenbasis as follows [Sarandy (2004)]:

$$|\Psi(t)> = \sum_{n=1}^{d} a_n(t)\exp(-i\phi_n(t))|e^{(n)}(t)>, \qquad (6.28)$$

$$\phi_n(t) \equiv \frac{1}{\hbar}\int_0^t E_n(t')dt' . \qquad (6.29)$$

$\phi_n(t)$ is the so-called dynamical phase; it generalizes the standard phase factor $-iE_n t/\hbar$ to the case where E_n is a function of time. Substituting this expansion into the Schrödinger equation, Eq. (6.21), one gets

$$\sum_n \hat{H}(t)a_n(t)\exp(-i\phi_n(t))|e^{(n)}(t)>= i\hbar\sum_n \exp(-i\phi_n(t))$$
$$\times\left(\dot{a}_n|e^{(n)}(t)> +a_n|\dot{e}^{(n)}(t)> +a_n(-iE_n/\hbar)|e^{(n)}(t)>\right). \quad (6.30)$$

Multiplying from left with $< e^{(k)}|\exp(i\phi_k(t))$ one obtains

$$\dot{a}_k = -\sum_n a_n < e^{(k)}|\dot{e}^{(n)} > \exp(-i\phi_n(t) + i\phi_k(t)). \quad (6.31)$$

This relation can further be transformed. For this purpose we take the time derivative of Eq. (6.25):

$$\dot{\hat{H}}|e^{(n)}(t)> +\hat{H}|\dot{e}^{(n)}>= \dot{E}_n|e^{(n)}> +E_n|\dot{e}^{(n)}> . \quad (6.32)$$

Multiplying from left with $< e^{(k)}|$ with $k \neq n$ one gets

$$< e^{(k)}|\dot{\hat{H}}|e^{(n)}> =< e^{(k)}|\dot{e}^{(n)}> g_{nk}. \quad (6.33)$$

Based on this identification Eq. (6.31) can be rewritten as

$$\dot{a}_k = -a_k < e^{(k)}|\dot{e}^{(k)}> -\sum_{n\neq k} a_n \frac{< e^{(k)}|\dot{\hat{H}}|e^{(n)}>}{g_{nk}}\exp(-i\phi_n(t) + i\phi_k(t)). \quad (6.34)$$

Adiabatic evolution requires that each a_k must evolve independently from any other coefficient. Each individual contribution under the sum should thus be negligible. This requirement implies the proposition. □

Remark 6.1 Lower bound for evolution time *A rough lower bound for the evolution time can be obtained from Eq. (6.27):*
Writing $\Delta\hat{H} = \dot{\hat{H}}\Delta t$ *and* $\epsilon_{nk} \equiv < e^{(k)}(t)|\Delta\hat{H}|e^{(n)}(t) >$ *one finds, cf. [Messiah (1990)],*

$$\hbar\frac{\epsilon_{nk}}{g_{nk}^2} \ll \Delta t. \quad (6.35)$$

Δt *thus scales as* ΔE_{min}^{-2} *for given (n, k). If the evolution time becomes too short, the probability for inter-level transitions increases. For a two-level model this so-called "diabatic" transition probability is conveniently estimated from the Landau–Zener formula; for a simple derivation see [Vutha (2010)].*

Remark 6.2 *Using Eq. (6.33) the condition Eq. (6.27) can alternatively be written as*

$$\hbar \left| \frac{< e^{(k)} | \dot{e}^{(n)} >}{g_{kn}(t)} \right| \ll 1 \,. \tag{6.36}$$

The result can be generalized: In the case of degeneracies the adiabatic condition applies to the respective finite-dimensional eigenspaces rather than to single eigenstates.

The general applicability of Eq. (6.27) has been challenged [Amin (2009)]. However, reported inconsistencies concerning its use can be avoided if the evolution time Δt fulfills a more sophisticated inequality than Eq. (6.35).

Remark 6.3 Adiabatic process in thermodynamics *A thermodynamic process is said to be adiabatic, if there is no gain or loss of heat. According to the second law, Eq. (4.110), this implies constant entropy under reversible condition.*

The quantum dynamics of closed systems is reversible and keeps the (von Neumann) entropy, whether or not the control dynamics is adiabatic in the above (microscopic) sense.

Example 6.1 Driven spin *For the Hamiltonian*

$$\hat{H}(t) = \frac{1}{2}(H_0 \hat{1} + H_3(t)\hat{\sigma}_3) \tag{6.37}$$

the eigenvalues depend on time, but not the eigenfunctions. As a consequence adiabaticity always applies, independent of the function $H_3(t)$. The time dependence is typically implemented by means of a varying external magnetic field in three directions, $B_3(t)$, that is, $H_3 = \hbar\Gamma B_3(t)$.

This example is an instant of partial quantum control, Definition 4.7, where the uncontrolled (first part) and controlled (second part) always commute.

Example 6.2 Parametric quantum oscillator *The parametric harmonic oscillator is defined as (cf. Example 5.10)*

$$\hat{H}(t) = \frac{1}{2m_0}\hat{p}^2 + \frac{1}{2}m_0\omega^2(1 + f(t))\hat{x}^2 \,. \tag{6.38}$$

We consider the finite process $0 \leq t \leq \Delta t$ *with*

$$\omega = \begin{cases} \omega_0 & for \quad t = 0 \\ \omega_1 & for \quad t = \Delta t \,. \end{cases} \tag{6.39}$$

If the oscillator is initially in the thermal state $\beta = 1/(k_B T)$ *the energy expectation value at* $t = \Delta t$ *is given by [Abah (2012); Deffner (2010)]*

$$< H(\Delta t) >= \frac{\hbar\omega_1}{2} Q^* \coth\left(\beta\hbar\omega_0/2\right) \tag{6.40}$$

with the measure of non-adiabaticity $Q^* \geq 1$. *This measure has first been introduced by [Husumi (1953)]. Adiabatic evolution requires* $Q^* = 1$, *in which case the transition matrix* $K_{nm}(\Delta t)$ *reduces to the Kronecker delta:*

$$K_{nm}(\Delta t) = |\int dx \int dx_0 \phi^*(\Delta t, x) U(x, x_0, \Delta t) \phi_n(0, x_0)|^2 \rightarrow \delta_{nm} \,. \tag{6.41}$$

In the above equation U is the propagator for the parametric oscillator in real space. The zero-temperature limit of Eq. (6.40) gives the adiabatic constant already found by Einstein:

$$< H(\Delta t) > /\hbar\omega_1 = const. \tag{6.42}$$

Example 6.3 Spin in a rotating external field *The situation would, of course, change once we decided to rotate the magnetic field, keeping its modulus constant, for example. In this case we had to include the vector character of field and spin. If the two vectors are not aligned—as it must happen during finite-speed rotation of* $\vec{B}$*—a precession would be induced, which perturbs the simple following-up of the spin state. This effect becomes increasingly dominant as the rotation speed is increased.*

Definition 6.3 Non-adiabaticity measures *The parameter* $Q^* > 1$ *indicates non-adiabatic behavior. Alternatively, one may use the so-called mixing (or energy-) entropy, cf. Definition 3.84, now defined in the instantaneous energy basis. Contrary to the von Neumann entropy this entropy is not conserved; its increase also points to non-adiabaticity [Rezek (2010)]. This measure can rather easily be calculated.*

6.2.2 *Adiabaticity in Open Systems*

Closed systems, even if non-autonomous, evolve unitarily. Open systems can approximately be described by master equations of the Lindblad type, cf. Eq. (3.368). Here we generalize to an explicitly time-dependent system Hamiltonian and/or to a time-dependent environment (bath).[a] For a detailed discussion focused on periodic time dependence see [Alicki (2006)]. One notes that the validity of the Lindblad equation involves an adiabaticity criterion.

$$\frac{d}{dt}\hat{\rho} = \hat{\mathcal{L}}_{coh}(t) + \hat{\mathcal{L}}_{incoh}(t)\,, \tag{6.43}$$

$$\hat{\mathcal{L}}_{incoh}(t) = \sum_q W_q(t)\hat{G}_q(t)\hat{\rho}\hat{G}_q^\dagger(t) - \frac{1}{2}\sum_q W_q(t)(\hat{\rho}\hat{G}_q^\dagger(t)\hat{G}_q(t) + \hat{G}_q^\dagger(t)\hat{G}_q(t)\hat{\rho}). \tag{6.44}$$

The Lindblad equation can be rewritten in a form resembling the Schrödinger equation by introducing the "superoperator"[b] $\mathbb{L}$ with

$$\frac{d}{dt}\hat{\rho} = \mathbb{L}\hat{\rho}\,. \tag{6.45}$$

However, the analogy of (instantaneous) Hamiltonian eigenstates is lost, as this superoperator is non-Hermitian and cannot in general be diagonalized. It has been suggested to consider instead the so-called adiabatic Jordan blocks [Sarandy (2004)]. In this rather involved language a sufficient condition for open quantum system adiabatic dynamics can be formulated. Reliable applications have not yet appeared so far. In the adiabatic limit approximate results in terms of effective time-dependent environment operators $\sqrt{W_q}\hat{G}_q(t)$ have been considered in [Cai (2010)].

A much simpler approach can be based on the moving attractor model.

Definition 6.4 Moving attractor *In relaxation time approximation the Lindblad equation reads, cf. [Birjukov (2008)]:*

$$\frac{d}{dt}\hat{\rho} = -\frac{i}{\hbar}[\hat{H}(t), \hat{\rho}] - \frac{1}{\tau_R}(\hat{\rho} - \hat{\rho}_{attr}(t))\,. \tag{6.46}$$

---

[a] For thermodynamic cycles we typically have to alternate between time-dependent Hamiltonians and time-dependent baths.

[b] The action of the superoperator is to transform one operator into a new one—while a standard (quantum) operator transforms one state into a new state.

Here we have allowed for a time-dependent attractor state, $\hat{\rho}_{attr}(t)$. Typically, the attractor state will be controlled by few time-dependent parameters. We expect that in the adiabatic limit, that is, when the motion of the attractor state in its parameter space is slow enough, the actual state $\hat{\rho}$ will follow suit and remain close to the instantaneous attractor.

This model will be taken up in Section 7.3.

6.2.3 *Adiabatic Quantum Computing*

Many computationally hard problems can be recast into optimization problems. The quantum adiabatic evolution provides an attractive approach to optimization problems. According to the quantum adiabatic theorem a quantum system remains in its instantaneous eigenstate, if the system Hamiltonian varies slowly enough and if there is a gap between this eigenvalue and the set of other eigenvalues. The adiabatic scheme discussed here involves the ground state only [Farhi (2001)].

Remark 6.4 *Adiabatic quantum computation is equivalent to the standard circuit model [Aharonov (2007)]; contrary to the latter it lacks schemes for fault-tolerant operation.*

Typically, the system intended to implement the adiabatic quantum computing will interact with its environment. The use of the ground state already indicates that the temperature of the open system must be small compared to the minimum gap. Indeed, it turns out that this is a rather general requirement for the robustness of adiabatic quantum computation with respect to decoherence [Childs (2002)]. Note that the ground state is a special thermal state; it avoids serious problems with non-equilibrium states needed for other quantum computer schemes.

Definition 6.5 Quantum adiabatic evolution *Suppose a quantum system starts in an initial Hamiltonian $\hat{H}_0$, the ground state of which, $|\psi_g(0)>$, is known. Let us consider the time-dependent interpolation*

$$\hat{H}(t) = (1 - s(t))\hat{H}_0 + s(t)\hat{H}_P\,. \qquad (6.47)$$

The function s(t) slowly varies from 0 (at $t = 0$) to 1 (at $t = t_f$). Then the original ground state goes over into the ground state of $\hat{H}_P$, the so-

called problem Hamiltonian. This ground state is assumed to encode the answer to the optimization problem.

Factorization has received much attention after P. Shor published his now famous quantum algorithm [Nielsen (2000)]. Adiabatic quantum computing presents an interesting alternative; it requires less working qubits.

Example 6.4 Factorization as optimization *The factorization problem can be expressed by* $N = p \times q$*, where N is the known product, and p and q are the prime factors to be found. Define the cost function,*

$$f(x, y) = (N - xy)^2 \geq 0\,. \tag{6.48}$$

Then the minimal value of f(x,y) provides the solution $f(p, q) = 0$*.*

Proposition 6.6 Problem Hamiltonian for factorization *Based on the cost function the corresponding problem Hamiltonian can be written as*

$$\hat{H}_P = (N\hat{1} - \hat{x}\hat{y})^2\,. \tag{6.49}$$

The eigenvalues of both the operators, $\hat{x}, \hat{y}$*, should be discrete and given by natural numbers. A convenient representation is binary,*

$$x = \sum_{j=0}^{n_x-1} 2^j n_j \quad n_j = 0, 1\,. \tag{6.50}$$

The corresponding equation holds for y; n_x *and* n_y *are the respective bit widths. Space for the two binary numbers can be provided via* $n = n_x + n_y$ *spins: Indeed, the eigenvalues of the operator for spin j,*

$$\hat{b}(j) \equiv \frac{\hat{1}(j) - \hat{\sigma}_3(j)}{2}\,, \tag{6.51}$$

are = 0, 1. The appropriate sum operators can thus be formulated as

$$\hat{x} = \sum_{i=0}^{n_x-1} 2^i \hat{b}(i)\,, \tag{6.52}$$

$$\hat{y} = \sum_{k=0}^{n_y-1} 2^k \hat{b}(n_x + k)\,. \tag{6.53}$$

Example 6.5 Factorization of N = 21 *A direct application of the above formalism has been realized experimentally by [Peng (2008)].*

Restricting oneself to N being odd, also the factors x and y must be odd, their last bit thus equal 1. This last bit can be omitted from the computation, that is, the number of qubits needed is reduced by 2. Choosing $x < y$ *and N = 21 it thus suffices to include* $n_x = 1$ *and* $n_y = 2$ *qubits, that is, a total of three spins. Such a three-qubit nuclear magnetic resonance (NMR) quantum processor has successfully been investigated.*

Example 6.6 Factorization of N = 143 *Xu et al. [Xu (2012)] used a somewhat more efficient scheme to factorize 143 on merely four qubits of a liquid crystal NMR quantum processor with dipole–dipole couplings.*

Example 6.7 Determination of Ramsey numbers *Some special so-called two-color Ramsey numbers have been computed based on up to 128 qubits [Bian (2012)]. This is so far the largest array of qubits successfully used for computation; the solution is based on quantum adiabatic evolution.*

Remark 6.5 Special purpose machine *It should be clear from the above examples that the factorization of a given number N would require a properly adapted problem-Hamiltonian (specifically selected Hamiltonian parameters). Such problem-Hamiltonians are not freely programmable. It is not clear to what extent such a hardware adaptation could be carried out in an efficient and reliable way for increasing problem size N. Furthermore, the efficiency (speed) of the adiabatic quantum computation goes down with the eigenvalue gap, which decreases as N increases. Both these limitations pose serious practical problems; they are variants of the unavoidable scaling problems to be faced by any serious quantum computer implementation.*

6.3 Beyond Adiabaticity: Jarzynski Relation

The Jarzynski relation has originally been discussed in an entirely classical context. This fluctuation theorem has found numerous applications in the nanodomain [Ritort (2003)]. For a recent review see [Seifert (2012)]. Here we restrict ourselves to the quantum version [Mukamel (2003)], which is inherently based on aspects of observational quantum thermodynamics, Section 5.5, related to a (rapid) change of the underlying Hamiltonian. Of prime interest will be measurement-induced fluctuations of work, an extensive thermodynamic variable.

6.3.1 *Jarzynski Relation for Unitary Motion*

We consider an adiabatic process (in the sense of thermodynamics, cf. Remark 6.3) imposed on a system originally in a canonical equilibrium state. As it will turn out, the final state does not need to be an equilibrium state. Work as related to a two-time measurement of energy has been introduced in Definition 5.22,

$$W_{mn}(t, 0) \equiv E_m(t) - E_n(0)\,. \tag{6.54}$$

Instead of considering the average of the work directly, we focus here on

$$X_{mn}(t, 0) \equiv \exp\left(-\beta W_{mn}(t, 0)\right), \tag{6.55}$$

and take for the initial probability distribution the canonical result, Eq. (4.36),

$$P_n(0) = \frac{1}{Z_0} \exp\left(-\beta E_n(0)\right). \tag{6.56}$$

The corresponding average then is

$$\begin{aligned} << X(t, 0) >> &= \sum_{mn} X_{mn}(t, 0) K_{mn}(t) P_n(0) \\ &= \frac{1}{Z_0} \sum_{mn} K_{mn}(t) \exp\left(-\beta E_m(t)\right). \end{aligned} \tag{6.57}$$

Applying the sum rule, Eq. (6.16), and the partition sum,

$$Z(t) = \sum_{m} \exp\left(-\beta E_m(t)\right) = \exp\left(-\beta F(t)\right), \tag{6.58}$$

we end up with the quantum Jarzynski relation [Mukamel (2003)]

$$<< \exp\left(-\beta W(t, 0)\right) >>= \exp\left(-\beta \Delta F(t)\right). \tag{6.59}$$

Here we have introduced the change in free energy, $\Delta F(t) = F(t) - F(0)$, cf. Definition 4.29. This thermodynamic property can thus be expressed as

$$\boxed{\Delta F(t) = \Delta F^{JR} \equiv -\frac{1}{\beta} \ln << \exp\left(-\beta W(t, 0)\right) >>.} \tag{6.60}$$

ΔF^{JR} is the Jarzynski estimator for the free energy change; in the present case the estimator is identical with the true $\Delta F(t)$. This identity holds, if the initial state is a canonical equilibrium state, if the external control generates a purely unitary evolution, and if the resulting process is characterized by two correlated energy measurements. Contrary to the classical Jarzynski relation there are no paths quantum mechanically!

Remarkably, the RHS of the Jarzynski relation will, in general, contain highly non-adiabatic and non-equilibrium features; nevertheless, this special average allows to estimate the change of a thermodynamic equilibrium property, $\Delta F(t)$.

Furthermore, the Jarzynski relation contains the second law in the following sense:

Proposition 6.7 Jarzynski relation and Second law *We first note that the above relation can be restated as*

$$<< \exp(-\beta W(t,0) + \beta \Delta F(t)) >>= 1\,. \tag{6.61}$$

Using Jensen's inequality [Ritort (2003)], $\overline{\exp y} \geq \exp \overline{y}$, *where the overline indicates any kind of ensemble averaging, one finds*

$$<< W(t,0) >> -\Delta F(t) \geq 0\,. \tag{6.62}$$

Identify $W \;=<<\; W(t,0) \;>>$. *Then the dissipated work (under isothermal condition) must be non-negative,*

$$W_{diss} = W - \Delta F \geq 0\,, \tag{6.63}$$

as required by the second law, cf. Eq. (4.140).

The quantum Jarzynski relation has to presuppose a closed system (unitary evolution). This is rather restrictive; numerous attempts have been made to relax this condition.

Remark 6.6 Approximate Jarzynski relation for subsystems *Consider a bipartite system composed of the system in a strict sense, A, and a "classical" (large) bath. The system A is subject to an external time-dependent control. The total system is closed so that the Jarzynski relation applies to A + B. By definition the bath B can only exchange heat,* $\Delta Q_{mn}(B) = -\Delta Q_{mn}(A)$. *Assuming weak coupling (no energy* *contributions due to interaction),*

$$W_{mn}(t) = \Delta E_{mn}(t) = \Delta W_{mn}(A) + \Delta Q_{mn}(A) \\ +\Delta Q_{mn}(B) = \Delta W_{mn}(A), \quad (6.64)$$

$$\Delta F(AB) = \Delta F(A) + \Delta F(B) \approx \Delta F(A), \quad (6.65)$$

the Jarzynski relation for the total system boils down to a Jarzynski relation for subsystem A only [Campisi (2009)]. (In the last equation it has been assumed that the free energy of the large bath B would not change. This is not quite convincing: While the relative change of B may, indeed, be small, this change could well be of comparable order with respect to $\Delta F(A)$.)

6.3.2 *Jarzynski Relation for Non-Unitary Evolution*

For non-unitary evolution the sum rule Eq. (6.16) does no longer hold. For such cases we define

$$\sum_i K_{fi}(t) \equiv s(f), \quad (6.66)$$

that is, the sum is allowed to depend on the final state f. Then the generalized Jarzynski relation reads

$$<< \exp(-\beta W_{fi}) >> = \frac{Z(t)}{Z_0}\bar{s}, \quad (6.67)$$

$$\bar{s} \equiv \frac{1}{Z(t)} \sum_f \exp(-\beta E_f(t)) s(f). \quad (6.68)$$

$\bar{s}$ appears as a thermal average over the s(f). The ratio of the partition sums is

$$\frac{Z(t)}{Z_0} = \exp(-\beta \Delta F), \quad (6.69)$$

as before. But now there is a systematic difference between the Jarzynski estimator and the true $\Delta F(t)$:

$$\Delta F^{JR}(t) - \Delta F(t) = -\frac{1}{\beta} \ln \bar{s}, \quad (6.70)$$

$$\Delta_{rel}^{JR} \equiv |\frac{\ln \bar{s}}{\beta \Delta F}|. \quad (6.71)$$

Δ_{rel}^{JR} is the relative deviation.

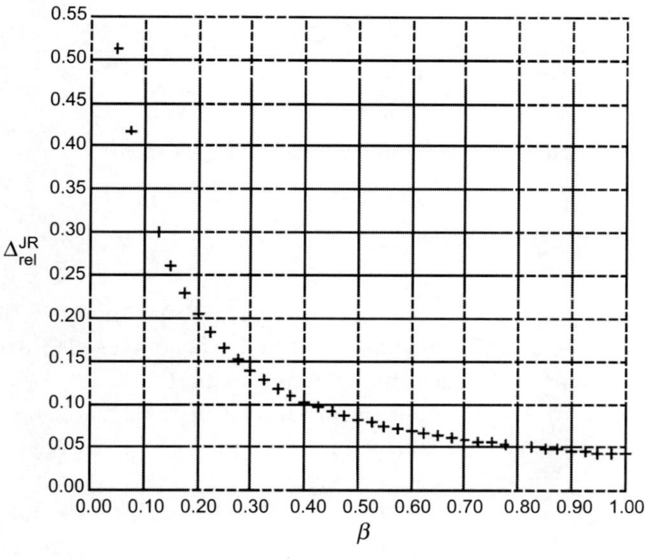

Figure 6.1 Boundary switching process: Relative error of the Jarzynski estimator as a function of β.

There are various reasons for non-unitary evolution. A standard case is the effective dynamics reflecting the behavior of open systems (see, e.g., the Lindblad master equation, Eq. (3.368)). Less well known is the fact that even the mechanical control itself might give rise to non-unitarity and thus violate the standard Jarzynski relation:

Example 6.8 Boundary switching process *We consider a particle in a box under instantaneous expansion of the box length, $L_0 \to L_1$ with $r = L_1/L_0 \geq 1$. The eigenfunctions of the potential well with different widths are not defined on the same Hilbert space; the functions of the broader well cannot be expanded into those of the smaller one. The resulting change is thus non-unitary, and we expect a deviation from the Jarzynski relation [Teifel (2010)].*

According to Eq. (6.18) the transition matrix under instantaneous change, $t \to 0^+$, reads

$$K_{fi}(0^+) = | < e^{(f)}(0^+)|e^{(i)}(0) > |^2 , \tag{6.72}$$

and the resulting sum rule, indeed, depends on f:

$$s(f) = \sum_i K_{fi}(0^+) = \frac{1}{r} - \frac{\sin(2f\pi/r)}{2f\pi}. \qquad (6.73)$$

Numerical results for $\Delta^{J\,R}_{rel}$ are shown for r = 1.31 in Fig. 6.1.

6.4 Feedback Control

The paradigm example for a self-regulating machinery is the so-called "Watt governor" (flyball governor). This device stabilizes the speed of a classical steam engine under fluctuating loads.

A more recent example (laser control of molecules) has been discussed in Section 2.2.7: in this case the respective adaption is enhanced by means of a genetic algorithm.

The notorious Maxwell's demon also belongs to the feedback category. The demon is supposed to selectively manipulate the final states of gas molecules via measurement-based actions ("closed-loop control"): A kind of rectification of thermal fluctuations.

In any case the general aim is optimization.

Definition 6.6 Open-loop versus closed-loop control *Control theory attempts to influence the behavior of dynamical systems. The control is called "open loop," if no measurement of the system output is used to alter the control; this is typical for the control underlying thermodynamic processes. In a "closed-loop" situation a sensor monitors the system and uses the data to modify the controller by means of a transfer function (= feedback loop). In the quantum domain such a feedback loop requires quantum measurements; their inherent stochastic nature means noise.*

Example 6.9 Feedback cooling *Consider a tripartite system ABC, where A is the system under control (linear harmonic oscillator)*

$$\hat{H}(A) = \frac{\hat{p}^2(A)}{2m} + \frac{1}{2}m\omega_0^2\hat{x}^2(A), \qquad (6.74)$$

and B is an environment operating as an external force, cf. Example 5.8,

$$\hat{H}_{eff}(A) = \hat{H}(A) - \hat{x}(A)F(B). \qquad (6.75)$$

Subsystem C interacts with A and will be measured to get (approximately) instantaneous information about A in the form of $< \hat{x}(A) >$ *and* $< \hat{p}(A) >$. *This information is fed back into B [Hopkins (2003)]:*

$$F(B) = -\gamma(m\omega_0 < \hat{x}(A) > + < \hat{p}(A) >)\,. \qquad (6.76)$$

This $F(B)$ *turns out to be an optimal choice for minimizing the average energy of A and thus to cool the system. Based on this feedback force* $\hat{H}_{eff}(A)$ *has become a non-linear Hamiltonian. The effective dynamics for A is governed by*

$$d\hat{\rho} = -\frac{i}{\hbar}[\hat{H}_{eff}(A), \hat{\rho}]dt + (\mathrm{noiseterms})\,. \qquad (6.77)$$

The noise terms include shot noise, back-action noise, and "technical noise"; it unavoidably limits the performance of this feedback device. (There is no universal result available for the efficiency.) Continuous feedback loop operations that permanently stabilize quantum systems around target states are being realized only recently [Sayrin (2011)].)

Note the difference to refrigerator cooling: The latter is controlled by thermodynamic processes, with no direct Hamilton dynamics whatsoever; simple thermodynamic constraints apply in this case.

Example 6.10 Jarzynski scenario with feedback *A classical Jarzynski scenario has also been studied with respect to feedback [Toyabe (2010)]. Subject to thermal fluctuations a particle jumps between steps of a spiral staircase. The step size is of order* $k_B T$. *On average the particle falls down unless pushed up from the outside. Measuring the particle position at regular intervals one can place a block behind the particle, if it was found to have moved up. Feedback renders the effective dynamics non-unitary.*

The corresponding value $\overline{s}$ *in the generalized Jarzynski relation, Eq. (6.67), is related here to the efficacy by which that very information has been exploited.*

Remark 6.7 Generalized thermodynamics? The case of hidden costs I *Again we are confronted with the idea that information might directly serve as a physical input, thus calling for a "generalized" thermodynamics, cf. Remark 4.16 on Brillouin's negentropy. Indeed, for fixed temperature T the second law, Eq. (6.63), has been postulated to read now [Sagawa (2008)],*

$$\Delta F - \Delta W \le k_B T S_I\,, \qquad (6.78)$$

where $S_I \geq 0$ is the mutual information, cf. Definition 2.19, obtained by the measurement device about the subsystem to be controlled. These measurements tend to introduce non-equilibrium features. Apparently, a violation of the second law,

$$\Delta F - \Delta W \leq 0\,, \tag{6.79}$$

might result, see Eq. (4.140). However, Eq. (6.78) is an effective description, based on a certain partition of the total scenario, cf. Example 4.11.

It is reminiscent of Maxwell's demon:. In a "complete" picture, as we know, the standard second law cannot be violated by the demon—if its physical constraints due to cyclic operation are properly taken into account, cf. Section 4.2.6 and Proposition 4.6. There is no reason to expect that this finding might fail in other realizations like those of Example 6.10.

The general proviso applies: ". . . it is not information that can be converted to work but heat that can be converted into work by use of information." [Levitin (2011)]. And the gain of information is a costly physical process, cf. also Section 7.4.3.

Example 6.11 Coherent quantum feedback *Contrary to the cases mentioned so far the controller does not need to be classical. A simple example is as follows [Lloyd (2000)]: Consider the task of taking a spin (A) in the unknown pure state*

$$|\psi(A) >= \alpha|0> + \beta|1> \tag{6.80}$$

into the state $|1>$. This could be done by measurement and classical feedback: After finding state $|0>$ with probability $P_0 = |\alpha|^2$ apply an electromagnetic pulse to flip this state. Otherwise, after measuring $|1>$, do nothing. Alternatively add a "controller bit" (B) in state $|1>$. The SWAP gate, Definition 3.22, interchanges the states between A and B so that the system spin is then in the state $|1>$, as desired. (This process is unitary and thus reversible.)

Note, however, that for each subsequent control step (possibly required due to ongoing perturbations of the system spin) we would need a fresh controller spin!

Remark 6.8 Modes of quantum control *As indicated by the above example: "Measurements can always be replaced considering* *instead the joint evolution of two systems, a primary system and an auxiliary" [Jacobs (2013)]. There are fundamental differences, though. In measurement-based feedback the information is amplified and transformed to classical numbers, which can be processed by a classical computer; this processing can be extremely sophisticated. The auxiliary, on the other hand, tends to be a simple system the possible effects of which on the primary system are necessarily very limited.*

Example 6.12 Quantum error correction *The basic idea of quantum error correction is to consider a special subspace of states only. If the initial state belongs to this subspace and if not too many errors have occurred, there exists a quantum operation that is able to restore the initial state. This is a closed-loop feedback scenario.*

6.5 Quantum Gates and Algorithmic Cooling

In an informal sense, an algorithm is a set of rules that defines a sequence of operations. Typically, such instructions can be implemented by a computer. Just like information, algorithms can be expressed in many different forms. A pertinent example is given by the Turing machine, cf. Definition 2.21.

Thermodynamic machines are typically characterized by their operation in control space, cf. Section 7.2; this defines a procedure on the respective level of description. Of prime interest are energy currents and their interconnects. Nevertheless, in the quasi-stationary limit (infinite baths, infinite work reservoirs) the states of the various subsystems do not change.

Efficient cooling mechanisms are of practical importance, even down to the nano-level. Quantum refrigerators should thus be among the first real-world applications of quantum thermodynamics.

For finite systems ("nano-thermodynamics") cooling, for example, implies change of state of the target subsystem: reduction of entropy. In a closed system this can only work at the expense of an increased entropy somewhere else (the total entropy is conserved.) The process is finite. It is equivalent to a state transformation. Its description defines a procedure (algorithm), but now on the micro-level. Algorithmic cooling in this sense becomes a subfield of gate-based quantum computation.

Quantum gates allow a simplified description of the unitary evolution of non-autonomous systems.

Algorithmic cooling procedures have first been introduced by [Boykin (2002)]. Below we give a basic description in terms of a non-interacting three-spin model [Rempp (2007)].

$$\hat{H} = \sum_{\mu=A;B;C} \Delta E \hat{\sigma}_3(\mu) . \tag{6.81}$$

Step 1: Each spin is prepared in a canonical state with temperature $k_B T(\mu) = 1/\beta_0$:

$$\hat{\rho}_0 = \hat{\rho}(A) \otimes \hat{\rho}(B) \otimes \hat{\rho}(C) , \tag{6.82}$$

$$\hat{\rho}(\mu) = \frac{1}{2}\hat{1}(\mu) + \frac{\epsilon_0}{2}\hat{\sigma}_3(\mu) , \tag{6.83}$$

$$\epsilon_0 = \frac{1 - \exp(-\beta_0 \Delta E)}{1 + \exp(-\beta_0 \Delta E)} . \tag{6.84}$$

Step 2: We apply the unitary transformation

$$\hat{U} = \hat{U}_{CSWAP}(ABC) \otimes \hat{U}_{CNOT}(AB) . \tag{6.85}$$

The CNOT gate has been defined in Example 3.23. The CSWAP (controlled SWAP) is a three qbit gate. If the central qbit (B) is in state $|e^{(1)}(B) >$ a SWAP operation according to Example 3.22 is carried out between (A) and (C); otherwise no action. The new state is

$$\hat{\rho}_1 = = \hat{U} \hat{\rho}_0 \hat{U}^{-1} , \tag{6.86}$$

$$\hat{\rho}_1(A) = \mathrm{Tr}_{BC}\{\hat{\rho}_1\} . \tag{6.87}$$

Notably, the effective temperature of (A) is reduced:

$$\beta_1(A) = \frac{3}{2}\beta_0 - \frac{\ln\left(3 - \frac{8}{3+\exp(\beta_0 \Delta E)}\right)}{\Delta E} . \tag{6.88}$$

For small β_0 we thus get

$$\frac{\beta_1(A)}{\beta_0} \approx \frac{3}{2} . \tag{6.89}$$

The process can be iterated: Perform this (first order) procedure with a total of three separate three-spin systems and combine from each set the cold spin into the second-order set. Then $\beta_2(A) \approx (3/2)^2\beta_0$. One can thus bring down the temperature as far as one wishes; however, in agreement with the third law, the procedure is inefficient. For N repetitions one would need a total of 3^N qubits, that is, the resources increase exponentially!

Chapter 7

Control Cycles

Jonas Burgert: In the Cauldron, 2010, based on photography by Andreas Wunschirs [Reprinted with permission from Schreiber and Wipplinger (2011)].

Quantum Thermodynamic Processes: Energy and Information Flow at the Nanoscale
Günter Mahler

ISBN 978-981-4463-73-7 (Hardcover), 978-981-4463-74-4 (eBook)
www.panstanford.com

7.1 Nodes and Subsystems

Definition 7.1 Thermodynamic machine *Functionality (cf. Section 1.4) requires a basic structural design, which constraints the allowed paths of operation. In most cases the design calls for a multipartite system.*

For thermodynamic machines[a] *the architecture typically consists of a "central node," classically the "working medium." This node provides various "contacts." Subject to external or internal control, each contact may be open or closed; when open, it connects to a specific subsystem, a heat or a work reservoir. By definition energy exchanged with a work reservoir is work, energy exchanged with a heat reservoir is heat. Reservoirs are often taken to be infinite, so that—despite of receiving or supplying energy—their state does not change. The machine operation is stationary—which ceases to hold for finite subsystems. In the quantum domain the subsystems are no longer guaranteed to function as simple heat or work reservoirs, and the node may be replaced by a direct many-body interaction, cf. Section 8.2.2.*

The net effect is the conversion of energy between the various contacts. Such conversion processes play an important role not only in engineering but also in "natural" or even biological systems. While the nature of subsystems and nodes may completely change, the underlying thermodynamic features tend to remain almost the same ("universality").

For continuous operation some source of non-equilibrium suffices to drive the system. Cyclic operation requires, in addition, time-dependent external control in some parameter space. This space must be at least two-dimensional. The central node has to be in the same state after completion of each cycle: $\Delta U(node) = 0$, $\Delta S(node) = 0$ *(cyclic operation). This does not hold for the machine as a whole.*

Definition 7.2 Sign convention *The various energy contributions per subsystem have to be added over one complete cycle. The following*

[a]Non-thermodynamic quantum motors have also been suggested [Bustos-Martin (2013)]. They would, for example, convert electrical into mechanical energy up to the ideal efficiency $\eta = 1$.

sign-convention applies: Energy packages entering the node are counted positive, those leaving the node are counted negative.

Definition 7.3 Efficiency *The main purpose of thermodynamic machines is to convert heat into work and vice versa. In the ideal case the conservation of total energy applies: The sum of all input currents must be equal to the sum of all output currents. A measure for performance is the ratio between desired output and the input (or cost) required. This ratio need not be 1, because only part of the various energy flows enter. An example is the efficiency η for the heat engine,*

$$\eta = \frac{\textit{negative work output}}{\textit{positive heat input}} \geq 0\,. \tag{7.1}$$

For multistroke machines (typically there are four strokes) the contributions per stroke have to be considered separately. They may differ by sign and origin (reservoirs). It is assumed here that any negative heat term (i.e., leaving the central node) is lost, that is, cannot be used again, implying $\eta < 1$. This is intuitive as long as this backflow does not re-enter the same reservoir, from which the positive heat is being drawn.

It will turn out that the definition of efficiency is not always straightforward but has to be adjusted when appropriate.

Remark 7.1 Levels of description *For any such a typical thermodynamic machine there are various levels of description. These refer to the individual parts as follows:*

(1) All parts treated classically ("classical thermodynamics"),
(2) central node treated as a quantum model,
(3) heat baths and central node treated as quantum,
(4) central node and work reservoir treated as quantum (there is no explicit time dependence of the Hamiltonian to control the contacts),
(5) all parts treated fully quantum. Note that, in general, quantum subsystems per se are neither heat nor work reservoirs: The type of energy exchange they support has to be clarified from case to case.

For the standard case (1) the (classical) node would be the working medium in a strict sense, for example, a gas contained in a cylinder with moving walls. This node is under time-dependent external control and connects with classical heat and work reservoirs. In the following we will mainly be concerned with case (2). The quantum node is typically a few-level system described by a Hamiltonian system, subject to mechanical control (i.e., change of model parameters). The respective micro-state is an equilibrium state, subject to statistical control. Both these control types give rise to a parametrized quantum thermodynamic description.

Note that for this description to become effective, appropriate (changeable) environments are needed, see Chapter 5. Before we can meaningfully speak about the gradual change of equilibrium states, their mere existence must be guaranteed. In the following these pre-conditions are not modeled in any detail.

7.2 Control Space: Parametrization of Quantum Thermodynamics

Quantum thermodynamics is constrained here to equilibrium states, which are stationary for given model parameters. Dynamical features thus require the gradual change of model parameters, which serve as control parameters. There are two classes: mechanical and statistical control.

The mechanical control derives from the access to Hamiltonian parameters and the statistical control from the access to the momentary equilibrium state. Such control features are non-trivial, but we know from experience that specific realizations are feasible.

Definition 7.4 Mechanical control *We take up the parametrization sketched already in Section 4.3.3, see also [Birjukov (2008)]. For this purpose we start from the Hamiltonian model with the eigenvalue equation*

$$\hat{H}\,|e^{(i)}>= E_i|e^{(i)}> \; . \tag{7.2}$$

The details of this model are irrelevant; in fact, it could be a single molecule, a particle in a box, or an interacting many-particle system. In any case, it will involve a set of model parameters. For the following it is assumed that at least some of these parameters can be modified *from the outside such that the resulting spectrum changes. This dependence is taken to be a function of a single control parameter γ in the form:*

$$E_i(\gamma) = g(\gamma)\epsilon_i \quad i = 1, 2, \ldots d\,. \tag{7.3}$$

d is the Hilbert space dimension, which may be infinite. This spectral deformation (a simple re-scaling of the spectrum, see Eq. (4.7)) defines the "mechanical" control; $g(\gamma)$ is a non-negative monotonous function of γ, the ϵ_i are characteristic constants in units of energy.[a]

Definition 7.5 Statistical control *The microscopic equilibrium state is stationary and thus diagonal in the energy representation. It is a function of the spectrum, E_i. We assume an additional parameter dependence of this function via α, the "statistical" control. It is essential that this parameter can be modified from the outside.*

The equilibrium state is usually derived by some standard methods. For the present purpose of thermodynamic machines the concrete form of this function is irrelevant, though. For example, there could be conditions under which the nearly uniform distribution

$$\rho_{ii}^{uniform}(\alpha) = \begin{cases} 1-\alpha & if \quad i = 1 \\ \dfrac{\alpha}{1-d} & if \quad i = 1, 2, \ldots d \end{cases} \tag{7.4}$$

would be applicable—though the physical justification is of no concern here.

The canonical attractor state is much more standard and reads in the respective energy representation

$$\rho_{ii}^{can}(\alpha) = < e^{(i)}|\hat{\rho}(\alpha)|e^{(i)} >= \frac{1}{Z_{can}} \exp\left(-\alpha\epsilon_i\right), \tag{7.5}$$

$$Z_{can} = \sum_i \exp\left(-\alpha\epsilon_i\right), \tag{7.6}$$

$$\alpha = g(\gamma)\beta\,. \tag{7.7}$$

In any case, this two-dimensional control space (α, γ) suffices to derive a complete description of any conventional thermodynamic process and any machine cycle. Here we assume a quasi-static evolution: The parameter changes are slow enough so that the system

---

[a]The physical nature of this control has to be left open here; concrete examples are given in Example 4.1.

always remains in the instantaneous thermodynamic equilibrium state.

The description is closed: All intensive as well as extensive thermodynamic variables can be re-expressed in terms of those two control parameters. Beyond this control the environment enters indirectly via certain constraints like constant entropy, constant temperature, etc.

Definition 7.6 Parametrized thermodynamic state functions
The internal energy and the von Neumann entropy, respectively, now read:

$$U(\alpha, \gamma) = g(\gamma) \sum_i \epsilon_i \rho_{ii}(\alpha) \equiv g(\gamma) h(\alpha), \tag{7.8}$$

$$S_{vN}(\alpha) = -\sum_i \ln(\rho_{ii}(\alpha)) \rho_{ii}(\alpha). \tag{7.9}$$

To guarantee uniqueness we assume both functions, $g(\gamma)$, $h(\alpha)$, to be strictly monotonous. These parametrized state functions characterize all thermodynamic properties of the system.

Relevant intensive parameters are the thermodynamic temperature T and the generalized pressure $\tilde{p}$.

Definition 7.7 Temperature in control space

$$T = \left(\frac{\partial U}{\partial S}\right)_\gamma = \frac{1}{k_B} \left(\frac{\partial U}{\partial \alpha}\right)_\gamma \left(\frac{dS_{vN}}{d\alpha}\right)^{-1}. \tag{7.10}$$

The latter expression can be made more explicit:

$$\begin{aligned}\frac{dS_{vN}}{d\alpha} &= -\sum_i \frac{1}{\rho_{ii}} \frac{d\rho_{ii}}{d\alpha} \rho_{ii} - \sum_i \ln \rho_{ii} \frac{d\rho_{ii}}{d\alpha} \\ &= -\sum_i \ln \rho_{ii} \frac{d\rho_{ii}}{d\alpha}. \end{aligned} \tag{7.11}$$

We thus obtain

$$\boxed{\begin{aligned} &T(\alpha, \gamma) = g(\gamma) \Theta(\alpha)^{-1}, \\ &\Theta(\alpha) \equiv -k_B \textstyle\sum_i \ln \rho_{ii} \dfrac{d\rho_{ii}}{d\alpha} \left(\dfrac{dh}{d\alpha}\right)^{-1} = -k_B \dfrac{\sum_i \ln \rho_{ii} \frac{d\rho_{ii}}{d\alpha}}{\sum_i \epsilon_i \frac{d\rho_{ii}}{d\alpha}}. \end{aligned}} \tag{7.12}$$

Example 7.1 *In the canonical case we have with* $\ln \rho_{ii} = -\epsilon_i \alpha - \ln \sum_i \exp(-\epsilon_i \alpha)$

$$\Theta(\alpha) = k_B \alpha\,, \tag{7.13}$$

in agreement with Eq. (7.7). The term $dh/d\alpha$ is negative.

Just for curiosity we also consider Eq. (7.4). In this case

$$\frac{dh}{d\alpha} = \frac{1}{d-1} \sum_{i=2}^{d} \epsilon_i - \epsilon_1 > 0\,, \tag{7.14}$$

$$k_B T = -g(\gamma) \frac{dh}{d\alpha} \frac{1}{\sum_i \ln \rho_{ii} \frac{d\rho_{ii}}{d\alpha}}\,, \tag{7.15}$$

$$\sum_i \ln \rho_{ii} \frac{d\rho_{ii}}{d\alpha} = \ln \frac{\alpha}{(d-1)(1-\alpha)} < 0\,. \tag{7.16}$$

This gives for

$$\Theta(\alpha) = k_B \left(\frac{dh}{d\alpha} \right)^{-1} \ln \frac{(d-1)(1-\alpha)}{\alpha}\,. \tag{7.17}$$

For $T > 0$ we have to require $\Theta > 0$ and thus $\alpha < 1 - d^{-1}$.

We see that for any combination of control parameters α, γ there is a corresponding absolute temperature T, whether the attractor is canonical or not. Note that this would no longer be the case for spectral deformations violating Eq. (4.7); we would then easily get non-equilibrium states—even under quasi-static evolution. In the following we will restrict ourselves to the canonical attractor.

Definition 7.8 Generalized pressure in control space *In standard thermodynamics pressure is defined as the conjugate variable to volume V. According to Eq. (4.84) the generalized pressure, $\tilde{p}$, is introduced as the variable conjugate to γ:*

$$\tilde{p}(\alpha, \gamma) = -\left(\frac{\partial U}{\partial \gamma} \right)_{\alpha} = -\frac{dg(\gamma)}{d\gamma} h(\alpha)\,. \tag{7.18}$$

The physical significance of this $\tilde{p}$ will depend on the physical significance of γ, see Example 4.1. With γ being dimensionless, $\tilde{p}$ has the same dimension as U.

Definition 7.9 First law in control space

$$
\boxed{
\begin{aligned}
dU(\alpha,\gamma) &= đQ + đW\,,\\
đQ(\alpha,\gamma) &= \left(\frac{\partial U}{\partial\alpha}\right)_{\gamma} d\alpha = g(\gamma)\frac{dh}{d\alpha}d\alpha = T(\alpha,\gamma)dS\,,\\
đW(\alpha,\gamma) &= \left(\frac{\partial U}{\partial\gamma}\right)_{\alpha} d\gamma = \frac{dg}{d\gamma}h(\alpha)d\gamma = -\tilde{p}d\gamma\,.
\end{aligned}
}
\tag{7.19}
$$

This is Gibbs fundamental form, cf. Eq. (4.102), which includes the second law. It comes as a generalization of Eq. (4.86).

Based on Eq. (7.12), we find

$$dS = \frac{đQ(\alpha,\gamma)}{T(\alpha,\gamma)} = \Theta(\alpha)\frac{dh}{d\alpha}d\alpha \tag{7.20}$$

For a cyclic process with $\Delta U = 0$ it holds that

$$\Delta W = -\oint T\,dS\,. \tag{7.21}$$

Remark 7.2 Universality and its limits *It is well-known that basic thermodynamic features, in particular those relating to thermodynamic machines, do not depend on details of the thermodynamic system under consideration, for example, its thermal equation of state. This important aspect carries over to the quantum regime: in fact, the two-dimensional parametrization is independent of details of the Hamilton model and independent even of the nature of the attractor state. (The attractor state need not be canonical.)*

However, just as it happens in the classical domain, this universality is lost once dynamical aspects in a strict sense (relating to non-equilibrium and irreversibility) have to be taken into account. In this case the behavior of the thermodynamic system is influenced also by its "eigen dynamics" as described by the respective master equation. Model-dependent features cannot be avoided.

7.3 Quasi-Static Cycles

Definition 7.10 Basic types of processes *In the control plane a quasi-static process is specified by a line and thus requires one additional condition. We follow here [Birjukov (2008)]; similar* *ideas have been brought forward, for example, in [Kieu (2004)] and [Quan (2007)].*

Elementary examples are:

(1) *Adiabatic (isentropic):* $\alpha = const.$
(2) *Isochoric:* $\gamma = const.$
(3) *Isothermal:* $T(\alpha, \gamma) = const.$
(4) *Isobaric:* $\tilde{p}(\alpha, \gamma) = const.$

Definition 7.11 Basic cycles *Any closed line in the two-dimensional control space may count as a possible cycle. Analytical results are obtained for more special cycles, which consist of four "strokes" made up from two different pairs of the above basic processes:*

- *Process (1) + (2) + (1) + (2): Otto cycle*
- *Process (1) + (3) + (1) + (3): Carnot cycle*
- *Process (1) + (4) + (1) + (4): Brayton cycle*
- *Process (2) + (3) + (2) + (3): Stirling cycle*
- *Process (3) + (4) + (3) + (4): Ericsson cycle*

There are also cycles composed of three different basic process types. The (ideal) Diesel cycle, for example, consists of the process sequence (1) + (4) + (1) + (2).

Definition 7.12 Sense of rotation *Given some control plane* (α, γ), *the actual performance of a cycle can be either clockwise (negative) or anti-clockwise (positive). The sense of rotation (together with the functions* $h(\alpha), g(\gamma)$*) controls the sign of the various energy contributions, see, for example, Table 7.1. A change of the sense of rotation is possible as each cycle is reversible.*

Remark 7.3 Pure state cycle *An entirely different quantum cycle has been suggested by [Bender (2000)].*

This cycle is entirely based on pure states, $|\psi>$*; it is a combination of two "adiabatic" and two "isoenergetic steps." The isoenergetic step is defined by* $U = <\psi|\hat{H}|\psi> = const.$ *(This definition could be generalized to mixed states.) As the von Neumann entropy is always zero here, adiabaticity cannot be interpreted in the thermodynamic sense, but in the sense of adiabatic Schrödinger dynamics only, see*

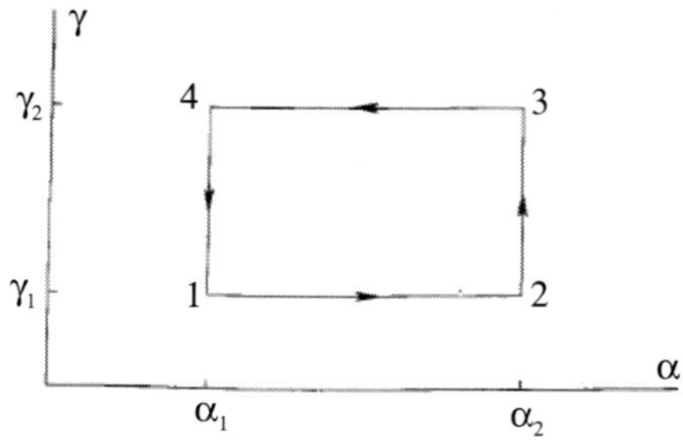

Figure 7.1 Otto cycle in control space.

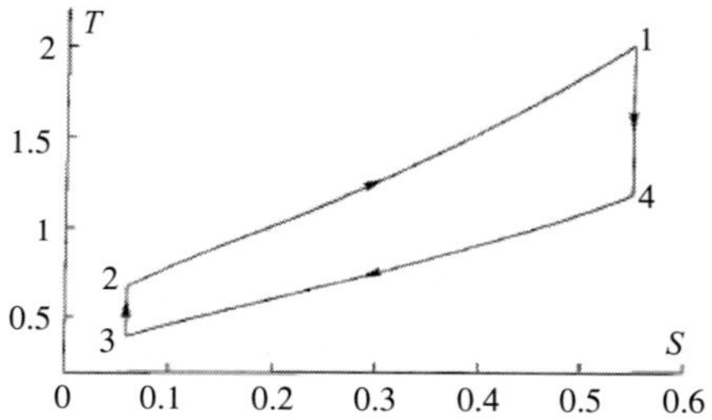

Figure 7.2 Otto cycle in T-S space for $\alpha_1 = 0.5, \alpha_2 = 1.5, \gamma_1 = 1, \gamma_2 = 1.3$. Model: Particle in a box.

Definition 6.2. Heat and work are formally identified via

$$dU = (< d\psi|\hat{U}\,\psi > + < \psi|\hat{H}\,|d\psi >) + < \psi|d\hat{H}\,|\psi > \equiv dQ^* + dW^*\,, \tag{7.22}$$

see [Abe (2011)]. This heat dQ^* has no relation whatsoever to entropy. The potential implementation of such a scenario—let alone its thermodynamic significance—remains unclear.

7.3.1 *Otto Cycle*

The Otto cycle consists of two adiabatic and two isochoric steps; seen from the control plane it is the most basic and direct cycle, see Fig. 7.1. As will be shown below, the resulting efficiency depends on some model details, which, however, can be summarized in the form of Eq. (7.33).

The cycle consists of the following four steps.

Step 1 $\to$ 2 (isochoric):

$$W_{12} = 0\,, \tag{7.23}$$

$$Q_{12} = \int_{\alpha_1}^{\alpha_2} g(\gamma_1)\frac{dh}{d\alpha}d\alpha = g(\gamma_1)(h(\alpha_2) - h(\alpha_1)) = -Q_{21}\,. \tag{7.24}$$

Step 2 $\to$ 3 (adiabatic):

$$W_{23} = \int_{\gamma_1}^{\gamma_2} h(\alpha_2)\frac{dg}{d\gamma}d\gamma = h(\alpha_2)(g(\gamma_2) - g(\gamma_1)) = -W_{32}\,, \tag{7.25}$$

$$Q_{23} = 0\,. \tag{7.26}$$

Step 3 $\to$ 4 (isochoric):

$$W_{34} = 0\,, \tag{7.27}$$

$$Q_{34} = \int_{\alpha_2}^{\alpha_1} g(\gamma_2)\frac{dh}{d\alpha}d\alpha = g(\gamma_2)(h(\alpha_1) - h(\alpha_2)) = -Q_{43}\,. \tag{7.28}$$

Step 4 $\to$ 1 (adiabatic):

$$W_{41} = \int_{\gamma_2}^{\gamma_1} h(\alpha_1)\frac{dg}{d\gamma}d\gamma = h(\alpha_1)(g(\gamma_1) - g(\gamma_2)) = -W_{14}\,, \tag{7.29}$$

$$Q_{41} = 0\,. \tag{7.30}$$

The present sense of rotation, see Fig. 7.1, is counter-clockwise, that is, positive, cf. Definition 7.12. In this case the total work per cycle is

$$\begin{aligned}\Delta W_{tot}^{(+)} &= W_{23} + W_{41} = (g(\gamma_2) - g(\gamma_1))(h(\alpha_2) - h(\alpha_1)) \quad (7.31)\\ &= \sum_i (E_i(\gamma_2) - E_i(\gamma_1))(\rho_{ii}(\alpha_2) - \rho_{ii}(\alpha_1))\\ &= -\Delta W_{tot}^{(-)}\,. \quad (7.32)\end{aligned}$$

For a heat engine the total work per cycle, ΔW_{tot}, must be negative (i.e., leaving the thermodynamic system for external use). This can happen for one specific sense of rotation only. We see from Table 7.1 that there is always just one step with positive heat contribution, that is, serving as a heat source. The respective efficiency is defined according to Eq. (7.1).

The signs of the energy terms are controlled by monotonous functions $h(\alpha)$, $g(\gamma)$, which may either increase or decrease with increasing control parameter. We thus have to distinguish four

Table 7.1 Efficiency η for Otto engine, depending on the behavior of $h(\alpha)$ and $g(\gamma)$

$h(\alpha)$	$g(\gamma)$	Work[a]	Heat 1	Heat 2	η
$\frac{dh}{d\alpha} < 0$	$\frac{dg}{d\gamma} > 0$	$\Delta W_{tot}^{(+)} < 0$	$Q_{34} > 0$	$Q_{12} < 0$	$\frac{-\Delta W_{tot}^{(+)}}{Q_{34}} = 1 - \frac{g(\gamma_1)}{g(\gamma_2)}$
$\frac{dh}{d\alpha} < 0$	$\frac{dg}{d\gamma} < 0$	$\Delta W_{tot}^{(-)} < 0$	$Q_{43} < 0$	$Q_{21} > 0$	$\frac{-\Delta W_{tot}^{(-)}}{Q_{21}} = 1 - \frac{g(\gamma_2)}{g(\gamma_1)}$
$\frac{dh}{d\alpha} > 0$	$\frac{dg}{d\gamma} > 0$	$\Delta W_{tot}^{(-)} < 0$	$Q_{43} > 0$	$Q_{21} < 0$	$\frac{-\Delta W_{tot}^{(-)}}{Q_{43}} = 1 - \frac{g(\gamma_1)}{g(\gamma_2)}$
$\frac{dh}{d\alpha} > 0$	$\frac{dg}{d\gamma} < 0$	$\Delta W_{tot}^{(+)} < 0$	$Q_{34} < 0$	$Q_{12} > 0$	$\frac{-\Delta W_{tot}^{(+)}}{Q_{12}} = 1 - \frac{g(\gamma_2)}{g(\gamma_1)}$

[a]Sense of rotation specified by $(\pm)$.

different cases, see Table 7.1. The results can be summarized as follows [Birjukov (2008)]:

$$\boxed{\eta_{otto} = \frac{-\Delta W_{tot}}{\text{positive heat from warm reservoir}} = 1 - \frac{\min g(\gamma)}{\max g(\gamma)}} \,. \tag{7.33}$$

Proposition 7.1 Entropy balance *The Otto cycle is reversible, that is, the total entropy change per cycle is zero. This is easily seen by analyzing the entropy change along the two isochoric steps. Based on Eq. (7.20) we find,*

$$S_{12} = \int_{\alpha_1}^{\alpha_2} \Theta(\alpha) \frac{dh}{d\alpha} d\alpha = -S_{34} \,, \tag{7.34}$$

and thus $\Delta S_{tot} = S_{12} + S_{34} = 0$.

Example 7.2 *For the canonical case we have* $dh/d\alpha < 0$*, for the spectral deformation with* $g(\gamma) = \gamma$ *(spin in a magnetic field or harmonic oscillator, cf. Example 4.1) it holds that* $dg/d\gamma > 0$*. The respective efficiency is*

$$\eta_{otto} = 1 - \frac{\gamma_1}{\gamma_2} \quad \gamma_2 > \gamma_1 \,. \tag{7.35}$$

For a particle in a D-dimensional box it holds that $g(\gamma) = \gamma^{-2/D}$*. The efficiency is*

$$\eta_{otto} = 1 - \left(\frac{\gamma_1}{\gamma_2}\right)^{2/D} \,. \tag{7.36}$$

Remark 7.4 Maximum/minimum temperatures *The Otto cycle does not explicitly refer to baths; they are not needed as means to keep* *temperatures constant. Nevertheless, temperatures $T(\alpha, \gamma)$ do exist at any point in control space. Of interest are the extremal temperatures T_c (cold) and T_h (hot). According to Eq. (7.12) these can be expressed as*

$$k_B T_c = \frac{\min g(\gamma)}{\max \Theta(\alpha)}, \tag{7.37}$$

$$k_B T_h = \frac{\max g(\gamma)}{\min \Theta(\alpha)}. \tag{7.38}$$

For the canonical case, $\Theta(\alpha) = k_B\alpha$, a control model with $g(\gamma) = \gamma$, and assuming $\alpha_1 < \alpha_2$, $\gamma_1 < \gamma_2$, we thus obtain

$$T_c = \gamma_1/\alpha_2 \tag{7.39}$$

$$T_h = \gamma_2/\alpha_1 \tag{7.40}$$

Remark 7.5 Carnot limit *The above relations allow us to re-express Eq. (7.33) as*

$$\eta_{otto} = 1 - \frac{T_c \max \Theta(\alpha)}{T_h \min \Theta(\alpha)} \leq 1 - \frac{T_c}{T_h} \equiv \eta_{carnot}. \tag{7.41}$$

The Carnot result, Eq. (7.62), functions as an upper bound. This bound could be reached for $\alpha_1 \to \alpha_2$ only, that is, when there is no work output at all.

Under the same control model one finds

$$\frac{Q_{12}}{\alpha_2 T_c} + \frac{Q_{34}}{\alpha_1 T_h} = 0, \tag{7.42}$$

which should be compared with Eq. (7.56). For the Otto cycle the total entropy change $\Delta S = 0$ can be expressed in terms of the two heat contributions over "effective temperatures" only.

Remark 7.6 Classical limit *The classical Otto efficiency depends on the so-called compression ratio V_1/V_2, where $V_2 > V_1$ are the limiting volumes of the cycle:*

$$\eta^{cl}_{otto} = 1 - \left(\frac{V_1}{V_2}\right)^{\kappa - 1}. \tag{7.43}$$

Here, $\kappa \equiv c_p/c_V = (f+2)/f$. Identifying the degrees of freedom per particle, f, with the dimension D of the box (cf. Eq. (4.72)), the classical and the quantum result are the same. For given f the efficiency increases with increasing compression ratio $V_2/V_1 \to \infty$.

Remark 7.7 Heat engine versus heat pump *The Otto heat engine becomes a heat pump, if the sense of rotation of the cycle is changed, that is, from positive to negative or vice versa. In this case all signs of energy contributions are flipped. The efficiency is*

$$\eta_{otto}^{pump} = \frac{-\text{heat to warm reservoir}}{\Delta W_{tot}} = \frac{1}{\eta_{otto}}. \tag{7.44}$$

For the spin in a magnetic field, Example 7.2, this would mean

$$\eta_{otto}^{pump} = \frac{\gamma_2}{\gamma_2 - \gamma_1}. \tag{7.45}$$

Remarkably, in the quantum case the sense of rotation is not the only way to change functionality. This is so, because the Otto process cannot work as a heat engine for all variable combinations $\gamma_1, \gamma_2, T_c, T_h$. Indeed, given three parameters, the fourth is constrained. For example, if

$$\boxed{\frac{\max g(\gamma)}{\min g(\gamma)} \geq \frac{T_h}{T_c}}, \tag{7.46}$$

the cycle operates as a heat pump (or refrigerator) rather than as a heat engine. This condition is most easily understood on the basis of the "virtual qubit," Section 8.1.

7.3.2 *Carnot Cycle*

In classical thermodynamics the Carnot cycle is known to have maximum efficiency for fixed heat reservoirs $T_c < T_h$. The cycle combines two adiabatic and two isothermal steps, see Fig. 7.3.

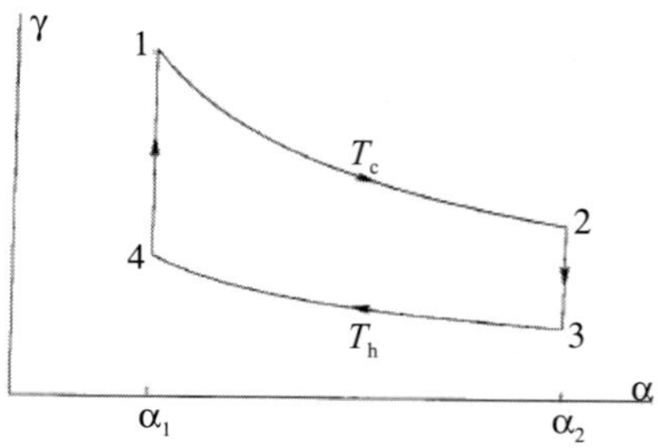

Figure 7.3 Carnot cycle in control space.

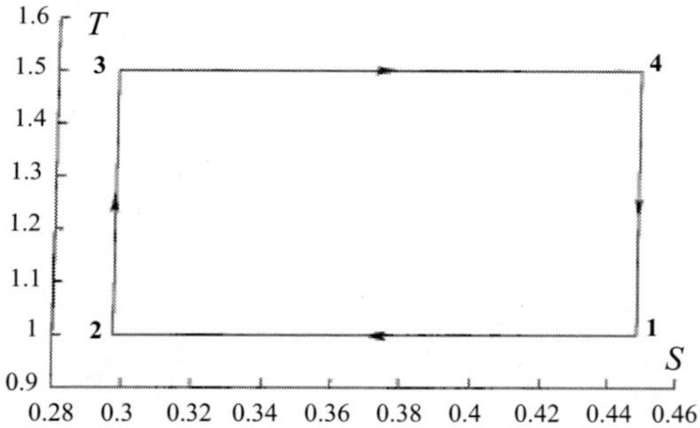

Figure 7.4 Carnot cycle in T–S space for $\alpha_1 = 0.6$, $\alpha_2 = 0.8$, $T_c = 1$, $T_h = 1.5$. Model: Particle in a box.

In the isothermal step the two control parameters are correlated in order to keep $T(\alpha, \gamma)$ constant. Taking the derivative of Eq. (7.12) one obtains

$$\begin{aligned} g(\gamma) &= \Theta(\alpha)T\,, \\ \frac{dg}{d\gamma}d\gamma &= T\frac{d\Theta}{d\alpha}d\alpha\,. \end{aligned} \tag{7.47}$$

The latter identity is convenient to change variables: Based on Eq. (7.19) we obtain for

Step 1 → 2 (isothermal at T_c):

$$W_{12} = \int_{\gamma_1}^{\gamma_2} h(\alpha)\frac{dg}{d\gamma}d\gamma = T_c\int_{\alpha_1}^{\alpha_2} h(\alpha)\frac{d\Theta}{d\alpha}d\alpha = -W_{21}\,, \tag{7.48}$$

$$Q_{12} = \int_{\gamma_1}^{\gamma_2} g(\gamma)\frac{dh}{d\alpha}d\alpha = T_c\int_{\alpha_1}^{\alpha_2} \Theta(\alpha)\frac{dh}{d\alpha}d\alpha = -Q_{21}\,. \tag{7.49}$$

Step 2 → 3 (adiabatic):

$$\begin{aligned} W_{23} &= \int_{\gamma_2}^{\gamma_3} h(\alpha_2)\frac{dg}{d\gamma}d\gamma = h(\alpha_2)(g(\gamma_3) - g(\gamma_2)) \\ &= h(\alpha)\Theta(\alpha_2)(T_h - T_c) = -W_{32}\,, \end{aligned} \tag{7.50}$$

$$Q_{23} = 0\,. \tag{7.51}$$

Step 3 → 4 (isothermal at T_h):

$$W_{34} = \int_{\gamma_3}^{\gamma_4} h(\alpha)\frac{dg}{d\gamma}d\gamma = T_h\int_{\alpha_2}^{\alpha_1} h(\alpha)\frac{d\Theta}{d\alpha}d\alpha = -W_{43}\,, \tag{7.52}$$

$$Q_{34} = \int_{\gamma_3}^{\gamma_4} g(\gamma)\frac{dh}{d\alpha}d\alpha = T_h\int_{\alpha_2}^{\alpha_1} \Theta(\alpha)\frac{dh}{d\alpha}d\alpha = -Q_{43}\,. \tag{7.53}$$

Step 4 → 1 (adiabatic):

$$W_{41} = \int_{\gamma_4}^{\gamma_1} h(\alpha_2)\frac{dg}{d\gamma}d\gamma = h(\alpha_2)(g(\gamma_4) - g(\gamma_1))$$
$$= h(\alpha)\Theta(\alpha_2)(T_h - T_c) = -W_{14}\,, \tag{7.54}$$
$$Q_{41} = 0\,. \tag{7.55}$$

The heat term Q_{12}, (Q_{21}) relates to the cold bath and Q_{34}, (Q_{43}) to the hot bath. We confirm that

$$\boxed{\frac{Q_{12}}{T_c} + \frac{Q_{34}}{T_h} = \Delta S = 0\,,} \tag{7.56}$$

that is, the entropy over one cycle stays constant, the cycle is reversible. Such a relation holds, of course, also for the opposite sense of rotation. The total work per cycle (transferred to or from the work reservoir) is

$$\Delta W_{tot}^{(+)} = W_{12} + W_{23} + W_{34} + W_{41} \tag{7.57}$$
$$= (T_h - T_c)\left(h(\alpha_2)\Theta(\alpha_2) - h(\alpha_1)\Theta(\alpha_1) - \int_{\alpha_1}^{\alpha_2} h(\alpha)\frac{d\Theta}{d\alpha}d\alpha\right)$$
$$= (T_h - T_c)\int_{\alpha_1}^{\alpha_2}\Theta(\alpha)\frac{dh}{d\alpha} = -\Delta W_{tot}^{(-)}\,. \tag{7.58}$$

The last line results via partial integration. Noting that according to Eq. (7.11)

$$\Theta(\alpha)\frac{dh}{d\alpha} = -k_B\sum_i \ln\rho_{ii}\frac{d\rho_{ii}}{d\alpha} = k_B\frac{dS_{vN}(\alpha)}{d\alpha} = \frac{dS}{d\alpha}\,, \tag{7.59}$$

the total work and the heat Q_{34} can be simplified as

$$\Delta W_{tot}^{(+)} = (T_h - T_c)(S(\alpha_2) - S(\alpha_1))\,, \tag{7.60}$$
$$Q_{34} = -T_h(S(\alpha_2) - S(\alpha_1))\,. \tag{7.61}$$

For $dS/d\alpha < 0$ (which is the case for the canonical attractor) the total work $\Delta W_{tot}^{(+)}$ is negative and Q_{34} is positive. The corresponding Carnot efficiency is

$$\boxed{\eta_{carnot} = \frac{-\Delta W_{tot}^{(+)}}{Q_{34}} = 1 - \frac{T_c}{T_h}\,.} \tag{7.62}$$

For $dS/d\alpha > 0$ the sense of rotation would have to be changed, but the same result obtains. Changing the sense of rotation from (+)

to (−) for $dS/d\alpha < 0$ all signs flip and one gets the heat pump functionality. The respective efficiency is

$$\eta_{carnot}^{pump} = \frac{1}{\eta_{carnot}} = \frac{T_h}{T_h - T_c}. \tag{7.63}$$

Obviously, this efficiency can be larger than 1. Again, the results are universal in the sense of Remark 7.2.

Finally, the refrigerator efficiency is

$$\eta_{carnot}^{fridge} = \frac{Q_{21}}{\Delta W_{tot}^{(-)}} = \frac{T_c}{T_h - T_c} \tag{7.64}$$

7.3.3 *Stirling Cycle*

The Stirling cycle is easily combined from the results of the Otto and the Carnot cycle, see Fig. 7.5.

For simplicity we restrict ourselves to $dh/d\alpha < 0$ and $dg/d\gamma < 0$; these conditions are fulfilled for the particle in a box and the canonical attractor. The sense of rotation in the control parameter plane (α, γ) is clockwise (negative), see Fig. 7.5. Note the sequential order of the α values; they are related by

$$\Theta(\alpha_1) = \frac{T_h}{T_c}\Theta(\alpha_4), \tag{7.65}$$

$$\Theta(\alpha_2) = \frac{T_h}{T_c}\Theta(\alpha_3). \tag{7.66}$$

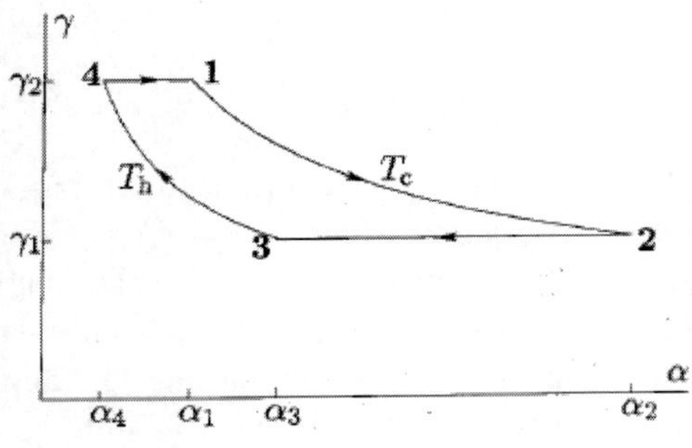

Figure 7.5 Stirling cycle in control space.

Step 1 → 2 (isothermal at T_c):

$$W_{12} = T_c \int_{\alpha_1}^{\alpha_2} h(\alpha) \frac{d\Theta}{d\alpha} d\alpha \,, \tag{7.67}$$

$$Q_{12} = T_c \int_{\alpha_1}^{\alpha_2} \Theta(\alpha) \frac{dh}{d\alpha} d\alpha < 0 \,. \tag{7.68}$$

Step 2 → 3 (isochoric):

$$W_{23} = 0 \,, \tag{7.69}$$

$$Q_{23} = \int_{\alpha_2}^{\alpha_3} g(\gamma_1) \frac{dh}{d\alpha} d\alpha = g(\gamma_1)(h(\alpha_3) - h(\alpha_2)) > 0 \,. \tag{7.70}$$

Step 3 → 4 (isothermal at T_h):

$$W_{34} = T_h \int_{\alpha_3}^{\alpha_4} h(\alpha) \frac{d\Theta}{d\alpha} d\alpha \,, \tag{7.71}$$

$$Q_{34} = T_h \int_{\alpha_3}^{\alpha_4} \Theta(\alpha) \frac{dh}{d\alpha} d\alpha > 0 \,. \tag{7.72}$$

Step 4 → 1 (isochoric):

$$W_{41} = 0 \,, \tag{7.73}$$

$$Q_{41} = \int_{\alpha_4}^{\alpha_1} g(\gamma_2) \frac{dh}{d\alpha} d\alpha = g(\gamma_2)(h(\alpha_3) - h(\alpha_2)) < 0 \,. \tag{7.74}$$

In the classical Stirling cycle the heat contributions along the two isochoric steps cancel. But here we have four non-zero heat contributions with pairwise opposite signs. We have to make sure that each pair with definite sign connects with one specific heat reservoir only, so that the inflow and outflow can physically be separated. Then the correct definition of the respective efficiency reads:

$$\eta_{stirling} = \frac{-\Delta W_{tot}^{(-)}}{Q_{23} + Q_{34}} \,. \tag{7.75}$$

If Q_{23} was neglected, the efficiency could become larger than 1, that is, larger than the Carnot limit, which would thus violate the second law. The concrete result for the efficiency is rather complicated.

The Stirling cycle is reversible: this results from the combination of the two isothermal entropy contributions (as defined for the Carnot cycle) and the two isochoric terms (as defined for the Otto cycle). The behavior in the entropy–temperature plane is shown in Fig. 7.6.

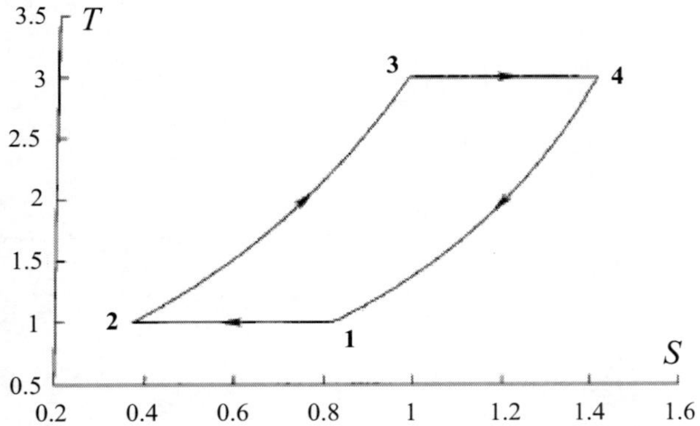

Figure 7.6 Stirling cycle in T–S space for $\gamma_1 = 1.2, \gamma_2 = 1.8, T_c = 1, T_h = 3$. Model: Particle in a box.

7.3.4 *Influence of Negative Temperature*

The absolute temperature according to Definition 7.7 does not require T to be positive. Indeed, negative temperatures may be imparted on quantum systems (A) by environments (B) with a bounded energy spectrum $E_{min} \leq E \leq E_{max}$, like for spin networks. In this case the entropy S(A) as a function of the "working point" (i.e., the respective band index j) within (B)—after having passed through its maximum value—approaches zero for the upper (just as for the lower) energy bound. The bands at both bounds are non-degenerate. The decreasing degeneracy as a function of increasing environmental energy implies a negative temperature for (A), cf. Example 5.3. Given such a situation, what are the potential consequences for thermodynamic cycles?

Remark 7.8 Hot and cold *For the following it is convenient to consider $\beta = 1/k_B T$ rather than T itself. The zeroth law, cf. Theorem 4.9, tells us that*

$$\frac{dS}{k_B} = (\beta(A) - \beta(B))\, đ Q(A) \geq 0\,. \tag{7.76}$$

By definition (A) is colder than (B) if heat flows from (B) to (A). This is an irreversible relaxation process. For subsystem A to receive heat, $đ Q(A) > 0$, it must hold that

$$\beta(A) > \beta(B)\,. \tag{7.77}$$

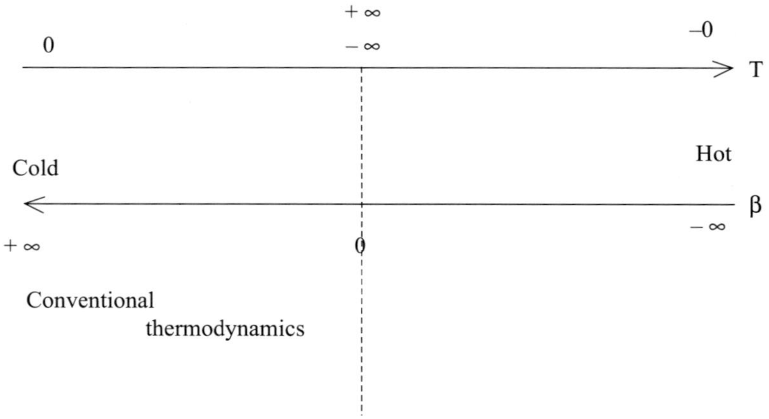

Figure 7.7 Hot and cold: The inverse temperature, $\beta = 1/k_B T$, goes smoothly from $+\infty$ to $-\infty$.

This is the case for any $\beta(A) > 0$, if $\beta(B) < 0$. In this sense negative temperatures are hotter than all positive temperatures. If both temperatures are negative, (A) is colder than (B) if $-\beta(A) < -\beta(B)$. On the linear β-scale temperatures get colder and colder as we move to the right (get hotter and hotter as we move to the left), irrespective of sign, see Fig. 7.7.

Proposition 7.2 Carnot cycle with negative temperatures *If we flip the sign for both baths, the results for the Carnot efficiency do not change at all.*

It is more interesting to consider the Carnot cycle under the condition $T_c > 0$ and $T_h < 0$. Equation (7.56) then reads:

$$\frac{Q_{12}}{T_c} - \frac{Q_{34}}{T_h} = (Q_{12}\beta_c - Q_{34}\beta_h)k_B = 0\,, \tag{7.78}$$

and the sign of both heat terms must always be identical, for a heat engine positive, for a heat pump negative. It makes no sense to restrict oneself to the input by one heat reservoir only. The efficiency should thus be

$$\eta_{carnot} = \frac{\Delta W_{tot}^{(+)}}{Q_{12} + Q_{34}} = 1\,. \tag{7.79}$$

The efficiency of the heat pump would also be 1. The need to modify the definition of efficiency has already been encountered for the Stirling cycle.

Remark 7.9 Extrapolation *The above result is not so surprising when considering the Carnot efficiency in the limit $\beta_h \to 0$ (for finite $\beta_c > 0$): We approach unity. A naïve extrapolation to negative β_h would lead to efficiencies larger than 1, a rather unphysical situation [Rapp (2010)]. According to the above consideration, the (modified) efficiency stays at 1.*

Remark 7.10 Relaxation *Remarkably, a single bath of negative temperature (connected via an appropriate node with a single work reservoir) might operate as a continuous heat engine: Indeed, with the heat change $đQ$ chosen as negative (i.e., leaving the bath), the entropy of the bath, $dS = -đQ/T_h$, nevertheless increases. This is a relaxation process (spontaneous and irreversible). If the bath is infinite, the process would be quasi-isothermal and stationary.*

The continuous Lasing model, Example 5.18, shows such a relaxation behavior; however, the output is not work in a strict sense (according to LEMBAS it is heat). It is an open question yet, whether a single bath heat engine as suggested above can be realized. Also the Carnot cycle involving negative temperatures still remains futuristic.

Occupation inversion is routinely achieved for lasing systems; however, the corresponding negative "virtual temperature" is a spectrally local feature and does not apply for the system as a whole.

Remark 7.11 Experimental realization of negative temperatures *Negative system temperatures have been realized recently in an ultra-cold quantum gas [Braun (2013)]. This multi-step experiment starts from a quantum gas in an optical lattice confined by a parabolic potential at positive temperature. After flipping the sign of the confining potential and the sign of the particle-interaction the energy gets bounded from above—a precondition for negative temperatures. The negative temperature state is eventually reached after relaxation; it is metastable at most.*

7.4 Dynamical Aspects

7.4.1 *Finite-Time Thermodynamics: The Curzon–Ahlborn (CA) Model*

The fundamental properties of thermodynamic machines follow under the condition of quasi-static evolution, that is, for infinite cycle times. Needless to say that this zero-power limit is of little interest as far as technical applications are concerned. One would like to replace the infinite time limit by a "finite time" scenario. Unfortunately, this will mean non-equilibrium and dissipation—making universal results (like the Carnot efficiency η_{carnot}) rather unlikely. Entropy production is the key physical concept underlying irreversible models of heat engines.

Surprisingly there are simple finite-time models, which still produce (seemingly) universal results. Such a (originally classical) model has been suggested in [Curzon (1975)] based on the approximation of "endoreversibility."

Definition 7.13 Endo-reversibility *The total model is formally split into a standard reversible kernel, the Carnot engine (T_c, T_h), and a dissipative embedding: the latter consists of two resistive leads, one to supply and one to remove heat by means of finite (positive) temperature differences, $T_{max} - T_h$ and $T_c - T_{min}$, cf. Fig. 7.8. T_{max} and T_{min} have to be provided from the outside.*

Rather than looking for the optimum efficiency for zero currents, η_{carnot}, we now aim at the efficiency under maximum power output. This leads to the CA result, $\eta_{CA} \leq \eta_{carnot}$. For a simple derivation we follow here [Bandyopadhyay (2006)].

The heat flows are written as

$$\frac{\Delta Q_1}{\Delta t} = K_1(T_{max} - T_h) > 0\,, \tag{7.80}$$

$$\frac{\Delta Q_2}{\Delta t} = K_2(T_{min} - T_c) < 0\,. \tag{7.81}$$

The material parameters K_j specify the heat conduction. The sign convention is positive terms go into system, negative terms leave system. For the endoreversible part the (thermodynamic)

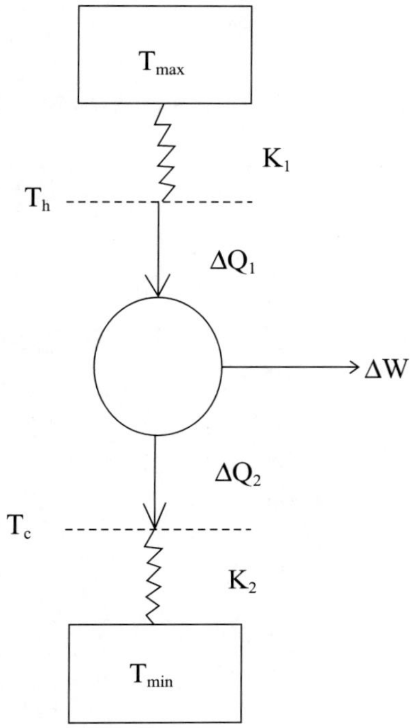

Figure 7.8 Endoreversible Carnot process. ΔQ_j heat exchange, ΔW work exchange. K_j are transport coefficients.

adiabaticity condition reads, cf. Eq. (7.56),

$$\frac{\Delta Q_1}{T_h} = -\frac{\Delta Q_2}{T_c} \,. \tag{7.82}$$

Based on the abbreviation

$$\tau \equiv \frac{T_c}{T_h} = \frac{-\Delta Q_2}{\Delta Q_1} \leq 1 \,, \tag{7.83}$$

we thus conclude

$$\frac{K_1}{T_h}(T_{max} - T_h) = \frac{K_2}{T_2}(T_h\tau - T_{min}) \,, \tag{7.84}$$

$$K_1 T_{max} - K_1 T_h = K_2 T_h - K_2 T_{min}/\tau \,, \tag{7.85}$$

$$T_h = \frac{K_1 T_{max} + K_2 T_{min}/\tau}{K_1 + K_2} \,. \tag{7.86}$$

The work output is then per cycle (according to the first law $\Delta U = 0$)

$$-\Delta W = \Delta Q_1 + \Delta Q_2 = \Delta Q_1 (1 - \tau)\,. \tag{7.87}$$

We use Eq. (7.86) to obtain for

$$\Delta Q_1 = K_1 (T_{max} - T_h) \Delta t \tag{7.88}$$

$$= K_1 K_2 \Delta t \frac{T_{max} - T_{min}/\tau}{K_1 + K_2}\,. \tag{7.89}$$

The time increment Δt is finite and proportional to the total cycle time of the machine. The power

$$\frac{|\Delta W(\tau)|}{\Delta t} = \frac{1 - \tau}{K_1^{-1} + K_2^{-1}} (T_{max} - T_{min}/\tau)\,, \tag{7.90}$$

is now maximized with respect to τ,

$$\frac{d}{d\tau} \frac{|\Delta W|}{\Delta t} = 0 \rightarrow \frac{d}{d\tau} (1 - \tau)(T_{max} - T_{min}/\tau) = 0\,, \tag{7.91}$$

implying

$$\tau^2 = \frac{T_{min}}{T_{max}}\,. \tag{7.92}$$

The maximum power output is

$$\left(\frac{|\Delta W|}{\Delta t} \right)_{max} = \frac{(\sqrt{T_{max}} - \sqrt{T_{min}})^2}{K_1^{-1} + K_2^{-1}}\,, \tag{7.93}$$

which is obtained under the so-called CA efficiency,

$$\boxed{\eta_{CA} = \frac{-\Delta W}{\Delta Q_1} = 1 - \tau = 1 - \sqrt{\frac{T_{min}}{T_{max}}}\,.} \tag{7.94}$$

Note that this surprisingly simple result is independent of K_1, K_2. A CA efficiency has also been derived for refrigerators [Yan (1990)]:

$$\eta_{CA}^{(fridge)} = \frac{1}{\sqrt{1 - T_{min}/T_{max}}} - 1\,. \tag{7.95}$$

Remark 7.12 Universality? *Despite its formal similarity with the fundamental Carnot result, η_{carnot}, the CA formula does not share the universal applicability of the former. A number of investigations in the quantum domain have shown this.*

The CA efficiency at maximum power can be rewritten in the form

$$\eta_{CA} = 1 - \sqrt{1 - \eta_{carnot}} = \frac{\eta_{carnot}}{2} + \frac{\eta_{carnot}^2}{8} + \cdots . \tag{7.96}$$

While the result as such is definitely not universal, the first two expansion coefficients appear to be universal [Esposito (2010)]. The coefficient $1/2$ *derives from the symmetry of the Onsager matrix (linear response).*

While the CA result is not universal in its original physical setting, it gains supplementary significance from other inquiries. Below there are two examples. Other cases (within the realm of energy converters) have also been suggested [Angulo-Brown (2010)]. All these findings would point toward a kind of "universality" in an entirely different sense.

Example 7.3 Work extremum principle *Let us consider the Hamiltonian [Allahverdyan (2008)]*

$$\hat{H}(t) = \hat{H}(A) + \hat{H}(B) + \hat{V}(t), \tag{7.97}$$

$$\hat{V} = 0 \quad for \quad t < 0 \quad and \quad t > \tau . \tag{7.98}$$

It is explicitly time dependent during the finite time increment τ*. The work extracted during this unitary step is*

$$\Delta W = \mathrm{Tr}\{\hat{H}(\hat{\rho}(0) - \hat{\rho}(\tau))\} . \tag{7.99}$$

Here it is assumed that for $t \leq 0$ *sub-system (A) has been in contact with a bath* $\beta(A)$ *and subsystem (B) with a bath* $\beta(B) > \beta(A)$ *(weak coupling limit). The resulting state is*

$$\hat{\rho}(0) = \hat{\rho}_{can}(A) \otimes \hat{\rho}_{can}(B) . \tag{7.100}$$

For $t > \tau$ *the two subsystems are decoupled and brought again in contact with their original baths. This two-step cycle can be repeated at will: bath contact, unitary evolution, bath contact, etc.*

For fixed bath temperatures the model is still dependent on the spectra for the two subsystems and on $\hat{V}$*. These dependencies are eventually optimized with respect to maximum work output. It is found that the efficiency at maximal work is bounded from below by* η_{CA}.

The following Bayesian approach, cf. Proposition 2.4, indicates that the CA result may be re-interpreted in terms of subjective control deficits.

For this purpose the original CA setting is substituted by a "statistical game" played by an outside observer [Johal (2010)]. The real physical uncertainty (due to tracing out part of the total system) is thus replaced by the indeterminacy within parameter space (subjective ignorance) characterizing the otherwise ideal machine (in the limit of infinite cycle time). To become operative, this subjective ignorance must eventually translate into limited control.

Example 7.4 Ensemble of Otto cycles *We consider a two-level system subject to the spectral deformation (cf. Eq. (4.6))*

$$E_n = \gamma \epsilon_n \quad n = 0, 1, \qquad \text{for} \quad \gamma_1 \leq \gamma \leq \gamma_2 . \tag{7.101}$$

The respective efficiency is, cf. Example 7.2,

$$\eta = 1 - \frac{\gamma_1}{\gamma_2} . \tag{7.102}$$

Defining $\epsilon_0 \equiv 0$ *the work per cycle can be written as*

$$\begin{aligned}\Delta W(\gamma_2, \eta) &= (\gamma_2 - \gamma_1)\epsilon_1(\rho_1^{(2)} - \rho_1^{(1)}) \\ &= \epsilon_1 \gamma_2 \eta \left(\frac{1}{1 + \exp(\gamma_2 \epsilon_1 / k_B T_h)} \right. \\ &\quad \left. - \frac{1}{1 + \exp(\gamma_2 \epsilon_1 (1 - \eta) / k_B T_c)} \right) .\end{aligned} \tag{7.103}$$

Here, $T_h > T_c$. *Now consider an ensemble of machines specified by the probability [Johal (2010)]*

$$\Pi(\gamma_2) = (\ln \gamma_{max}/\gamma_{min})^{-1} \frac{1}{\gamma_2} . \tag{7.104}$$

For fixed efficiency the ensemble averaged work per cycle is

$$\overline{\Delta W(\eta)} = \int_{\gamma_{min}}^{\gamma_{max}} \Delta W(\gamma_2, \eta) \Pi(\gamma_2) d\gamma_2 . \tag{7.105}$$

The maximum work as a function of η *is found requiring*

$$\frac{d\overline{\Delta W}}{d\eta} = 0 . \tag{7.106}$$

Under the additional conditions $\gamma_{min} \rightarrow 0$ *and* $\epsilon_1 \gamma_{max} \gg k_B T_h$ *one obtains for the efficiency, which maximizes the averaged work [Johal (2010)]:*

$$\eta^* \approx 1 - \sqrt{T_c/T_h} = \eta_{CA} . \tag{7.107}$$

Note that the CA efficiency here gives the maximum averaged work rather than the power! Of course, the choice of the prior distribution is debatable; similar results are obtained for a larger class of priors, but still within severe restrictions.

Lack of control—of whatever origin—tends to degrade performance. In a way the Bayesian model confirms that "objective" limitations (arising, e.g., due to friction) and subjective inabilities may have equivalent consequences.

7.4.2 *Non-Equilibrium and Friction*

Definition 7.14 Friction *"Friction is the price for moving too fast" [Rezek (2010)]. As such it will enter in one form or other any model within finite-time thermodynamics. It is useful to distinguish two types:*

(1) "External friction," due to dissipation of the system kinetic energy to its environment. This is most easily understood in terms of scattering: The forward/backward asymmetry in the co-moving frame tends to relax down toward the symmetric (isotropic) situation of the true rest frame.

(2) "Internal friction," due to the generation of internal excitations (non-adiabaticity). While the latter process by itself is reversible (because it is unitary), the increase of mixing entropy (cf. Definition 6.3) and the associated excess work to be provided from the outside indicates a "resistance" of the system against change. Indeed, as shown already by [Husumi (1953)] for the case of the harmonic oscillator, the energy of the system is increased on average under the action of a time-dependent ω.

Remark 7.13 Quantum friction *In its most elementary form quantum friction is expected to occur, when two flat surfaces separated by a vacuum gap are sliding relative to each other at zero temperature. The contact-free friction—related to the Casimir-effect—would be due to the relative motion of quantum fluctuations [Volokitin (2011)].*

Needless to say that in the present context we are far away from this limit. Nevertheless, could there still be genuine quantum effects in operation? Non-commutativity as related to partial control, cf. Example 4.7, has been proposed as a possible source along adiabatic steps [Feldmann (2003)], see below.

Influence of external friction For thermodynamic processes in the control space as defined in Section 7.2 non-equilibrium effects can approximately be included as follows: Let the state dynamics be described by the "moving attractor," cf. Definition 6.4. In the overdamped case (neglecting the coherent contribution) this equation of motion reduces to

$$\frac{d}{dt}\hat{\rho}^* = -\frac{1}{\tau_R}(\hat{\rho}^* - \hat{\rho}(\alpha(t))\,. \tag{7.108}$$

The speed of $\alpha(t)$ will control the momentary deviation of the non-equilibrium state $\hat{\rho}^*$ from the attractor, $\hat{\rho}(\alpha(t)$:

$$\hat{\rho}^*(t) \rightarrow \rho^*_{ii}(\alpha(t), \Delta\rho_i) = \rho_{ii}(\alpha(t)) + \Delta\rho_i(t)\,. \tag{7.109}$$

The deviations $\Delta\rho_i(t)$ cannot be controlled directly; they depend on history. However, for a stationary cycle the deviations will become implicit functions of the position in control space, $\Delta\rho_i \rightarrow \Delta\rho_i(\alpha, \gamma)$.

A form of "process thermodynamics" can now be introduced [Birjukov (2008)]:

$$U^*(\alpha, \gamma, \Delta\rho_i) = g(\gamma)\sum_i \epsilon_1 \rho^*_{ii}\,, \tag{7.110}$$

$$S^*(\alpha, \Delta\rho_i) = -\sum_i \ln \rho^*_{ii}(\alpha) \cdot \rho^*_{ii}\,. \tag{7.111}$$

The total differential of the internal energy (generalized first law) is

$$dU^* = đW^* + đQ^*\,, \tag{7.112}$$

$$đW^* = \left(\frac{\partial U^*}{\partial \gamma}\right)_{\alpha, \Delta\rho_i} d\gamma = \frac{dg}{d\gamma}\sum_i \epsilon_i \rho^*_{ii} d\gamma\,, \tag{7.113}$$

$$\begin{aligned} đQ^* &= \left(\frac{\partial U^*}{\partial \alpha}\right)_{\gamma, \Delta\rho_i} d\alpha + \sum_i \left(\frac{\partial U^*}{\partial \Delta\rho_i}\right)_{\gamma, \alpha} d(\Delta\rho_i) \\ &= g(\gamma)\sum_i \epsilon_i \left(\frac{d\rho_{ii}}{d\alpha} d\alpha + d(\Delta\rho_i)\right) . = g(\gamma)\sum_i \epsilon_i d\rho^*_{ii} \end{aligned} \tag{7.114}$$

We define the process temperature as the response to an entropy-change for fixed deviation $\Delta\rho_i$:

$$T^*(\alpha, \gamma, \Delta\rho_i) = \left(\frac{\partial U^*}{\partial \alpha}\right)_{\gamma, \Delta\rho_i} \left(\frac{\partial S^*}{\partial \alpha}\right)^{-1}_{\Delta\rho_i} . \qquad (7.115)$$

Observing that

$$dS^* = \left(\frac{\partial S^*}{\partial \alpha}\right)_{\Delta\rho_i} + \sum_i \left(\frac{\partial S^*}{\partial \Delta\rho_i}\right)_\alpha d(\Delta\rho_i), \qquad (7.116)$$

the heat term can be re-written as

$$đQ^* = T^* dS^* - đQ^*_{diss}, \qquad (7.117)$$

$$đQ^*_{diss} = T^* \sum_i \left(\frac{\partial S^*}{\partial \Delta\rho_i}\right)_\alpha d(\Delta\rho_i) - \sum_i \left(\frac{\partial U^*}{\partial \Delta\rho_i}\right)_{\gamma,\alpha} d(\Delta\rho_i). \qquad (7.118)$$

The analysis has now to be specialized for the various process steps; there is no relaxation along the adiabatic steps. For any cycle the work per cycle is

$$\Delta W^* = -\left(\oint T^* dS^* - \oint đQ^*_{diss}\right). \qquad (7.119)$$

External friction (dissipation) will thus lead to a velocity-dependent work output. Surprisingly, for the case of an Otto cycle the maximum efficiency is the same as in the quasi-static limit; the efficiency for the Carnot cycle is reduced, as might have been expected.

Influence of internal friction In the following we study two examples, both based on the two-spin Heisenberg model, Definition 3.9.

Example 7.5 *We consider the Hamiltonian:*

$$\hat{H}(t) = \gamma(t)\hat{H}_c + \hat{H}_{nc}, \qquad (7.120)$$

$$\hat{H}_c = \frac{\hbar\Gamma B_3}{2}(\hat{\sigma}_3(A) + \hat{\sigma}_3(B)), \qquad (7.121)$$

$$\hat{H}_{nc} = \frac{J}{2}(\hat{\sigma}_1(A) \otimes \hat{\sigma}_1(B) - \hat{\sigma}_2(A) \otimes \hat{\sigma}_2(B)). \qquad (7.122)$$

The first part is an instance of spectral deformation as introduced in Definition 4.6; it is controlled by the external magnetic field $\gamma(t) =$

$B(t)/B_3$. The second part is "uncontrolled"; it does not commute with the first part, cf. Definition 4.7. With the control function

$$\Omega^2(\gamma) \equiv (\gamma \hbar B_3 \Gamma)^2 + J^2 \geq \gamma \hbar B_3/, \tag{7.123}$$

the parameter-dependent spectrum of $\hat{H}$ is [Feldmann (2003)]

$$\boxed{E_1 = -\Omega(\gamma) \quad E_2 = 0 \quad E_3 = 0 \quad E_4 = +\Omega(\gamma)\,.} \tag{7.124}$$

Note that the control function has not the scaling form of the spectral deformation model, Definition 4.6. The thermodynamic Otto cycle is composed of four individual branches (isochoric, adiabatic). In [Feldmann (2003)] a detailed analysis has been provided in terms of the Lindblad master equation in the Heisenberg picture (see Eq. (3.375)). The work per time (power) is $\mathcal{P} = \dot{\gamma} < \hat{H}_c >$, the heat flow $\dot{Q}$ derives from dissipation, Eq. (3.375). The first law (energy balance) thus reads

$$\frac{dU}{dt} = < \frac{d\hat{H}}{dt} > = \mathcal{P} + \dot{Q}\,, \tag{7.125}$$

$$\dot{Q} \equiv < \hat{\mathcal{L}}^*(\hat{H}) > \,. \tag{7.126}$$

The numerical results can eventually be discussed in the $\{S, T\}$- or the $\{S_H, T_{dyn}\}$ - plane, where S_H is the mixing (or energy-) entropy, see Definition 3.84, and

$$T_{dyn} \equiv \left(\frac{\partial U}{\partial t}\right)_\gamma \cdot \left(\frac{\partial S_H}{\partial t}\right)_\gamma^{-1}\,, \tag{7.127}$$

is the associated temperature, called "dynamical temperature" [Feldmann (2003)].

The irreversibility is here indirectly caused by the finite time duration of the adiabatic steps; it is closely linked with the violation of the quantum adiabatic condition, see Proposition 6.5, that is, with the rapid change of the energy level structure.[a] The resulting increase of system energy is dissipated and must be supplied from the outside. But as for the case of extrinsic friction, the maximum efficiency is found to be the same as in the quasi-static limit (cf. Example 7.2):

$$\eta_{otto} = 1 - \frac{\Omega(\gamma_1)}{\Omega(\gamma_2)} \quad for\ \gamma_2 > \gamma_1\,. \tag{7.128}$$

[a]For the parametrized quantum oscillator similar results have been derived a long time ago by [Husumi (1953)], cf. Example 6.2.

Example 7.6 *We again start from the Heisenberg model, now in its isotropic variant, $J_3 = J$. Its eigenstates and eigenvalues are given in Eq. (3.118). For the following it is convenient to shift all the energy levels by $-J$, so that the parameter dependences of the eigenvalues come in separate groups:*

$$\boxed{E_1 = -\hbar\Gamma B_3 \quad E_2 = -4J \quad E_3 = 0 \quad E_4 = \hbar\Gamma B_3\,.} \tag{7.129}$$

Following [Thom (2011)] we keep $J > 0$ and vary the external magnetic field $B_3 > 0$ in the range $B_3^{(2)} \leq B_3 \leq B_3^{(1)}$. This is another instance of partial control, leading to internal friction. But contrary to Eq. (7.124) this spectrum violates the spectral deformation model, Definition 4.6, not by the control function itself, but by the (fixed) separate level $E_2 = -4J$.

Here we analyze the Otto cycle directly along the lines of Section 7.3.1. Note that the two adiabatic steps (change of spectrum, constant occupation) lead to non-equilibrium, as the spectrum gets seriously deformed, cf. Remark 4.10. For the two isochoric steps we immediately find:

Step $1 \to 2$: $\rho_{jj}(1) \to \rho_{jj}(2)$

$$W_{12} = 0\,, \tag{7.130}$$

$$Q_{12} = Q^* + \hbar\Gamma B_3^{(1)}(\rho_{11}(1) - \rho_{11}(2) + \rho_{44}(2) - \rho_{44}(1))\,, \tag{7.131}$$

$$Q^* \equiv 4J\,(\rho_{22}(1) - \rho_{22}(2))\,. \tag{7.132}$$

Step $3 \to 4$: $\rho_{jj}(2) \to \rho_{jj}(1)$

$$W_{34} = 0\,, \tag{7.133}$$

$$Q_{34} = -Q^* - \hbar\Gamma B_3^{(2)}(\rho_{11}(1) - \rho_{11}(2) + \rho_{44}(2) - \rho_{44}(1))\,. \tag{7.134}$$

The total work per cycle we simply infer from the conservation of energy,

$$-\Delta W_{tot} = Q_{12} + Q_{34} = \hbar\Gamma(B_3^{(1)} - B_3^{(2)}) \\ (\rho_{11}(1) - \rho_{11}(2) + \rho_{44}(2) - \rho_{44}(1))\,. \tag{7.135}$$

The efficiency (for the heat engine) is defined as usual:

$$\eta_{otto} = \frac{-\Delta W_{tot}}{Q_{12}}\,. \tag{7.136}$$

For zero interaction, $J = 0$, the heat term Q^* is zero, and the standard result, Eq. (7.33), is recovered,

$$\eta_{otto} = 1 - \frac{B_3^{(2)}}{B_3^{(1)}}, \tag{7.137}$$

confirming the findings of Example 7.5. In general, however, the efficiency deviates from this result. Remarkably, non-zero interaction may even enhance the efficiency. Indeed, if Q^* turns out to be negative, the total heat input Q_{12} is reduced. This is the case for $\rho_{22}(2) > \rho_{22}(1)$, which can happen, if J is small enough compared to $\hbar\Gamma B_3$. However, with or without increase, the efficiency always stays below the Carnot limit.

Also for the quantum oscillator model the non-adiabaticity modifies the Otto efficiency [Abah (2012)]: the respective efficiency might become larger or smaller than the quasi-static η_{otto}. Friction is an essential limiting property also for refrigeration: without it the cooling power could be optimized beyond any bounds thus violating the third law, cf. Section 7.5.1.

7.4.3 *Bath Correlations: Photo-Carnot Engine*

Let us ask: Can one improve the quantum Carnot engine by means of quantum correlations within the bath? Obviously, this question requires a more detailed study. The model below has first been suggested by [Scully (2003)] and clarified later on by [Dillenschneider (2009)].

The working medium (A) is taken to be a cavity photon gas at temperature $T(A)$. The Carnot scenario is realized by coupling this system to two reservoirs, R1 and R2, which do not only differ by temperature: Bath R2 is "standard," while R1 is a bath of specially correlated atom pairs.

We start by considering the non-standard bath R1 in more detail. It consists of pairs of two-level atoms (BC) with identical energy-splitting $\hbar\omega$. The atoms interact via a Heisenberg–XX coupling, cf. Example 3.9 for $J_3 = 0$:

$$\hat{H}(BC) = \frac{\hbar\omega}{2}(\hat{\sigma}_3(B) + \hat{\sigma}_3(C)) + \frac{J}{2}(\hat{\sigma}_+(B) \otimes \hat{\sigma}_-(C) + c.c). \tag{7.138}$$

The pair gas is assumed to be in a thermal equilibrium state with $\beta = 1/k_B T(BC)$. According to Eq. (3.118) this state is explicitly given by

$$\hat{\rho}(BC) = \frac{1}{Z}\left(\exp\left(\beta\hbar\omega\right)|\psi_1><\psi_1| + \exp\left(\beta 2J\right)|\psi_2><\psi_2| \right. \\ \left. + \exp\left(-\beta 2J\right)|\psi_3><\psi_3| + \exp\left(-\beta\hbar\omega\right)|\psi_4><\psi_4|\right). \tag{7.139}$$

Its concurrence C can be evaluated in closed form, cf. Definition 3.63; a positive value indicates entanglement between the two two-level atoms. It is non-zero (the thermal state being entangled) for sufficiently low temperatures and/or strong coupling J.

We now consider a beam of such correlated atom pairs, acting as the hot bath, $T(BC) \equiv T_h$. The interaction is deliberately restricted as follows: Only one atom out of each pair is brought in contact with the photon gas, but only once. After a while, that is, after a sequence of such interactions, a new stationary solution results, in which the cavity field is in a thermal state with increased temperature $T(A)' > T_h$. The zeroth law is violated; it would hold only if the total pair (BC) as such had interacted with the photon field. The individual atoms are not even in a thermal state, they serve as a kind of "high-performance fuel."

The second heat bath R2 is a conventional heat bath at temperature T_c. The photon gas (A) as the working medium and the two baths make up the photon Carnot cycle.

For low temperatures (large β) the effective efficiency is then found to be

$$\eta_{carnot}^{eff} = 1 - \frac{T_c}{T_h} + \frac{T_c}{T_h}\left(1 + \frac{1}{\beta\hbar\omega}\ln C\right) \geq \eta_{carnot}\,. \tag{7.140}$$

Note that because of $C \leq 1$ the last term in parenthesis, $\ln C$, is negative. This result is another instance of "generalized thermodynamics," in which information measures enter, cf. Eq. (6.78).

Remark 7.14 Generalized thermodynamics? The case of hidden costs II *As stressed by the authors, the potential violation of the second law is apparent only: Additional work has to be provided to prepare and stabilize the correlated bath to begin with. But this work does not enter the standard efficiency definition—though it should. The individual atoms generate a higher temperature than nominally* *included as T_h. Contrary to unjustified speculations this hypothetical device is of theoretical rather than practical interest only.*

On the other hand, it is quite remarkable to see how the "information language" is able to deal with a compact but incomplete physical description. Thermodynamics is an effective theory in the sense of Section 1.1; the use of information concepts allows a further reduction. This works even though information and its codification is essentially subjective, cf. Remark 4.23. Unfortunately, the cost of information is hidden—and thus easily leads to wrong predictions, cf. Remark 6.7.

7.5 Special Applications

7.5.1 *Otto Cycle: Cooling and the Third Law*

Refrigerator Let us assume $dh/d\alpha < 0$ and $dg/d\gamma > 0$, cf. Definition 7.6. The Otto cycle runs in the refrigerator mode as we change the sense of rotation $(+) \rightarrow (-)$, cf. Table 7.1. Both, the heat intake Q_{12} from the colder bath T_c and the work W_{tot} become positive. The corresponding efficiency is defined by

$$\eta_{otto}^{fridge} = \frac{Q_{12}}{W_{tot}^{(-)}} = \frac{\gamma_1}{\gamma_2 - \gamma_1} . \tag{7.141}$$

Based on Eq. (7.39), Eq. (7.40) we have $T_c = \gamma_1/\alpha_2$ and $T_h = \gamma_2/\alpha_1$. We conclude for $T_c \rightarrow 0$:

$$\eta_{otto}^{fridge} = \frac{T_c\alpha_2}{T_h\alpha_1 - T_c\alpha_2} \sim T_c . \tag{7.142}$$

The efficiency and thus the heat Q_{12} scales linearly with T_c. This result confirms the third law: We cannot reach absolute zero temperature, cf. 4.14.

Finite time scenario Instead of the heat extraction we consider the cooling rate,

$$R \equiv \frac{Q_{12}}{\Delta t} , \tag{7.143}$$

where Δt is the cycle time. This is a finite-time scenario. Friction does not allow us to make the cycle time as small as we might wish.

A rough estimate is as follows: The optimum efficiency is obtained in the Carnot limit, where we have with $\alpha_1 \approx \alpha_2$

$$\frac{\gamma_1}{\gamma_2} \sim \frac{T_c}{T_h} . \tag{7.144}$$

This means that the minimum energies reached for $\gamma = \gamma_1$ scale linearly with T_c. In order to avoid non-adiabaticity effects—which would lead to an increase of system energy and thus to dissipation, which eventually prevents further cooling—the evolution time should be, according to Eq. (6.35),

$$\Delta t \sim \text{minimal energies}^{-2} \sim T_c^{-2} . \tag{7.145}$$

As $Q_{12} \sim T_c$ this means for the cooling rate,

$$R \sim T_c^3 . \tag{7.146}$$

This gives a dynamical interpretation of the third law: With the cooling rate going to zero for $T_c \to 0$ the cycle time of the device would go to infinity. Taking the cycle time (and thus the total operation time needed) as a limiting resource, the approach to zero temperature becomes inefficient.

Remark 7.15 *Based on the quantum oscillator model it has been argued[Rezek (2009)] that the cycle time may be reduced to $T_c^{-1/2}$ so that the cooling rate would scale as $R \sim T_c^{3/2}$.*

7.5.2 *Autonomous Machines*

Any control, whether static or time-dependent, might eventually be traced back to some appropriate physical environment. Insisting to avoid infinite regress, this environment should, in turn, not be under the influence of some time-dependent control again. System and environment are then defined by an explicitly time-independent Hamiltonian, that is, as an autonomous system.

In this case the cycle of a thermodynamic machine has to derive from a time-dependent state of an otherwise time-independent "mechanical" environment. Also the selective coupling to different baths has to be under the control of the internal motion. Indeed, this is what happens for a classical steam engine. The momentary position (phase) of the piston is strictly correlated with the opening and closing of appropriate valves.

Unfortunately, in the quantum domain, in particular in the nano-limit, the pure coherent motion of a subsystem becomes rather problematic, as, in general, interactions lead to entanglement and entanglement leads to local decoherence, cf. Section 5.2.1.

Example 7.7 Spin-Oscillator model *Following [Tonner (2006)] we start from the spin-Boson model as defined in Eq. (3.129).*

$$\hat{H} = \hat{H}(A) + \hat{H}(B) + \hat{V}(AB), \tag{7.147}$$

$$\hat{H}(A) = \frac{\hbar\omega(A)}{2}\hat{\sigma}_j(A), \tag{7.148}$$

$$\hat{V}(AB) = L\hat{\sigma}_j(A)\hat{x}(B). \tag{7.149}$$

Taking now $j = 3$ *and assuming the factorization approximation, Proposition 5.11, to be applicable, we are led to the effective spin-Hamiltonian*

$$\hat{H}^{eff}(A;t) = \left(\frac{\hbar\omega(A)}{2} + L < \psi(B;t)|\hat{x}|\psi(B;t) >\right)\hat{\sigma}_3(A). \tag{7.150}$$

With the oscillator being prepared in the coherent state $|\alpha(B) >$ *with* $\alpha = |\alpha| \exp(i\Phi_0)$*, one thus gets the time-dependent effective energy splitting*

$$\Delta E(A;t) \equiv \hbar\omega(A) + 2\sqrt{2}L|\alpha| \cos(\omega(B)t + \Phi_0). \tag{7.151}$$

Of course, this oscillatory mechanical control for subsystem A only works on a time scale, on which the factorization is still valid.

To proceed toward a fully cyclic thermodynamic heat engine the mechanical control has now to be supplemented by a phase-dependent coupling to two external baths with the temperatures T_c, $T_h > T_c$*, see Fig. 7.9.*

The effective dynamics is then described by a Lindblad master equation, cf. Proposition 3.26. The baths are supposed to produce the following environment operators,

$$\hat{G}_q = W_q\hat{\sigma}_\pm(A) \otimes \Pi_q(B) \quad q = c, h. \tag{7.152}$$

Here, the $\Pi_q(B)$ *are "time slot operators." They are supposed to make the coupling of the spin to the baths depend on the momentary state of the oscillator: a kind of "quantum valve." This valve can be thought of as performing a fidelity test between the actual state and some properly chosen reference state of the oscillator; the spin-bath*

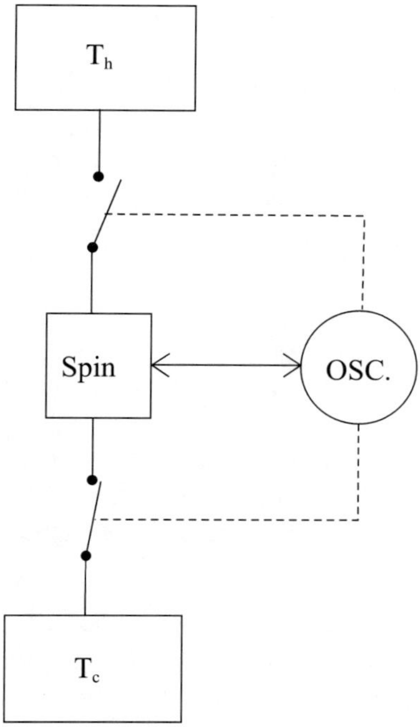

Figure 7.9 Autonomous spin-oscillator model. The oscillator represents the work reservoir and controls the bath couplings.

coupling is then taken to be a function of this momentary fidelity; for details see [Tonner (2006)]. The design is very special, requiring non-classical bath engineering. Presently it is meant as a proof of principle rather than suggesting a practical realization.

Numerical results indicate that this machine indeed shows thermodynamic cycles producing an increase of the mechanical energy of the oscillator (which operates also as the work reservoir). However, this behavior is transient only. Eventually, decoherence destroys this motion; it invalidates the factorization approximation and gives way to a quasi-stationary state of heat transport between the two baths (leakage)—a fundamental limitation and the main result of this investigation.

Remark 7.16 Quantum and classical limit *In the quantum limit in a strict sense, that is, a single spin and a low-intensity coherent* *field with* $|\alpha| \approx 1$*, any machine function tends to vanish. This means that there is a threshold for the amplifying behavior:* $|\alpha|$ *must be large enough to get started. But even then amplification does not go on for ever.*

For a full classical behavior we have to not only increase $|\alpha|$ *but also increase the number of spins, N. Coupling and transition operators are then replaced by their collective counterparts, for example,*

$$\hat{V}(AB) \rightarrow L\hat{\Sigma}_3\hat{x}(B), \tag{7.153}$$

$$\hat{\Sigma}_3 \equiv \sum_{\mu=1}^{N} \hat{\sigma}_3(\mu). \tag{7.154}$$

In the limit $N, |\alpha| \rightarrow \infty$ *decoherence disappears and the cycle does no longer deteriorate. (For unlimited operation, the mechanical energy in the oscillator mode had to be stabilized by driving an external appliance.)*

Example 7.8 Single-ion heat engine *The authors [Abah (2012)] consider a single ion of mass m within a trap specified by the pseudo-potential* $\hat{V}$*. The appropriate model description is*

$$\hat{H} = \frac{1}{2m}(\hat{p}_x^2 + \hat{p}_y^2 + \hat{p}_z^2) + \hat{V}(\hat{x}, \hat{y}, \hat{z}), \tag{7.155}$$

$$\hat{V}(\hat{x}, \hat{y}, \hat{z}) = \frac{m\omega_A^2}{2}\frac{(\hat{x}^2 + \hat{y}^2)r_0^4}{(1 + \hat{z}C/2m)^4} + \frac{m\omega_B^2}{2}\hat{z}^2, \tag{7.156}$$

$$C \equiv 2m\tan\theta/r_0. \tag{7.157}$$

Here θ *is the angle between the electrodes and the trap axis z, and* r_0 *the radial distance of the ion to the electrodes. We now exploit the fact that for small* $\hat{z}$ *the denominator of the first part in* $\hat{V}$ *can be expanded. The zero-order term gives the harmonic potential in terms of* $\hat{x}$ *and* $\hat{y}$*, the first-order term the linear coupling between radial (x,y) and axial (z) motion. This coupling term reads*

$$\hat{V}(AB) = -C\hat{z}\omega_A^2(\hat{x}^2 + \hat{y}^2). \tag{7.158}$$

The total model thus splits into two parts, subsystem A (a two-dimensional radial oscillator with ω_A *in terms of x, y) and subsystem B (a one-dimensional axial oscillator with* ω_B *in terms of z), with weak inter-subsystem coupling,* $\hat{V}(A, B)$*,*

$$\hat{H} \rightarrow \hat{H}(A) + \hat{H}(B) + \hat{V}(AB). \tag{7.159}$$

Assuming validity of the factorization approximation, cf. Proposition 5.11, we get the following effective time-dependent Hamiltonian for the axial oscillator,

$$\hat{H}^{eff}(B;t) = \hat{H}(B) - C\omega_A^2 \hat{z} < \psi(A;t)|(\hat{x}^2 + \hat{y}^2)|\psi(A;t) > . \tag{7.160}$$

As we know, see Remark 5.9, the state of (A) may be a general mixed state, in particular, a thermal state. Heating or cooling changes the effective potential for subsystem B and thus the net force (including the constant gravitational force), moving the ion either up or down in the z-direction. These thermally induced displacements (mechanical work) will turn out to represent the analogue to the controlled motion of the piston of a classical heat engine.

To incorporate this effect as a kind of self-control, the present model has to be extended: At each axial end point of the trap a separate bath is positioned: At the upper end point the hot bath of temperature T_h, at the lower point the cold bath of temperature T_c. With one of these baths the radial motion of the ion eventually comes in contact, as it moves up or down, see Fig. 7.10 The contact flips the radial thermal state of the ion, thus flipping the force on (B). This implies a coherent oscillatory axial motion.

Correspondingly, we get the following effective time-dependent Hamiltonian for the radial oscillator (A),

$$\hat{H}^{eff}(A;t) = \hat{H}(A) - C\omega_A^2(\hat{x}^2 + \hat{y}^2) < \psi(B;t)|\hat{z}|\psi(B;t) > . \tag{7.161}$$

The second term on the RHS serves as a correction to the frequency ω_A which depends on the position $< \hat{z} >$. For subsystem (A) the respective frequency ω_A is thus time-dependent, correlated with the axial ion motion inside the trap. As limiting values we define $\omega_1 \leq \omega(A) \leq \omega_2$. ω_1 is reached, when the ion is in contact with the cold bath, ω_2 when in contact with the hot bath, see Fig. 7.10.

In the heat engine mode the axial motion also serves as the work reservoir and would thus continuously be amplified. To keep the cycle parameters stationary we have to transfer the mechanical energy to another "mechanical" subsystem.

Remark 7.17 Efficiency for maximum work output *For each thermodynamic cycle the results of Section 7.3.1 apply with $g(\gamma_i) = \gamma_i = \omega_i/\omega_0$, $i = 1,2$; $\alpha_2 = \gamma_1\beta_c$, and $\alpha_1 = \gamma_2\beta_h$.*

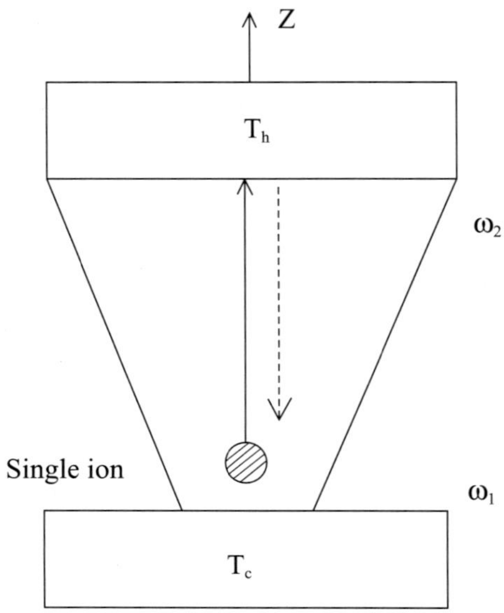

Figure 7.10 Autonomous single ion heat engine. The ion oscillates between the two baths.

Step $1 \rightarrow 2$ *(isochoric,* $\omega_1 = $ *const.): The oscillator state changes from* β_c *to* β_h*. (At upper end point of trap weak coupling to cold heat bath.)*

Step $2 \rightarrow 3$ *(adiabatic,* $\beta_h = $ *const.): The spectrum changes from* ω_1 *to* ω_2*. (Due to effective force ion moves down along z-axis.)*

Step $3 \rightarrow 4$ *(isochoric,* $\omega_2 = $ *const.): The oscillator state changes from* β_h *to* β_c*. (At lower end point of trap weak coupling to hot heat bath.)*

Step $4 \rightarrow 1$ *(adiabatic,* $\beta_c = $ *const.): The spectrum changes from* ω_2 *to* ω_1*. (Due to effective force ion moves up along z-axis.)*

Based on Eq. (4.45) we thus identify

$$\gamma_1 h(\alpha_2) = \frac{1}{2}\hbar\omega_1 \coth\left(\beta_c \hbar\omega_1/2\right) = U\left(\alpha_2, \gamma_1\right), \tag{7.162}$$

$$\gamma_2 h(\alpha_1) = \frac{1}{2}\hbar\omega_2 \coth\left(\beta_h \hbar\omega_2/2\right) = U\left(\alpha_1, \gamma_2\right), \tag{7.163}$$

$$\gamma_2 h(\alpha_2) = \frac{1}{2}\hbar\omega_2 \coth\left(\beta_c \hbar\omega_1/2\right) = U\left(\alpha_2, \gamma_2\right) = \frac{\omega_2}{\omega_1} U\left(\alpha_2, \gamma_1\right), \tag{7.164}$$

$$\gamma_1 h(\alpha_1) = \frac{1}{2}\hbar\omega_1 \coth\left(\beta_h \hbar\omega_2/2\right) = U\left(\alpha_1, \gamma_1\right) = \frac{\omega_1}{\omega_2} U\left(\alpha_1, \gamma_2\right). \tag{7.165}$$

The total work per cycle is according to Eq. (7.31)

$$\Delta W_{tot} = W_{23} + W_{41} = (\gamma_2 - \gamma_1)(h(\alpha_2) - h(\alpha_1)) \tag{7.166}$$

$$= \frac{1}{2}(\hbar\omega_2 - \hbar\omega_1)\coth\left(\beta_c \hbar\omega_1/2\right) + + \frac{1}{2}(\hbar\omega_1 - \hbar\omega_2)\coth\left(\beta_h \hbar\omega_2/2\right). \tag{7.167}$$

The total work in the high-temperature limit ($\beta\hbar\omega \ll 1$) is

$$\Delta W_{tot} \approx \frac{1}{\beta_c}\left(\frac{\omega_2}{\omega_1} - 1\right) + \frac{1}{\beta_h}\left(\frac{\omega_1}{\omega_2} - 1\right). \tag{7.168}$$

Assuming ω_1 to be fixed but ω_2 at our disposal the work output is maximized for

$$\frac{\omega_2}{\omega_1} = \sqrt{\beta_c/\beta_h}\,. \tag{7.169}$$

This gives for the standard result

$$\eta_{otto} = 1 - \frac{\omega_1}{\omega_2}. \tag{7.170}$$

the CA efficiency, Eq. (7.94). However, this result is here not related to finite-time aspects, and not related to maximum power.

For shorter cycle times non-equilibrium effects and/or quantum friction (non-adiabaticity) [Abah (2012)] has to be included, cf. Section 7.4.2.

Remark 7.18 Structural limits for ultra-small machines *The spatial separation of heat baths is rather challenging on small scales. Furthermore, the factorization approximation becomes a very problematic condition. The clear distinction between heat and work will no longer hold in the quantum domain. And finite (embedded) systems are expected to show a tendency toward stationary equilibrium, see also biological motors, Section 7.5.4.*

Are continuous machines more promising in the quantum domain? Confer the discussion on ultra-small refrigerators, Section 8.2.1.

7.5.3 *Entangling Machine*

Remark 7.19 Control of entanglement *What are sources and sinks of entanglement? Entanglement typically results from physical interactions between the subsystems involved; a notable exemption is entanglement swapping. (Here one can entangle particles that do not share any common past, cf. Definition 3.89).*

Uncontrolled interactions with the environment tend to destroy entanglement. In fact, the mathematical tool to incorporate such lack of control is to "trace out" the corresponding degrees of freedom. A procedure like this (leading to openness and noise) will typically be destructive, cannot help to improve control—unless the environment happens to be in its ground state.

Another possibility is "dynamical entanglement": Even if the above entanglement sink cannot be removed, its impact may (partly) be suppressed by means of some coherent driving of the system state, as will be discussed below.

For a long time the extent to which cyclic quantum machines could also produce useful non-classical effects in a strict sense was unclear; the standard thermodynamic functions of heating and cooling had been all too dominating.

On the other hand, entanglement has been identified as a resource (for quantum information processing tasks, but also for physical processes like energy transfer). However, entanglement is notoriously unstable with respect to decoherence via external baths. One remedy is active control via error correction, which, in turn depends on quite demanding algorithms and a large physical overhead [Nielsen (2000)]. Alternative approaches gain attention recently and are based on direct "physical design" (self-organization).

Given that the destructing effects due to the baths cannot directly be avoided, it is tempting to ask whether it may be possible to built a machine that could convert mechanical work (supplied from the outside) into entanglement stabilization. It appears that this might, indeed, be possible.

Definition 7.15 Model and entanglement *To understand the physical background, let us consider an interacting two-spin model (Heisenberg model) under external control [Guerreschi (2012)]. This external control is two-dimensional: there are two Hamilton parameters, J and $H_3 \equiv \hbar\Gamma B_3$. B_3 is the external magnetic field, Γ the gyromagnetic ratio, cf. Example 4.1.*

$$\hat{H}(t) = \frac{H_3(t)}{2}(\hat{\sigma}_3(A) + \hat{\sigma}_3(B)) + J(t)\hat{\sigma}_1(A) \otimes \hat{\sigma}_1(A). \tag{7.171}$$

Based on the abbreviations

$$\epsilon(t) = \sqrt{H_3^2(t) + J^2(t)}, \tag{7.172}$$

$$\eta(t) = \frac{\epsilon(t) - H_3(t)}{J(t)}, \tag{7.173}$$

one finds for the four eigenenergies and the eigenstates (expressed in terms of the product states = eigenstates for zero interaction $J = 0$, $|ij> \equiv |i(A)> \otimes |j(B)>$), $j = 1, \ldots 4$:

$$E_1(t) = -\epsilon(t) \quad |e^{(1)}(t)> = \frac{1}{\sqrt{1+\eta^2}}(-\eta|11> + |22>), \tag{7.174}$$

$$E_2(t) = -J(t) \quad |e^{(2)}> = \frac{1}{\sqrt{2}}(|12> - |21>), \tag{7.175}$$

$$E_3(t) = +J(t) \quad |e^{(3)}> = \frac{1}{\sqrt{2}}(|12> + |21>), \tag{7.176}$$

$$E_4(t) = +\epsilon(t) \quad |e^{(4)}(t)> = \frac{1}{\sqrt{1+\eta^2}}(|11> + \eta|22>). \tag{7.177}$$

The spin–spin entanglement in the ground state can be specified by the concurrence:

$$C = \frac{2\eta}{1+\eta^2}. \tag{7.178}$$

We consider two limiting cases: For $J \ll H_3$

$$\epsilon \to H_3 + J^2/H_3, \tag{7.179}$$

$$\eta \to J/H_3, \tag{7.180}$$

and C goes to zero, no entanglement, the ground state is $|e^{(1)}> = |22>$. For $J \gg H_3$

$$\epsilon \to J + H_3^2/(2J), \tag{7.181}$$

$$\eta \to 1 - H_3/J, \tag{7.182}$$

and C approaches 1, the ground state $|e^{(1)}>$ is maximal entangled.

Definition 7.16 Construction of work variable *The time dependence of the two Hamilton parameters $\{J, H_3\}$ is now assumed to derive from a single "distance" parameter $d(t)$:*

$$J(t) = \frac{J_0}{d(t)}, \tag{7.183}$$

$$H_3(t) = \hbar\Gamma B_3(t) = \hbar\Gamma(B_0 - B_1 \exp(-d^2(t)/4\sigma)) . \tag{7.184}$$

The spectrum E_j thus depends on this single parameter $d(t)$. However, this dependence has not the form of a simple spectral deformation, Eq. (4.7)—with the consequence that a thermal state prepared at some d would not remain thermal (in the absence of any bath coupling) for different d. Restricting ourselves to the ground state, this effect will be of no concern.

Definition 7.17 Adiabatic entanglement cycle *Let us suppose that a transition between those two entanglement limits, $C \approx 0$ and 1, respectively, is induced under the condition of adiabaticity, cf. Eq. (6.36). Starting in the ground state, the system entanglement can thus systematically be changed from zero to maximum and back. The function $C(d)$ is unique and reversible, that is, has no hysteresis.*

However, as we couple each spin of our interacting system to a separate heat bath at identical temperature T, the ground state entanglement C for given H_3 and J decreases with increasing T. It would disappear above a certain critical temperature.

It is now possible to show that the average entanglement $\overline{C}$ can be stabilized by means of (adiabatically slow) periodic control cycles in terms of d(t).

Remark 7.20 Stabilization of entanglement *How are the periodic control cycles able to stabilize entanglement? The qualitative idea behind is a kind of periodic "reset" as the ground state is forced into its product form. If the environment temperature is not too high, the system is preferably found in this state of zero entanglement. Under the change of the control parameter d the entanglement would built-up again (in the isolated case); but this transiently happens even in presence of the bath. However, to be operative the cycles have to be faster than the thermal relaxation time. Otherwise the decay of entanglement would have time to finish for each momentary control parameter value d. Non-equilibrium is essential, the process* *is dissipative and irreversible. The resulting entanglement is time-dependent, even erratic, so its practical use will require additional attention.*

Mechanical work (input) is thus used to suppress decoherence (as an output). Such a device is definitely non-classical. This function may be interpreted to define an entangling machine [Guerreschi (2012)].

Remark 7.21 Vibration-induced coherence and energy transport *A similar idea underlies [Chin (2013)]. The authors—having in mind a pigment–protein complex—study an exciton model coupled to a bath. The driving is taken to be "intrinsic," that is, to derive from semiclassical vibrational modes carried by the proteins. (The semiclassical treatment is essential.) On the one hand the emphasis of this paper is on advanced numerical methods. On the other hand the resulting exciton coherence is taken as a kind of resource for the energy transport along the molecular system. Coherent—as opposed to hopping transport—is known to have the potential of high efficiency.*

The system seems to operate in a "holistic way" in the sense that the entangling and the transport function are not modular, not separable from each other. From the engineering point of view this may turn out to be disadvantageous.

7.5.4 *Mechano-Chemical Control Space: Biological Motors*

Contrary to the devices discussed before the Carnot efficiency (based on thermal non-equilibrium, that is, temperature differences) does not apply here. Biological machines typically operate under isothermal conditions and in a two-dimensional "mechano-chemical" control space, cf. Definition 4.23. A convenient description is based on the Gibbs free energy, see Definition 4.31, under the condition of constant pressure and temperature, cf. Fig. 7.11,

$$dG(T, p, x, N_\nu) = -F\,dx + \sum_\nu \mu_\nu dN_\nu \qquad (7.185)$$

The first term is the mechanical energy, $\Delta_x U$, related to displacing a body by dx against an external force, F. The second part is the chemical energy, $\Delta_N U$; it is related to a chemical reaction

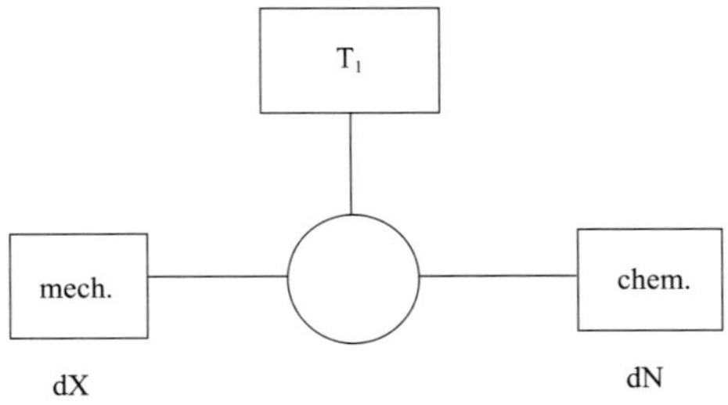

Figure 7.11 Mechano-chemical energy converter.

described by the overall change of particle number dN induced by the difference in chemical potential, $\Delta\mu$, cf. Eq. (4.147).

Definition 7.18 Cycles *An important biological function is to carry cargo to specific compartments within the cell. A significant part of this "cellular traffic" relies on motor proteins that move along filaments [Jülicher (1997)]. The filaments act as a kind of "highways." Various families of such machines have been identified.*

These devices are still based on thermodynamics insofar as they operate close to thermal equilibrium. And similar to the engineering-type thermodynamic machines individual cycles are clearly distinguishable. The total cycle can be characterized by the respective reaction rate, efficiency, and power output.

Definition 7.19 Mechano-chemical efficiency *The output is the mechanical energy $\Delta_x U$, the input the chemical energy $\Delta_N U$. For a system close to thermal equilibrium the pertinent thermodynamic efficiency thus reads*

$$\boxed{\eta_m = \frac{\Delta_x U/\Delta t}{\Delta_N U/\Delta t} \equiv -\frac{v \times F}{r\Delta\mu}.} \tag{7.186}$$

It is convenient to consider here the respective energy differences per time: $dN/dt = r$ is the reaction rate, $v = dx/dt$ the mechanical velocity. dt can be scaled down to the cycle time, the smallest meaningful time unit. The dissipation rate is

$$\dot{W}_{dis} = -v \times F + r\Delta\mu > 0.$$

Under the so-called stall-force condition (implying $v = 0$ and thus zero friction) the efficiency is zero. It holds that $|\eta_m| \leq 1$; dissipation

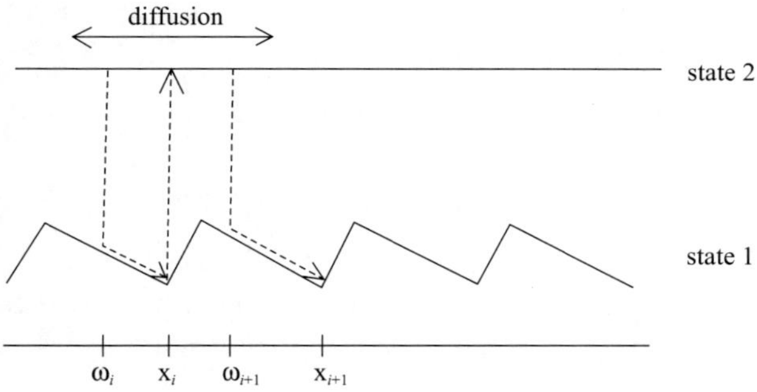

Figure 7.12 Ratchet model (four-stroke motor). For details see text.

may even help to approach the upper limit. Remarkably, for strict thermal equilibrium ($\vec{F} = 0$, $\Delta\mu = 0$) the efficiency cannot be defined [Jülicher (1997)].

The above thermodynamic analysis is abstract and does not yet specify a concrete device. Indeed, also for the Carnot or Otto cycle any implementation requires additional details. For the biological motor the molecular model of what is going on involves:

(i) the nature of translational motion,
(ii) the internal state dynamics within the motor protein,
(iii) the chemical reaction (e.g., hydrolysis) involving the environment, and the mutual relations between these sub-processes. Predominantly, the models are quasi-classical; nevertheless, on the individual molecule level the chemical process becomes discrete and delivers fixed energy quanta.

Definition 7.20 Flashing ratchet model *A simple flashing ratchet model is shown in Fig. 7.12. Here the effective potential changes between an asymmetric sawtooth (period L) and a flat potential, depending on the internal state of the motor [Jülicher (1997); Sasaki (2007)]. The switching between the two states is triggered by certain chemical changes within the motor: State 1: No nucleotide or ATP bound, state 2: ADP bound. The motion of the motor along the sawtooth potential is biased (uni-directional), the motion along the flat potential is diffusive. The net motion is biased on average.*

The whole process is cyclic (modulo the shift along the track). Qualitatively we distinguish four steps:

(1) *State 1; directed motion from position w_i to x_i.*
(2) *State change $1 \rightarrow 2$.*
(3) *State 2; diffusive motion to w_{i+1} or back to w_i.*
(4) *State change $2 \rightarrow 1$.*

Depending on the details of modeling there may be more than four steps within a cycle. Chemical energy is provided by

$$\mathrm{ATP} \rightarrow \mathrm{ADP} + \mathrm{P} + \Delta U_N\,, \tag{7.187}$$

implying $\Delta\mu = \mu_{ADP} + \mu_P - \mu_{ATP}$ and $dN = -dN_{ATP} = dN_{ADP} = dN_P$. This energy is converted into mechanical energy ΔU_N. In fact, this conversion is quantized, that is, comes in discrete steps: The consumption of a single ATP molecule leads to a single step forward, $d\vec{x}$, in the position of the motor. Typically the step size varies in multiples of the basis length L. It is assumed that the hydrolysis of ATP is catalyzed by a conformational change of the motor associated with its translational motion along the track. The chemical output is ADP and Phosphor P.

Remark 7.22 Forward/backward cycles *Cycles have to be physically implemented and are defined with respect to a corresponding control space. For simple cycles the forward and backward mode of operation are strictly interrelated (cf., e.g., the Carnot or Otto cycle).*

In the biological realm such relations can sometimes be violated calling for the use of qualitatively different cycles in the forward and backward direction [Klumpp (2012)].

Chapter 8

Continuous Operations

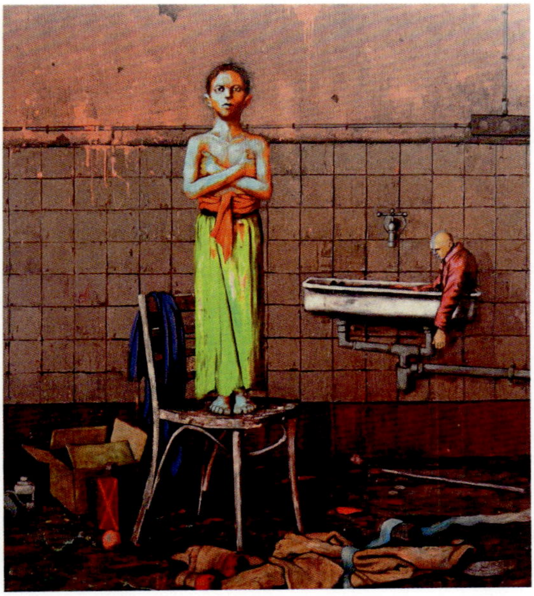

Jonas Burgert: Final Pride, 2007, based on photography by Lepkowski Studios [Reprinted with permission from Schreiber and Wipplinger (2011)].

Quantum Thermodynamic Processes: Energy and Information Flow at the Nanoscale
Günter Mahler

ISBN 978-981-4463-73-7 (Hardcover), 978-981-4463-74-4 (eBook)
www.panstanford.com

Cyclic thermodynamic machines have been characterized by the respective central node and the attached subsystems, see Definition 7.1. Typically we have to distinguish a total of three subsystems operating as work or heat reservoirs, respectively.

Thermodynamic functions realized by means of cyclic control can often be based also on continuous operation. In fact, continuous operation may be advantageous as no time-dependent mechanical control is needed ("no moving parts"). Probably the first such proposal in the quantum regime has been by [Scovil (1959)] addressing the three-level maser as a thermodynamic machine. The language was still borrowed from the cyclic counterpart. In general, however, the physics behind such devices differs significantly from its cyclic version. One may even say that knowing the architecture for cyclic control does not necessarily help to find the architecture for the continuous case.

Are there new overarching principles? It seems there are: In the quantum regime such continuous machines can often be understood based on three-body interactions (replacing a separate central node) and/or from the perspective of "virtual qubits" at "virtual temperatures" [Brunner (2012); Mahler (2012)]. The former concept allows to establish a kind of advanced "interaction logic"; the latter assumes the feasibility of a transition-selective thermal control (non-equilibrium). This is the basic source of non-equilibrium. While quite simple theoretically, its practical implementation may be challenging.

8.1 Virtual Qubit and Virtual Temperature

Definition 8.1 Effective two-level system: Virtual qubit *Consider a system composed of two non-interacting parts A, B with discrete spectra,* $E_i(A)$, $E_j(B)$. *Let each subsystem be in a thermal state with temperature* $T(A)$ *and* $T(B)$, *respectively, cf. Fig. 8.1. The respective occupation probabilities are* $p_i(A)$ *and* $p_j(B)$. *Then pick two states from A with energies* $E_g(A) < E_e(A) \equiv E_g(A) + \Delta E(A)$ *and two states form B with energies* $E_g(B) < E_e(B) \equiv E_g(B) + \Delta E(B)$.

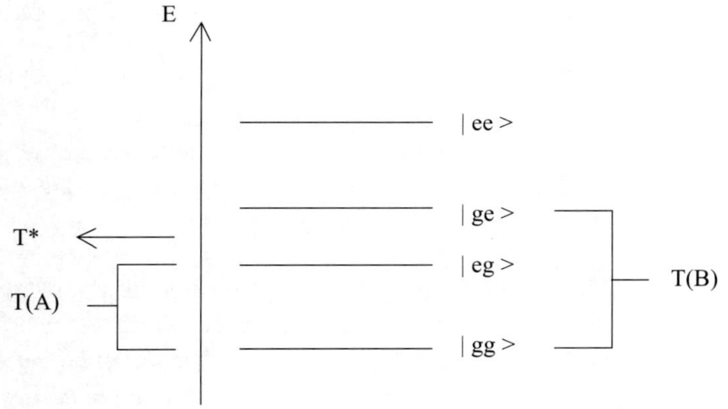

Figure 8.1 Virtual qubit: A and B taken as two-level systems. Brackets indicate imposed bath couplings. Virtual temperature T^* as "output" between $|ge>$ and $|eg>$.

Out of these select the two product states

$$|\psi_1> = |e(A), g(B)> \qquad E_1 = E_g(A) + E_g(B) + \Delta E(A), \quad (8.1)$$

$$|\psi_2> = |g(A), e(B)> \qquad E_2 = E_g(A) + E_g(B) + \Delta E(B). \quad (8.2)$$

Their respective occupation probabilities are $p_1 = p_e(A)p_g(B)$ *and* $p_2 = p_g(A)p_e(B)$. *Note that* $p_1 + p_2 < 1$. *These two selected states with energy splitting* $E^* \equiv E_2 - E_1 = \Delta E(B) - \Delta E(A) > 0$ *are now said to define our "virtual qubit,"*

$$\exp(-E^*/k_B T^*) \equiv \frac{p_2}{p_1} = \frac{\exp(-E_2/k_B T(B))}{\exp(-E_1/k_B T(A))}. \quad (8.3)$$

The "virtual temperature" T^*,

$$\boxed{T^* = \frac{E_2 - E_1}{E_2/k_B T(B) - E_1/k_B T(A)}} \quad (8.4)$$

is the temperature of an external bath that would be needed to generate those relative occupation probabilities directly. It can be positive or negative. T^* *is negative, if*

$$1 < \frac{E_2}{E_1} < \frac{T(B)}{T(A)} \equiv \frac{T_h}{T_c}. \quad (8.5)$$

Up to this point, however, this description is nothing but an ad hoc theoretical exercise. For it to become operative "in the real world," the virtual qubit has to be addressable, that is, one should be able to selectively couple it to other subsystems or baths. This is non-trivial, as it relates to actual design requirements. But only then can this concept become a powerful tool for implementing advanced machine functions.

Remark 8.1 "Effective walls" *In the macro-domain the origin of non-equilibrium has been described in terms of "walls" allowing for multiple state variables, cf. Definition 4.26. The virtual qubit concept may be considered a kind of analogue in the nano-domain, directly operating in frequency space.*

As shown below, the sign of virtual temperature controls the type of machine operation: Negative T^* will turn out to be associated with the heat engine mode, positive T^* with the refrigerator mode. It has been suggested [Brunner (2012)] to distinguish between work (as related to "negative virtual temperatures") and heat (as related to "positive virtual temperatures") [Brunner (2012)]: This is incorrect. Inverting probabilities amounts to characterizing a state, but work as well as heat has to be characterized by a process, not just a state [Mahler (2012)]. Thermalization, whether involving positive or negative temperatures, is always an irreversible process and not related to work in a strict sense.

In the examples to follow two or more selective bath couplings will be *assumed* to occur, so the underlying virtual qubit becomes operative—by definition. The concept may help to understand at first sight counterintuitive results.

Remark 8.2 Spectral temperature *For non-equilibrium states (which are still diagonal in the energy basis) a generalized temperature concept has been proposed [Gemmer (2009)]. It is based on the observation that for each neighboring pair of levels an effective temperature can be introduced—quite similar to virtual temperatures. For a true canonical state all these would be identical; in general, an overall characteristic temperature can be defined via appropriate averaging.*

8.1.1 *Two-Spin Node: Heat Engine (lasing)*

The simplest realization of a virtual qubit is via two "real" spins (A,B) with energy splittings $\Delta E(A) < \Delta E(B)$. Each spin is coupled to a different heat bath, RA, RB with temperature T(A) and T(B), respectively. The selective coupling is facilitated by the fact that the two spins may be spatially separated at opposing ends of a cavity. This cavity serves as the carrier of a single field mode (C), described as a quantum oscillator of frequency $\Omega(C)$. The field mode is resonantly coupled to the virtual transition $E^* = \Delta E(B) - \Delta E(A)$:

$$\hat{H}(ABC) = \frac{\Delta E(A)}{2}\hat{\sigma}_3(A) + \frac{\Delta E(B)}{2}\hat{\sigma}_3(B) + \hbar\Omega(C)\hat{a}^\dagger\hat{a}$$
$$+\hbar\lambda(\hat{\sigma}^-(A)\hat{\sigma}^+(B)\hat{a}^\dagger + \hat{\sigma}^+(A)\hat{\sigma}^-(B)\hat{a}) . \quad (8.6)$$

The ladder operators have been defined in Eq. (3.114). A negative virtual temperature T^* obtains for

$$1 < \frac{\Delta E(B)}{\Delta E(A)} < \frac{T(B)}{T(A)} \equiv \frac{T_h}{T_c} . \quad (8.7)$$

One may expect that under this condition an amplification of the oscillator state (a lasing process of power $\mathcal{P}(C)$) should result; the (infinite) energy resource would have to be provided by the two external heat baths RA, RB, that is, by the heat currents $\dot{Q}(A)$, $\dot{Q}(B)$; obviously there is no work done on the system (ABC). Under reversible operation (see Eq. (7.56)),

$$\frac{\dot{Q}(B)}{T(B)} + \frac{\dot{Q}(A)}{T(A)} = 0 , \quad (8.8)$$

and observing the energy conservation under stationary conditions,

$$\mathcal{P}(C) + \dot{Q}(A) + \dot{Q}(B) = 0 , \quad (8.9)$$

we get for the efficiency

$$\eta_{carnot} \equiv \frac{\mathcal{P}(C)}{\dot{Q}(B)} = 1 - \frac{T(A)}{T(B)} . \quad (8.10)$$

However, this simple analysis does not confirm the nature of $\mathcal{P}(C)$ as work W per time. A separate investigation is in order [Youssef (2009)]. We apply the LEMBAS principle, Section 5.2.4, for subsystem (C), the field mode. We convince ourselves that

$$\hat{H}_0^{eff}(C) = \hbar\Omega(C)\hat{a}^\dagger\hat{a}\,, \quad (8.11)$$

$$\hat{H}_1^{eff}(C) = \hbar\lambda(<\hat{\sigma}^-(A)\hat{\sigma}^+(B)>\hat{a}^\dagger + <\hat{\sigma}^+(A)\hat{\sigma}^-(B)>\hat{a})\,. \quad (8.12)$$

The respective flow of work exchanged between (AB) and (C) is thus

$$\dot{W}(C) = -\frac{i}{\hbar}\mathrm{Tr}\{[\hat{H}_0^{eff}(C);t),\,\hat{H}_1^{eff}(C);t)]\hat{\rho}(C)\}\,. \quad (8.13)$$

Observing that $[\hat{a}^\dagger\hat{a},\,\hat{a}^\dagger] = \hat{a}^\dagger$ and $[\hat{a}^\dagger\hat{a},\,\hat{a}] = -\hat{a}$ we find according to Eq. (5.134)

$$\begin{aligned}\dot{W}(C) &= -\frac{i}{\hbar}\mathrm{Tr}_C\{[\hat{H}_0^{eff}(C),\,\hat{H}_1^{eff}(C)]\}\\ &= -i\hbar\Omega\lambda(<\hat{\sigma}^-(A)\hat{\sigma}^+(B)><\hat{a}^\dagger>\\ &\quad -<\hat{\sigma}^+(A)\hat{\sigma}^-(B)><\hat{a}>)\,. \quad (8.14)\end{aligned}$$

This term is obviously zero, unless the oscillator is in a local coherent state. As it turns out [Youssef (2009)] the interaction between (C) and (AB) leads to entanglement and thus destroys local coherences. The energy gain $\mathcal{P}(C)$ is thus heat, not work. In fact, the field is in a phase-diffused Glauber state (Poisson statistics, cf. Definition 3.36) with continuously increasing mean photon number. This finding is in agreement with the closed relaxation model, Example 5.18. In this latter example the phase-diffused Glauber state shows up as a transient non-equilibrium state only, which asymptotically reaches the appropriate standard equilibrium. In the present model the non-equilibrium is frozen-in due to the infinite baths of different temperature, that is, due to infinite energy supply.

Remark 8.3 Three-level system + oscillator mode *Alternatively, the two-spin system (AB) can be replaced by a three-level system* (A^*) *with the two separate bath contacts between levels* $0-2(T_h)$ *and* $0-1(T_c)$. *The oscillator* (C) *is resonantly coupled to the virtual qubit defined by* $E^* = E_{21}$ *(see Fig. 8.2). The requirement for heat engine operation then is*

$$\frac{\Delta E_{20}}{\Delta E_{10}} < \frac{T_h}{T_c}\,. \quad (8.15)$$

The physical properties are essentially the same as for the two-spin model. In this case, based on a different definition (which does

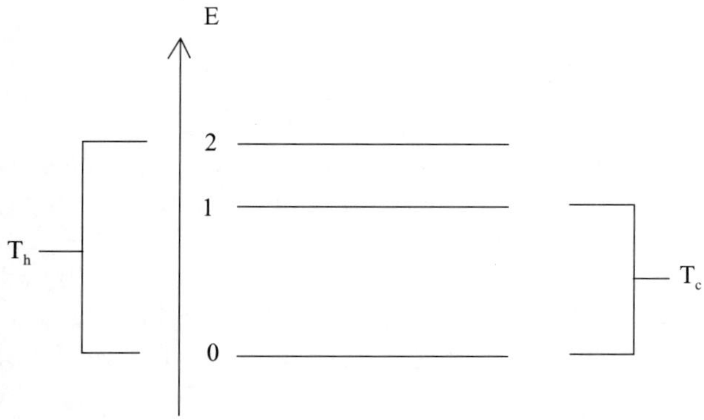

Figure 8.2 Three-level system with two separate baths. Selective coupling indicated as brackets.

not insist work to be associated with constant entropy), $\mathcal{P}(C)$ has been interpreted as work in [Boukobza (2006)]. This assignment is questionable, though.

8.1.2 *Effective Three-Level Node: Refrigerator*

Reversible cyclic machines can always be run in reverse sense of rotation to change the signs of all energy fluxes. This is not possible for continuous machines; nevertheless, the same machines running as heat engines (e.g., the two-spin or the three-level model) can also be used as refrigerator (or heat pump). One simply has to adjust the parameters such that the virtual temperature T^* is no longer negative. This means for the three-level model:

$$\frac{\Delta E_{20}}{\Delta E_{10}} > \frac{T_h}{T_c} \,. \tag{8.16}$$

Along these lines a simple quantum refrigerator has been discussed by [Kosloff (2000)]. Its efficiency is bounded from above by the Carnot result, Eq. (7.64).

A different architecture has been suggested by [Kolar (2012)]. Here the central node (A) is taken to be a single spin subject to a classical driving (classical work input). This driving results in a frequency modulation described by $f(t)$,

$$\hat{H}(A;t) = \frac{\hbar(\omega_0 + f(t))}{2}\hat{\sigma}_3(A), \tag{8.17}$$

Assuming a strictly periodic driving two side bands result, $E_{20} \equiv \hbar(\omega_0 + \Delta\omega)$ and $E_{10} \equiv \hbar(\omega_0 - \Delta\omega)$, respectively. Together with the ground state $E_0 \equiv 0$ this constitutes an effective three-level system ("Floquet representation").

As before, this three-level system is permanently coupled to two heat baths, (B) and (C). But these are both treated fully quantum now. The hot bath (B) of temperature T_h is selectively coupled to the transition $2-0$ via operator $\hat{b}(B)$ and is an infinite reservoir; the cold bath (C) of temperature T_c is coupled to the transition $1-0$ via operator $\hat{b}(C)$ and is a finite system. The respective coupling strengths are $\gamma(\nu)$, $\nu = B, C$. The total Hamiltonian takes the form of a generalized spin-Boson type model, cf. Example 3.13,

$$\hat{H}(ABC) = \hat{H}(A;t) + \hat{H}(B) + \hat{H}(C) + \hat{V}_1(AB) + \hat{V}_1(AC), \tag{8.18}$$

$$\hat{V}_j(A\nu) = \frac{\gamma_\nu}{2}\hat{\sigma}_j(A)(\hat{b}(\nu) + \hat{b}^\dagger(\nu)), \tag{8.19}$$

$$\hat{b}(\nu) \sim \sum_k g_k(\nu)\hat{a}_k(\nu). \tag{8.20}$$

Each Boson mode is decomposed into a large set of eigen modes to account for a specific physical bath. The heat pump or refrigerator condition is the same as before, that is, specified by Eq. (8.16):

$$\frac{\omega_0 + \Delta\omega}{\omega_0 - \Delta\omega} > \frac{T_h}{T_c} \tag{8.21}$$

which can alternatively be written as

$$\Delta\omega > \Delta\omega_c \equiv \omega_0\frac{T_h - T_c}{T_h + T_c} \tag{8.22}$$

The occupation of the Floquet levels is controlled by the respective non-Markovian master equation (including the baths). The bath (B) is assumed to be constant. As the cold bath (C) is finite, the heat flow $\dot{Q}(C)$ out of the cold bath will lower its temperature as a function of time,

$$C_V(T_c)\dot{T}_c = \dot{Q}_c < 0. \tag{8.23}$$

With the heat capacity $C_V(T_c) \sim T_c^\delta$, where $\delta \geq 1$ (c.f. Theorem 4.14 on the static version of the third law) and $\dot{Q}_c \sim -T_c^\epsilon$ we find

$$\dot{T}_c \sim -T_c^{\epsilon-\delta}. \tag{8.24}$$

For $(\epsilon - \delta) > 0$ the third law would dynamically be satisfied, that is, $T_c = 0$ cannot be reached within finite time. The authors claim that for specific coupling models $g_k(C)$ the parameter ϵ might become smaller than δ so that we could reach $T_c = 0$ within finite time.

However, the applicability of the standard thermodynamic relation Eq. (8.23) becomes doubtful here; in fact, weak coupling requires the interaction energy to be negligible with respect to any subsystem energy, so γ_C has to go to zero as T_c (and thus the energy of (C)) approaches zero; otherwise the finite coupling γ_C would necessarily lead to a local mixed state for (C), that is, to a non-zero effective temperature. But with γ_C the cooling rate also goes to zero and the third law is reinforced, cf. Section 8.3.1.

As expected, by changing the modulation rate to $\Delta\omega < \Delta\omega_c$ the machine can alternatively be run as a heat engine [Gelbwasser-Klimowsky (2013)].

8.1.3 *Cooling by Heating*

It has been noted early on that cooling does not necessarily require the input of work. However, in the classical domain pertinent scenarios require thermal systems with additional "non-mechanical" degrees of freedom.

Example 8.1 Classical version: Absorber refrigerator *The absorber refrigerator is based on an invertible chemical reaction $X \leftrightarrow Y$, the direction of which is controlled by the respective bath temperature (for more details, see, e.g., [Srikhirin (2001)]):*

$$T_1 \; on \; X \longrightarrow Y + \; waste \; heat \; T_2 \,, \tag{8.25}$$

$$T_3 \; on \; Y \longrightarrow X + \; waste \; heat \; T_2 \,, \tag{8.26}$$

with $T_3 > T_2 > T_1$. The node consists of a continuous cyclic flow, so that a given material cross-section passes alternatively the high- and low-temperature regions, see Fig. 8.3. To maintain this flow, some external pumping is needed (which can also provided by heat only). The respective cooling efficiency is $\eta^{fridge} = \dot{Q}_1/\dot{Q}_3$.

What kind of chemical reactions can be used? A typical implementation is based on NH_3 as the refrigerant, water as the absorbent. The concentration of the refrigerant serves as a thermodynamic variable

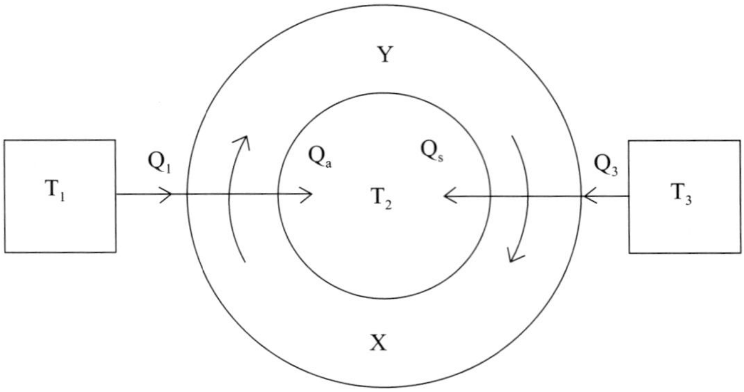

Figure 8.3 Classical absorber refrigerator: Simplified scheme. $X \rightarrow Y$ and $Y \rightarrow X$ denote chemical reactions, Q_i are heat currents. The node requires a circular material current.

in addition to pressure and temperature. In the absorption process we go from X (NH_3) to Y (NH_3 in aqueous solution); the inverse is the so-called separation process, induced by high temperature.

Surprisingly, in the quantum domain "work-less" refrigerators can be obtained in a much simpler fashion.

Example 8.2 Quantum version: Qutrit node *Consider a three-level system (eigenvalues $E_0 \equiv 0 < E_1 < E_2$) with the transition $0 \leftrightarrow 1$ selectively coupled to bath $j = 1$ with T_1 and transition $0 \leftrightarrow 2$ coupled to bath $j = 2$ with $T_2 > T_1$. The virtual temperature for transition $2 \leftrightarrow 1$ then is*

$$1/\beta^* = \frac{\omega_2 - \omega_1}{\beta_2\omega_2 - \beta_1\omega_1}\,. \tag{8.27}$$

In the following we suppose β^* to be positive. When we couple this very transition to a third bath $j = 3$ with temperature T_3, a conflict results: In the steady state heat currents flow, constrained by

$$\sum_{j=1}^{3} \dot{Q}_j = 0 \tag{8.28}$$

which is connected with the entropy production rate

$$\dot{S} - \sum_{j=1}^{3} \dot{Q}_j/T_j > 0 \tag{8.29}$$

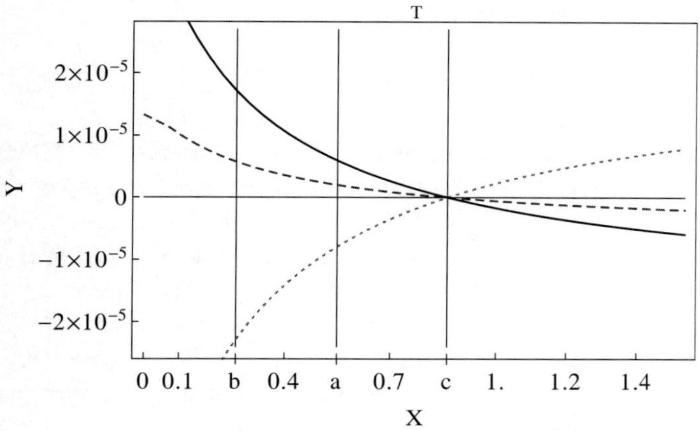

Figure 8.4 Stationary state heat fluxes, $Y = \{\dot{Q}_3, \dot{Q}_1, \dot{Q}_2\}$, as a function of the third bath temperature $X = T_3$ with $a = T_2$ (hot), $b = T_1$ (cold), $c = T^*$. For $T_3 < T^*$ the current $\dot{Q}_3$ starts largest, then $\dot{Q}_1$ and below (negative) $\dot{Q}_2$.

For $T_3 > T^*$ heat flows from the cold bath 1 to the hot bath 2, driven by the heat drawn from bath 3. The efficiency is

$$\eta^{fridge} = \frac{\dot{Q}_1}{\dot{Q}_3} = \frac{\omega_1}{\omega_2 - \omega_1} > 0 \tag{8.30}$$

At $T_3 = T^*$ all currents vanish (equilibrium), at $T_3 < T^*$ all currents switch sign, cf. Fig. 8.4.

Exactly along these lines (but replacing the three-level system by two qubits (A,B)) a “solar refrigerator” has been proposed [Cleuren (2012)]. In this case heat is pumped (associated with electrons) between two metallic leads, representing the cold bath 1 and the hot bath 2, respectively. A quantum dot (qubit) A with level-splitting $\hbar\omega(A)$ is connected with the cold bath 1, qubit B with level-splitting $\hbar\omega(B) > \hbar\omega(A)$ is connected with the hot bath 2. The solar photons of average energy $\hbar\omega = \hbar(\omega(B) - \omega(A))$ play the role of bath $j = 3$, that is, the external (thermal) energy source. Again, for $T_3 > T^*$ heat flows from the cold bath 1 to the hot bath with efficiency

$$\eta^{fridge} = \frac{\omega(A)}{\omega(B) - \omega(A)} > 0 \tag{8.31}$$

8.2 Three-Body Interactions

Definition 8.2 Non-linearity *Quantum mechanics is linear, see Remark 3.10). One speaks of non-linearity in Hamilton models, if three (or more) different subsystems are directly involved in a given interaction term rather than two (pairwise interactions). For phonon modes, for example, interactions of this type would specify anharmonicity. It has been argued that such a "non-linearity" was essential for a quantum absorber refrigerator to work [Martinez (2013)].*

Considering the simple three-bath qutrit model, Example 8.2 or the substitute model according to Remark 8.4, this statement must be taken with a grain of salt. Nevertheless, non-linearity is useful, indeed, as will be shown in the two examples to follow.

8.2.1 *Three-Spin Model*

The cooling of infinite systems can be characterized by a stationary heat current only, not by the actual lowering of temperature. In the following model we are concerned with the cooling of the smallest possible system, a single qubit (in the following spin (A)). Rather than looking at the detailed relaxation dynamics we focus on the asymptotic limit. The underlying mechanism combines cooling by heating (i.e., heat as the only resource) with SWAP-like interactions typical for algorithmic cooling, see Section 6.5. Let us consider the three-spin model [Linden (2010)]

$$\hat{H}(A, B, C) = \sum_{\nu} \hat{H}_0(\nu) + \hat{V}(ABC), \tag{8.32}$$

$$\hat{H}_0(\nu) = \frac{\hbar\omega(\nu)}{2}\hat{\sigma}_3(\nu), \tag{8.33}$$

$$\hat{V}(ABC) = \kappa(|010><101| + |101><010|). \tag{8.34}$$

Here the local spin states are written as $|0>; |1>$, the interaction $\hat{V}$ is in product state representation. Typical interactions we encountered so far had been bilinear in the form $\hat{V}(\mu\nu)$. The resonant (weak) three-body interaction $\hat{V}(ABC)$ with

$$\omega(B) = \omega(A) + \omega(C), \tag{8.35}$$

connects energetically degenerate states. It may be seen as implementing a generalized SWAP gate, cf. Example 3.22.

The three spins are now selectively coupled to two different baths: Spin A and B to a bath with temperature T_c and spin C to a bath with temperature $T_h > T_c$. Without interaction each subsystem would thus approach its thermal equilibrium state $\hat{\rho}_{can}(\nu)$. In relaxation-time approximation, see Definition 6.4, the subsystem dynamics is controlled by

$$\frac{d}{dt}\hat{\rho}(\nu) = -\frac{i}{\hbar}[\hat{H}_0(\nu), \hat{\rho}(\nu)] - \frac{1}{\tau(\nu)}(\hat{\rho}(\nu) - \hat{\rho}_{can}(\nu)) . \tag{8.36}$$

The stationary state of the total system is a simple product.

This state is modified once we include the interaction $\hat{V}$, which induces energy flows within the system: Spin B can receive per unit time energy $\hbar\omega(A)$ from A only, if accompanied by an energy transfer $\hbar\omega(C)$ from C. This process is more likely than its reverse, as spin C is in contact with a hot bath. On average, spin B thus gains energy, the other two spins loose energy. In the stationary limit the internal loss/gain of energy has locally to be counterbalanced by non-zero heat currents between the spins and their respective baths, cf. Eq. (3.374),

$$\begin{aligned} \dot{Q}(\nu) &= \mathrm{Tr}\{\hat{H}_0(\nu)\frac{d}{dt}\hat{\rho}(\nu)\} \\ &= -\frac{1}{\tau(\nu)}\mathrm{Tr}\{\hat{H}_0(\nu)(\hat{\rho}(\nu) - \hat{\rho}_{can}(\nu))\} , \end{aligned} \tag{8.37}$$

Defining for the resulting non-equilibrium state,

$$\dot{Q}(A) \equiv \dot{Q}_c , \tag{8.38}$$

$$\dot{Q}(B) = -\dot{Q}_c - \dot{Q}_h , \tag{8.39}$$

$$\dot{Q}(C) \equiv \dot{Q}_h . \tag{8.40}$$

consistency requires

$$\frac{\dot{Q}_c}{\dot{Q}_h} = \frac{\omega(A)}{\omega(C)} . \tag{8.41}$$

As heat flows from warmer to colder, spin A must have been cooled down with respect to its bath T_c, spin B heated up. The node (B, C) continuously extracts heat from A. This is in agreement with the virtual qubit analysis: With $E^* = \hbar(\omega(B) - \omega(C)) > 0$ the virtual temperature T^* of node (B,C) is positive,

$$\frac{\omega(B)}{T_c} > \frac{\omega(C)}{T_h} \tag{8.42}$$

As is common for the cyclic thermal machines, efficiencies are introduced as ratios of respective energy currents (energy per cycle) in the stationary limit. Along these lines the cooling efficiency of the device with respect to spin A may be defined as

$$\eta^{fridge} \equiv \frac{\dot{Q}(A)}{\dot{Q}(C)} = \frac{\omega(A)}{\omega(C)} . \tag{8.43}$$

Spin (C) extracts heat from the hot reservoir. This simple qualitative analysis does not provide the local effective temperatures $T_{eff}(A) < T_c$ or $T_{eff}(B) > T_c$ and their dependence on the various input parameters; for details see [Linden (2010); Popescu (2009)]. It is shown there that for $T_{eff}(A) \to 0$ one would have to let $\omega(C) \to \infty$. With the spin A approaching its ground state, the weak coupling limit would require to have the coupling parameter κ in $\hat{V}$ to go to zero—resulting in a cooling rate going to zero, in agreement with the third law.

Remark 8.4 Alternative model *The two-spin node (B,C) can be substituted by a single qutrit (B^*) (coupled to two different baths). Again there is qubit (A) to be cooled; in this way one avoids the use of a formal three-body interaction, that is, non-linearity [Linden (2010)].*

8.2.2 *Three-Oscillator Model*

The following "no-node model" by [Mari (2012)] (see also [Levy (2012)]) is also based on a direct three-body interaction: The respective optomechanical system consists of two optical field oscillators, A and B, and one mechanical oscillator C. The total Hamiltonian is

$$\hat{H}(ABC) = \sum_{\nu=A,B,C} \hbar\omega(\nu)\hat{a}^\dagger(\nu)\hat{a}(\nu) + \hat{V}(ABC) , \tag{8.44}$$

$$\hat{V}(ABC) = \hbar g(\hat{a}(A) + \hat{a}(B))^\dagger(\hat{a}(A) + \hat{a}(B))(\hat{a}(C) + \hat{a}^\dagger(C)) , \tag{8.45}$$

$$\omega(A) - \omega(B) = \omega(C) . \tag{8.46}$$

(C) thus couples to the virtual qubit $E^* = \hbar(\omega(A) - \omega(B)) > 0$. $\hat{V}(ABC)$ represents the non-linearity, cf. Section 8.2.1. Together with the resonance condition it establishes certain "logical constraints" in the sense of one particular transition to happen only together with some other transition. There is no separate node to connect the three fields.

In addition to coherent dynamics the system is subject to damping (local baths). As usual, their influence is described by the Lindblad master equation,

$$\frac{d}{dt}\hat{\rho} = -\frac{i}{\hbar}[\hat{H}(ABC), \hat{\rho}] + (\hat{\mathcal{L}}_{inc}(A) + \hat{\mathcal{L}}_{inc}(B) + \hat{\mathcal{L}}_{inc}(C))\hat{\rho}\,. \quad (8.47)$$

Now, mode (A) is assumed to be not driven and to stay close to its ground state, that is, $T(A) \approx 0$, while field (B) is taken to be prepared in an incoherent thermal state of temperature $T(B) > T(A)$. The virtual temperature is thus positive,

$$\frac{\omega(A)}{T(A)} > \frac{\omega(B)}{T(B)} \quad (8.48)$$

and we expect to be in the refrigerator mode with respect to the field (C). Indeed, detailed analysis shows that this "machinery" may lead to a reduction of the average occupation number for the mechanical mode (C), that is, to cooling (with respect to the state with zero coupling g).

8.3 Special Applications

We close by two applications: Both have to do with the control of entanglement.

8.3.1 *Dynamical Aspects of the Third Law Revisited*

For time-dependent Hamiltonians (classical driving) deviations from adiabaticity is an issue; it leads to friction and by that validates the third law. In the continuous case there is no explicit time dependence; limitations derive from the finite coupling strength to the environment instead: Inter-system coupling leads to entanglement (a quantum effect) and by that to local uncertainty, which is incompatible with zero temperature.

In [Cleuren (2012)], the authors came up with the claim that the rate of heat extraction from the cold bath would scale like

$$\dot{Q}_c \sim T_c^{\epsilon} \quad \epsilon = 1\,. \quad (8.49)$$

With this result put into Eq. (8.24) the term $\dot{T}_c$ should be at least (negative) constant for $T_c \rightarrow 0$ or might even diverge. This would allow to reach zero temperature in finite time, thus violating the dynamic version of the third law, similar to the case studied above.

Not surprisingly, this claim has roused controversy. Unfortunately, the debate cannot avoid to touch upon the validity and applicability of models to decide on fundamental issues. Models are always ambiguous. In order to proceed it is helpful to introduce general features characterizing the limiting behavior $T_c \rightarrow 0$: One should distinguish "circumstantial limits" and "dysfunctional" limits, as suggested in [Allahverdyan (2012)].

- Circumstantial limits: The set of model properties is in some sense accidental; it results from simplifying the scenario, thus discarding unimportant details. If seen as such an oversimplification a "more realistic" model may loose the desired function in the limit under investigation. Nevertheless, this consequence could appear *preliminary*, pending further technological sophistication, say. In fact, such kind of idealizations could thus become reachable even in practice.
- Dysfunctional limits: The set of model properties, while being essential for the desired function, would (asymptotically) destroy the very functionality of the device. It thus indicates an *internal conflict*: The property needed for the device operation will eventually suppress functionality in the limit of interest. Obviously, this conflict *cannot be removed* by means of future technology. The apparent violation of the third law (as a fundamental issue) should thus be removed with the help of a dysfunctional rather than a circumstantial argument.

[Levy (2012)] try to cure the violation of the third law by means of a "circumstantial argument": They propose to include an additional term in the original model by Cleuren et al. [Cleuren (2012)]. This amendment would do the job on the premise that any technological progress allowing to suppress this term would remain beyond reach.

Alternatively, [Allahverdyan (2012)] provide a "dysfunctional argument": For $T_c \to 0$ the system to be cooled has to find itself in its ground state, which is impossible provided that both the system–environment interaction and its commutator with the full Hamiltonian stay finite. This feature corroborates the entirely statistical findings of Example 3.14. But if the coupling is made progressively smaller the operation time of the cooling device goes to infinity (clearly a dysfunctional limit), thus enforcing the third law.

8.3.2 *Entangling Operation*

In the opposite limit we may wish to stabilize entanglement—in the presence of decoherence. We have discussed the entangling machine based on cyclic operation, cf. Section 7.5.3. This machine is based on non-equilibrium established by running the respective control cycle fast enough (though still slow enough to remain adiabatic). Obviously, such an effect cannot be transcribed into a continuous (stationary) situation. The following model exploits, instead, the so-called beam-splitter transformation.

Example 8.3 Beam-splitter scenario *Let us consider a thermal photon input field (A), which is so strongly damped that it is (approximately) reduced to the mixed state*

$$\hat{\rho}(A) \approx \rho_0 |0(A) >< 0(A)| + \rho_1 |1(A) >< 1(A)| \,. \tag{8.50}$$

The states $|j(A) >$ are eigenfunctions of the free field, that is, photon number states. If state $|1(A) >$ is fed into a beam splitter, one finds, cf. Definition 3.66,

$$|1(A) > \to t|1(C)0(D) > + r^*|0(C)1(D) > , \tag{8.51}$$

where r^* and t are complex parameters specific for the beam splitter and C,D denote the two output modes. If, instead, the above mixed state $\hat{\rho}(A)$ is used as input, a mixture of the vacuum state, $|0(C)0(D) >$ and the triplet state $|\psi_T(CD) >= |1(C)0(D) > + |0(C)1(D) > /\sqrt{2}$ results. This mixed state is always entangled. With a constant inflow of photons the entangling would operate continuously.

An (effective) implementation of such a beam-splitter scenario has been proposed by [Plenio (2002)]. Physically it consists of a two-level system S and two photon modes C, D. The two-level system is incoherently driven by a finite temperature thermal system RS of temperature $T(RS)$. Each photon mode is damped by a (zero-temperature) bath system RC and RD, respectively. The latter three environmental systems are not treated explicitly but only via their effect on the tripartite system (S,C,D), that is, via a master equation for the density matrix, $\hat{\rho}(SCD)$. The thermal system (RS) is modeled to have two channels (one for the spontaneous and induced decay, one for the incoherent excitation). The (infinite) baths RC, RD, are approximated as single decay channels, cf. Example 3.31. We summarize the key arguments.

The two-level system S (with spectrum $E_2 = \hbar\omega$, $E_1 = 0$) resonantly interacts with two photon modes (C,D) of identical frequencies located in their respective cavities:

$$\hat{H} = \hat{H}_0 + \hat{V}(S, C, D)\,, \tag{8.52}$$

$$\hat{H}_0 = \hbar\omega|2><2| + \hbar\omega\hat{b}^\dagger(C)\hat{b}(C) + \hbar\omega\hat{b}^\dagger(D)\hat{b}(D)\,. \tag{8.53}$$

Based on the beam-splitter transformation discussed above (taking the parameters to be real) the two "physical" modes (C,D) can be replaced by the effective modes (A,B):

$$\hat{a}(A) = t\hat{b}(C) + r\hat{b}(D)\,, \tag{8.54}$$

$$\hat{a}(B) = r\hat{b}(C) - t\hat{b}(D)\,, \tag{8.55}$$

so that the interaction term simplifies to

$$\hat{V}_{eff}(S, A) = g(\hat{a}(A)\hat{P}_{21} + \hat{a}^\dagger(A)\hat{P}_{12})\,. \tag{8.56}$$

Note that the second (transformed) mode $\hat{a}(B)$ does not appear in the transformed coupling Hamiltonian; this mode can thus be disregarded, if it is initially not occupied—as will be assumed now. The total dynamics—also transformed by means of the beam-splitter transformation—is described by the Lindblad master equation

$$\frac{d}{dt}\hat{\rho}(SA) = -\frac{i}{\hbar}[H(\hat{S}A), \hat{\rho}(SA)] + \hat{\mathcal{L}}_{inc}(A) + \hat{\mathcal{L}}_{driv}(S)\,. \tag{8.57}$$

$\hat{\mathcal{L}}_{inc}(A)$ is the damping of the cavity mode (A) with damping parameter κ, the corresponding term for mode (B), also proportional to κ, has been disregarded. The last term denotes incoherent driving of the two-level system (S), c.f. Remark 3.33. For given strength of this incoherent driving we consider the following cases:

(i) $\kappa = 0$ (perfect cavities): The stationary state for the effective mode (A) as well as for the physical modes (C,D) are thermal distributions. There is no entanglement between the physical modes.

(ii) Increase κ: The photon states with higher photon number are preferably damped away. The effective mode thus settles for the mixed state as specified in Eq. (8.50). The beam-splitter transformation back into the physical modes eventually leads to the entangled state

$$\begin{aligned}\hat{\rho}(C, D) \approx \rho_{00}|0(C)0(D) > < 0(C)0(D)| \\ + \rho_T |\psi_T(C D) > < \psi_T(C D)|, \end{aligned} \tag{8.58}$$

where $|\psi_T(C D) >$ is the triplet state as given above.

Remark 8.5 *From the "entanglement competition," cf. Remark 3.7, applied on system (S,C,D) embedded in the three infinite environmental systems (RS, RC, RD) one would expect that no entanglement should survive within the former. Indeed, the driving is incoherent, that is, cannot serve as a source of entanglement. Even without further damping of the cavity modes the effective mode (A) would be found to be thermal, and there would be no entanglement between the back-transformed physical modes (C,D) either. But with strong enough damping the effective mode (A) relaxes into a two-component mixed state, Eq. (8.50). In this case the beam-splitter transformation makes this state look like an entangled state with respect to the original physical modes (C,D).*

While the essential pre-steps are implemented by damping, the decisive role is played by the beam-splitter transformation applied on the appropriate mixed state.

This effective model is a demonstration of principle; it will hardly serve as a competitive entangling device.

Chapter 9

Outlook

Jonas Burgert: Cold Run, 2010, based on photography by Lepkowski Studios [Reprinted with permission from Schreiber and Wipplinger (2011)].

Quantum Thermodynamic Processes: Energy and Information Flow at the Nanoscale
Günter Mahler

ISBN 978-981-4463-73-7 (Hardcover), 978-981-4463-74-4 (eBook)
www.panstanford.com

9.1 On the Nature of the Quantum

There is the famous dictum by Feynman: "I think I can safely say that nobody understands quantum mechanics" [Feynman (1965)]. This pessimistic view (cf. Chapter 3) has apparently been taken as an invitation to uncritically transfer such a supposedly "unintelligible" scheme into other fields of research, even outside physics. Preferably those fields which appear to match that very reputation of being "strange." However, quantum theory is in fact not that mysterious at all.

In many respects, it is an extremely stringent and even logical framework for describing (certain fundamental aspects of) the world, cf. Chapter 3.

Why? It does not allow you to make implicit assumptions: If something cannot be decided, cannot be distinguished, it has to be left so! If it is impossible to decide through which of two slits a photon has passed, both paths have to be considered ("superimposed"). If two particles are identical, I should not have any means to address one and not the other; no model, no interaction, literally nothing is allowed to exist which would violate this symmetry. If I want to measure the angular momentum of a particle in a specific direction (like in the famous Stern–Gerlach apparatus for the spin, see Example 3.24) I have to apply a gradient of the magnetic field in that very direction. There is no way to do so in two directions "at the same time": The co-existence of the respective measurement outcomes cannot be ascertained, unless we are able to prove that the second measurement has no impact on the result of the previous one.

A logical consequence of incompatibility is undecidability. This, in turn, leads to uncertainty dictated by the inherent nature of concepts involved rather than subjective ignorance. And in order to characterize uncertainty statistics and probability theory should come in handy. However, the "total picture" faces serious limitations: While sequences of incompatible events can be combined to form so-called quantum "histories," such histories obey the Kolmogorov rules only if taken from a "consistent" set [Dowker (1995)].

Classical modeling tends to gloss over such subtleties—and for good practical reason. Even if we do not know how a macro-object has arrived in a room with two doors, we are save to assume it came through one of them—even if we are ignorant, there will be traces left in the environment (or just search for a possible witness). In short: There is always a way to "look closer."

Quantum behavior is not necessarily connected with the small—and classical behavior not necessarily with the large system limit, cf. Section 3.3.8. But what should count as a "true" non-classical phenomenon? Phases need not be quantum (there are classical waves). Sequential order effects are not necessarily quantum either (see your construction manual for putting together a piece of furniture).

9.1.1 *Quantum Phenomena*

Definition 9.1 Trivial quantum effects *Down on appropriate microscopic levels we will always need quantum theory as the only reliable tool for analysis (cf., e.g., the stability of atoms or molecules). Such basic features often serve as a kind of "black box" on a higher hierarchical level of description. An example would be the classical kinetic treatment of a (dilute) gas, where the stability of the individual particles is taken for granted. Crystals and the geometric shape of molecules are further examples. On the respective higher level of description the quantum nature of the basic constituents or concepts need no longer be of any concern: The quantum aspects are "encapsulated," they become "trivial," cf. Remark 5.7.*

Genuine quantum features typically do not survive in the macroscopic limit. Scaling up known quantum phenomena like interference (cf. Section 3.2.10; [Leggett (2002)]) thus requires a rapidly increasing expenditure in terms of physical resources; the practical impossibility to stabilize a large-scale quantum-computer points in the same direction. Larger and larger systems are increasingly hard to isolate, and uncontrolled interactions are detrimental for any explicit quantum features. In fact, this trend underlies quantum thermodynamics as the overwhelmingly stable alternative.

There are exemptions from the rule, though:

Definition 9.2 Macroscopic quantum phenomena *Phonons or plasmons as collective excitations within a condensed matter system may be taken as an example: millions and millions of subsystems participate. Nevertheless, such excitations contribute to electron–electron scattering, say, by means of quantized energy packages.*

Other macroscopic quantum phenomena, like superfluidity or Bose–Einstein condensation (BEC), are even more spectacular. The BEC is a gas of Bosons that are in the same quantum state; the effective description is therefore a single-particle Schrödinger equation with the particle–particle interaction included in an approximate way. This so-called Gross–Pitaevskii equation [Leggett (2001)] thus defines a sort of "self-similarity": The same type of equation is valid for phenomena on different hierarchical levels.[a]

Such descriptions have been proposed as blueprints for quantum features even within biology (cf., e.g., the so-called "Fröhlich condensate" [Hameroff (1996)]).

In addition, certain quantum features may occur—on a purely phenomenological level—based on a different mathematical background:

Definition 9.3 Higher-level quantum phenomena *Entanglement is identified as a typical quantum effect. In its full fledged form it cannot be transported into the classical domain. There are aspects, though, which allow for analogies. One typical feature of quantum entanglement is the fact that a pure total state will exhibit mixed state behavior locally. Such a phenomenon may, indeed, occur in the classical domain under the restriction of a coarse-grained (effective) description. Concrete (classical) examples have been analyzed under this constraint: two Brownian particles [Allahverdyan (2005b)], and the case of two coupled harmonic oscillators [beim Graben (2012)].*

"Looking closer" is deliberately inhibited here so that we can expect some quantum analogue features even on a higher hierarchical level of description, that is, away from the most fundamental one. Undecidability emerges as a kind of operational limitation.

These scenarios are remarkable in the sense that certain features known from a fundamental (quantum) level may re-emerge on a much

---

[a]Other examples range from non-linear optics to the theory of finance.

higher (somewhat phenomenological) level. But it is still all physics; nothing so far can be said about analogies outside the immediate physical realm.

9.1.2 *Is Quantum Mechanics Useful?*

Remark 9.1 A tension between "quantum" and "use" *A word of caution: Applications necessarily require some sort of "embedding with the rest of the world" (including the user); but open systems, more often than not, loose their quantum features, which they certainly would have, if isolated! Applications tend to make the applied system look classical—in agreement with our standard everyday experience. This feature points toward a kind of tension between quantum effects and their potential "use." Or is this merely a consequence of the "wrong perspective," inappropriate partitions?*

It is generally believed that classical physics emerges from quantum physics under appropriate conditions (see Section 3.3.8). Given that classical physics abounds with useful applications, the above question—is quantum mechanics useful?—can only mean, whether or not physics is useful already in the quantum regime.

It was Rolf Landauer [Landauer (1995)] who addressed this question in the context of information processing. At that time he was not so much concerned with quantum computing in a strict sense, but rather with limits due to unavoidable manufacturing defects: "smaller devices are more delicate." For example, interacting arrays of semiconductor quantum dots, suggested for use as a computer architecture, arose his strong criticism. Indeed, this proposal did not live up to expectations.

Definition 9.4 Quantum machine *Machines are characterized by their function, see Section 1.4. A quantum machine would thus be a device the function of which cannot satisfactorily be explained without recourse to quantum mechanics. In addition, one might even demand that the respective function could not (alternatively) be provided by a classical machine—at least, not under the given constraints of size (scaling). The quantumness of a machine may be defined in relation to* *a classical reference system, in which quantum features like coherence are suppressed.*

Remark 9.2 Modularity *Typical devices, including computers, are modular, that is, are systematically composed of independent parts. This is a tremendous engineering advantage. The modularity may easily be lost in the quantum domain, where holistic aspects reign, cf. Remark 7.21. If so, the design of composite quantum machines will present new challenges.*

Quantum features in technical devices have originally been based on the discreteness of energy levels. Dynamical features and specially prepared quantum states are finding niche applications only recently.

Remark 9.3 Real-world applications of quantum mechanics *The following list is not intended to be comprehensive:*

(1) Laser: This light amplification device is driven (pumped) from the outside to produce inversion in an (atomic) three-level system. The amplified light modes are selected by means of a so-called resonator.
(2) Transistor: This is a semiconductor device, essentially based on so-called p–n junctions. The level of description is semiconductor band theory (as such a quantum effect) combined with transport theory (kinetics).
(3) Ultraprecise clocks: Optical clocks exploit atomic transitions with extremely low linewidths. Recent versions are based on so-called "quantum logic spectroscopy" [Chou (2010)].
(4) Quantum key distribution (quantum cryptography): This technology is based on the delicacies of the quantum measurement process. Information gain (by the eavesdropper) implies disturbance [Nielsen (2000)], which can be detected by the legal communicators.
(5) Quantum random number generator: A pertinent implementation can be based on the vacuum field fluctuations [Symul (2011)]. This quantum noise is believed to be perfectly random over all frequencies.
(6) Quantum sensors: By using maximally entangled photon states the shot-noise limited sensitivity of an interferometer can be made

to scale like $1/N$ rather than $1/\sqrt{N}$, where N is the number of photons [Bollinger (1996)]. Quantum correlations can also be used to improve optical lithography. In a similar vain atomic magnetometers can be shown to benefit, for example, from spin squeezing.

(7) Quantum simulators: This proposal is based on multicomponent entangled states, combined with the phenomenon of interference [Nielsen (2000)]. The certification of correct working can be very challenging, though [Gogolin (2013)]. Small-scale versions have been realized; further progress continues to be slow.

Quantum miracles or quantum limitations? The spectrum of quantum applications in a strict sense is still somewhat limited. It is difficult to see how and where future technological developments should be expected. And even in cases where quantum effects do show up, it can become a quite subtle point to show that these effects are "essential" for the overall functions considered, cf. [Smolin (2013)].

Quantum aspects might show up also in a somewhat more "conservative" setting: Preliminary applications from quantum thermodynamics (thermodynamic machines) demonstrate that the well-established thermodynamic laws hold down to the extreme quantum limit—and thus for architectures that show little if any similarity with their macroscopic counterpart.

Finally, quantum ideas may contribute also to a better understanding of certain scientific problems, even beyond physics. This is still uncharted and unsafe territory, a brief overview should suffice as a closing of this book.

9.2 Quantum Processes Everywhere?

9.2.1 *Quantum Chemistry*

Quantum chemistry, as a branch of chemistry, is concerned with properties of chemical systems based on quantum modeling. Typical calculations involve various forms of approximations like the Born–Oppenheimer approximation, see Example 3.8. Also of prime interest is chemical processes, originally part of thermodynamic approaches. More recently, such processes have been manipulated via Laser pulses and the use of inherent superpositions of pathways.

Equilibrium structures are one further area of research. The chemical structure (of molecules), for example, appears as an entirely classical concept. However, this is not necessarily the case. There could be (energetically) equivalent structural alternatives, giving rise to resonance interaction between those, cf. Example 3.13. The interplay between resonance interaction and electron–phonon coupling then turns out to be essential for the resulting structure. For benzene the electron–phonon coupling is insufficient to overpower the resonance stabilization and so the fully delocalized structure results, the so-called "aromatic" nature of benzene. For ammonia, on the other hand, the electron–phonon coupling does overpower the resonance interaction and so two mirror image pyramidal classical equilibrium structures result [McKemmish (2009)].

In fact, "all biological and chemical charge transport processes are now known to be controlled by analogous interplays between vibrational motions and resonance couplings" [McKemmish (2009)].

9.2.2 *Quantum Biology*

Genuine quantum machines might be expected to occur in biological systems, cf. Definition 9.4, which would then provide examples for so-called "non-trivial" quantum effects [Wiseman (2007)]. Many researchers remain skeptical, though: Comparatively high temperatures, mesoscopic scales, and strong mutual interactions between the various parts could render the pertinent description essentially classical—except for those "trivial" (i.e., encapsulated) quantum effects.

Remark 9.4 Non-modularity of biological function? *The focus on "non-trivial" quantum effects in biological systems [Wiseman (2007)] would be misguided, though, if biological functions were typically the result of non-modular (holistic) architectures, in which smaller and smaller functional subunits (eventually based on quantum features) do not exist. In fact, if biological function (bio-computing) involves all length scales "at the same time," quantum features would always* *have to be taken into account.*[a] *In fact, coherence of any individual subsystem should be irrelevant, what counts is the coherence on the same structural level on which the function is defined. Effectively, this leads to size limitations, reminescent of quantum computation.*

Anyway, as argued by M. Conrad, the importance of the substrate architecture for such bio-computing tasks defines a serious challenge for further progress. On the other hand, it might open up new classes of artificial machines that would be more biology-like than present day machines [Conrad (1988)].

Example 9.1 Molecular motors *Molecular motors [Jülicher (1997)] are candidates to bridge the gap between classical and non-classical "energy converters," cf. Section 7.5.4. Up to now their modeling is rather phenomenological, though.*

On the microscopic scale the machinery that keeps biological cells working involves chemical reactions that are controlled and triggered by enzymes. Understanding enzyme catalysis is thus essential [Glowacki (2012)]. New models for reactions of that kind may require quantum tunneling.

The following examples have recently gained some popularity. Their microscopic level of description would certainly allow for quantum features in a strict sense.

Example 9.2 Energy harvesting: photosynthesis *Energy transfer within molecular networks is often described by semi-classical models in terns of "hopping" between discrete energy levels [Engel (2007); Sarovar (2010)]. Recent experimental data indicate that wave-like phenomena may well play a role. Even small amounts of entanglement seem to be present. This is not that surprising at all as we are concerned here with the micro-description of energy transport processes, cf. Section 9.2.1, and under laser excitation. In vivo experiments are not yet available. What is more surprising is the highly speculative (and rather superficial) interpretation in terms of "molecular quantum computations" [Engel (2007)], cf. Definition 2.2.*

---

[a] On the other hand, if it is true that modularity has evolutionary advantages [Clune (2013)], holistic architectures should be rather improbable.

Example 9.3 The avian compass *This example goes beyond the purely micro-level of description. In fact, knowledge about the workings of this navigation system is primarily derived from various behavioral experiments on the respective migratory birds. The microscopic correlate of this behavior is believed to be some photoreceptor pigment located in the bird's retina [Ritz (2000); Bandyopadhyay (2012)]. It is assumed that the brain receives a cumulative signal from (independent) radical pairs. The geomagnetic-dependent dynamics of spins of unpaired electrons in each radical pair is modeled by a Hamiltonian that consists of two electron spins (both interacting with the external magnetic field), while one of those spins interacts, in addition, with a nuclear spin. The radical pair has a finite lifetime. The set of parameters are adjusted to reproduce the observed behavior—and to be in rough agreement with other microscopic data. Noise appears to improve the functioning of the compass.*

Does this imply that the sensing of orientation is quantum? This is still open to debate. One should recall that our eyes also work on the quantum limit [Brunner (2008)]: We then experience flashes, which hardly provide any useful optical information—unless we consider ourselves as a kind of photon detector. Our everyday optical perception, nevertheless, appears to work rather classically.

Example 9.4 Vibration model of olfaction *Odorants are small molecules, rarely more than a few tens of atoms. While selective docking of such molecules is considered important by all current theories, it is doubtful whether docking alone could provide enough discrimination power [Brookes (2007)]. Luca Turin [Turin (2006)] suggested inelastic electron tunneling as a specific mechanism to incorporate the impact of odorant vibrations. The idea is that tunneling would occur between certain receptor states only when energy is conserved by the emission of an odorant phonon of the right frequency. In this sense olfaction would come rather close to the basic processes underlying optical sensing. In both these cases the origin of possible quantum features are fairly obvious: The pertinent mechanisms are basically molecular in nature.*

Turin's dynamical model is still controversial, but experimental evidence seems to be growing.

It goes without saying that the notorious "qualia problem" cannot be addressed by any such sensory models: How the experience of smell or colors can arise from its physico-chemical substrate remains a mystery.

9.2.3 *Quantum Evolution*

Organisms are selected to increase the amount of information they encode about their environment. One may thus focus on mechanisms that optimize means to reach these goals. Selection is exercised also by Maxwell's demon (cf. Theorem 4.19). This analogy might help to shed some new light on evolution from the point of view of statistical physics [Krakauer (2011)].

Definition 9.5 Darwin's demon *Natural selection can be modeled to result from a demon that is able to detect, memorize, and act upon variation in one generation, in order to induce an adaptive distribution of genotypes and thus organisms in the next generation. Bayesian updating (cf. Proposition 2.4) may provide an intuitive picture: "It is fairly natural to think of organisms as reified hypotheses about the state of the environment, and selection as a natural experiment that either supports (through survival) or refutes (through morbidity) a given genotype and attendant phenotype. Prior beliefs are encoded in the variation of heritable states of the genome encoding regular features of past environments" [Krakauer (2011)].*

Even in the context of evolution the potential "use" of quantum physics may be contemplated: If advantageous enough, why should such features not have been exploited? Well, most likely there are further constraints in terms of reliability and stability: Optimal solutions based on quantum dynamics may turn out to be too costly.

We have realized that Maxwell's demon cannot really violate the second law. Also Darwin's demon has to act subject to the laws and constraints of physics, cf. Remark 4.5. But what constraints? Maxwell's demon has been considered also in the quantum limit. Could this be a viable option also for Darwin's demon?

There is much speculation going on about the underlying "micro-dynamics" [Davies (2008)]; for Davies the "secret of life" lies with its remarkable information processing and replicating abilities. It is suggested—without providing any evidence—that living organisms might process information quantum mechanically. It is further claimed that "the transition from non-life to life was a quantum-mediated process." In order for this to happen "an element of teleology remains," though.

Similar ideas have been brought forward by McFadden in his book on "quantum evolution" [McFadden (2000)]. He believes that quantum mechanical rules would provide "... another way to overcome the huge improbability of the first self-replicator." In this work it is assumed that unitary dynamics would be applicable even to the entire biological cell and even on biological relevant time scales. In particular, the self-replication of proteins is suggested to be based on the so-called inverse quantum Zeno effect, cf. Proposition 3.25, by which a series of measurement projections is used to guide a quantum state into some target state, independent of the underlying Hamiltonian. In the context of adaptive mutation superpositions of many particle states are postulated to exist on macroscopic time scales (up to 100 seconds!). These quantum features would, he suggests, reduce the odds against the spontaneous generation of life.

However, all our experience from standard quantum physics tells us that the exploitation of such peculiarities would require suppression of decoherence and thus call for extreme isolation—requiring very special design (i.e., careful and elaborate experimentation). In the biological realm there is no convincing evidence that such (costly!) design requirements would be in effect.

So should we, nevertheless, expect "non-trivial" quantum effects in evolution? Darwin's demon differs from Maxwell's demon in a number of important aspects. One of those is adaptive feedback: Selected structures by themselves modify the constraints for future evolutionary steps. Such classical features should be considered a much more promising source of complexity.

Indeed, recent experimental findings indicate that evolution may follow a predictable genetic pattern, far from being random. The central issue, adaption, is still not well understood [Dobler (2012); Zhen (2012)]. Evolutionary degrees of freedom seem to be limited by physiological constraints.

9.2.4 *Quantum Brain*

Microtubuli have been identified to provide intracellular scaffolding as well as a delicate network of "highways," which carry traffic by so-called molecular motors [Jülicher (1997)].

In the Penrose–Hameroff proposal [Hameroff (1996); Hagan (2002)] quantum superpositions are assumed to occur in the tubulin dimers composing those microtubules. In fact, two alternative conformational states are taken to represent a formal qubit. This phase of quantum superpositions (supporting some kind of quantum computation!) is associated with a "preconscious process." Those coherent superpositions are taken to be inherently unstable and subject to "self-collapse." The latter is meant to be "objective," that is, not due to the standard environmental effects, but rather due to a (yet to be clarified) gravitational impact. In turn, each self-collapse event is claimed to correspond to an "instantaneous moment of conscious experience." Consciousness, it is argued, requires "non-computability." This is the basis of the so-called "orchestrated objective reduction" (Orch OR) model.

Definition 9.6 Orch OR model *The model may be seen as a four-step quantum machine:*

Step (1) *Superpositions within the formal qubits are being built up.*
Step (2) *The coherent motion starts to encompass large regions of the microtubule and its neighbors.*
Step (3) *The superposition reaches a "critical mass" inducing objective reduction.*
Step (4) *After reduction the ensemble of tubulins contains the quantum computational result.*

The Orch OR model is extremely speculative and controversial. The ad hoc modified quantum theory would require a "new physics," which is not in sight. Is consciousness a "non-trivial quantum effect?" And if so, is it nothing more than that? This would imply to dispose of the so-called qualia problem—without further justification.

Up to now there are no convincing arguments, let alone experimental data, supporting this view. Obviously, little can be said scientifically about the claimed relationship between brain function

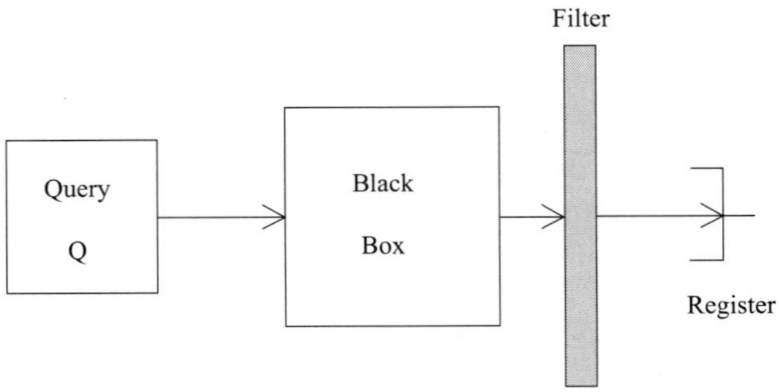

Figure 9.1 Human participant as "black box," subject to testing.

and its material substrate. All that can be done presently is to focus on the preconditions of this model, the question of macroscopic quantum coherence in biological system. Also the brain is an open system. Is it possible to overcome the notorious decoherence problem? While no final conclusions have been reached so far, most researchers consider the Orch OR model to be biologically unfeasible [McKemmish (2009)].

9.2.5 *Quantum Cognition*

Cognitive science focuses on the study of the mind and its underlying processes. It consists of various research disciplines, including psychology, neuroscience, and linguistics.

A subfield based in psychology is primarily concerned with "processes of thought" as initiated by "queries." To put "questions to nature" underlies all empirical natural sciences; sets of queries presented to human participants in a given psychological experiment may thus count as a reasonable analogy. In both cases we superficially deal with responses to stimuli (a kind of "pump and probe" scenario; see Fig. 9.1).

Remark 9.5 Rationality *Classical models of rational thinking are based on Boolean logic [Boole (2009)]. While limited in scope there is hardly any viable substitute [Primas (2007)]. Attempts to replace classical by quantum logic, for example, have met with limited success, if any.*

Remark 9.6 Frame in communication *The analogy between questions to nature and queries to human participants is quite superficial, though, if not misleading. In fact, the so-called "frame in communication" defined as the specific wording conveyed to the subject may be full of psychological and implicit meaning; different responses despite the "sameness of information" in a logical sense need not violate rationality [Sher (2011)]: The artificial query environment itself might give rise to apparent irrationality within the respective response, which would be called adaptive and well-founded in a more "natural setting".*[a] *Such a caveat has to be kept in mind in the following.*

Definition 9.7 Coarse-grained versus elementary queries *It has been suggested that any state-space of quantum theory could be based on basic binary alternatives, which Weizsäcker called "Ur" objects [Lyre (1997)]. In a similar vain one may expect that a query* $\tilde{Q}$ *as part of a rational discourse between human beings could be decomposed into independent elementary alternatives,* $Q(j)$; $j = 1, 2, \ldots,$ *each requiring a simple yes/no answer,* $A(j) = \pm 1$ *(= 1 bit of information). The original* $\tilde{Q}$ *might then be considered a "coarse-grained" query, the associated set* $\{Q(j)\}$ *a kind of detailed checklist. Of course, this would present an idealized situation.*

Quantum cognition postulates quantum-like behavior to occur in the cognition process, but denies the relevance of quantum effects in a strict sense, as assumed in the Orch OR model discussed above. The origin of such an apparent non-classical behavior has to remain open; it might result from coarse-graining, cf. Definition 9.3, or from implicit psychological connotations.

Most actual queries $\tilde{Q}$ will be coarse-grained—from a strictly logical point of view, ill-defined in an informational sense. Being forced to deal with such a $\tilde{Q}$ anyway, rather than with the full

[a]Your medical doctor may tell you that an operation leads to complications with a probability of 5% or that in 95% of cases there are no problems. Patients tend to feel better with the second statement—despite the logical equivalence. They take the wording as an implicit judgment by the doctor, that is, as an additional source of information. This is clever, not irrational within that very context—but not in a laboratory experiment.

set $\{Q(j)\}$ will leave room for implicit interpretations, context-dependencies, and biases by the addressee of the query.

But even under this condition a binary response can be enforced from the outside: one merely has to filter out (by means of the questionnaire) the large set of possible judgments induced by query $\tilde{Q}$ down to two alternative reactions: yes or no, passed or failed, and select or discard: $\tilde{A} = \pm 1$. Certainly, there is no reason to expect that such a rule of game would capture any fundamental features of the internal cognitive process as such. Rather it is the simplest possible output description, which—by means of this crude coarse-graining—suppresses any more balanced types of response (such as allowing three options including "do not know," or responses on a scale of 5 or 10 steps, etc.).

Definition 9.8 Incompatible queries *Contrary to the elementary $Q(j)$, two coarse-grained queries, $\tilde{Q}(1)$, $\tilde{Q}(2)$ need not be independent of each other. Such a mutual conflict is described as "incompatible." Contrary to physics this incompatibility is not due to the conflicting operational requirements in the attempt to measure incompatible observables; rather it is due to the lack of a clear operational procedure (i.e., the lack of a well-defined checklist). Nevertheless, coarse-graining may induce quantum-like features within an otherwise classical context, cf. Section 9.1.*

Proposition 9.1 Cognitive state *The resulting statistical behavior of the human participant is seen as a "propensity" encapsulated in his "cognitive state." This claim is taken quite literally: The probability P for $\tilde{A} = \pm 1$ is supposed to result from an algorithm with the cognitive state as an input. This algorithm is borrowed from the quantum theory of spins, the state becomes a normalized vector $|\psi>$ in the respective Hilbert space.*

Proposition 9.2 Quantum spin analogy *It has been suggested [Pothos (2012); Aerts (2011); Bruza (2009)] that each compatible coarse-grained query $\tilde{Q}(j)$ with enforced binary response can be associated with an effective quantum spin component $\hat{\sigma}_3(j)$. The basic answers $\tilde{A}(j) = \pm 1$ are taken to be eigenvalues associated with the orthogonal unit vectors $|a_{-1}(j) >= |e^{(1)}(j) >$, $|a_{+1}(j) >= |e^{(2)}(j) >$, of this spin j, see Eq. (3.67). If more than one "compatible"* *observable (query) needs to be included, one has to enlarge the state space correspondingly, for example, by postulating a spin-tensor space,* $|a_{\pm 1}(1)> \otimes |a_{\pm 1}(2)> \otimes \cdots$, *cf. Definition 3.55.*

A query $\tilde{Q}(2)$ *incompatible with* $\tilde{Q}(1)$ *would be associated with different spin components of the same effective spin as for query* $j = 1$*: Maximum incompatibility would thus be obtained for* $\hat{\sigma}_1(j)$, $\hat{\sigma}_3(j)$. *For* $\hat{\sigma}_1(j)$ *the eigenfunctions* $|+(j)>= 1/\sqrt{(2)}(|e^{(1)}(j)> +|e^{(2)}(j)>)$ *are rotated by the angle* $\theta_{12} = \pi/4$ *with respect to the original basis. Different angles allow for a reduced incompatibility.*

Proposition 9.3 Effective quantum projection *Cognition is a process. The query* $\tilde{Q}(j)$ *is taken to induce a kind of unitary pre-dynamics, which transforms the initial ("neutral") cognitive state into the respective "response state"* $|\psi(j)>$.

In a second and final step, this $|\psi(j)>$ *needs to be "broken down" (adapted) to the allowed responses, the yes/no-alternatives. This step is modeled like a quantum projection:*

$$\hat{P}_{\pm}(j) = |a_{\pm 1}(j)> < a_{\pm 1}(j)|, \qquad (9.1)$$

$$\hat{P}_{\pm}(j)|\psi(j)> = |a_{\pm 1}(j)> < a_{\pm 1}(j)|\psi(j)> . \qquad (9.2)$$

The Born rule is assumed to apply,

$$P(\tilde{A}(j) = \pm 1||\psi(j)>) = |< a_{\pm 1}(j)|\psi(j)> |^2 . \qquad (9.3)$$

In this way the probability P *is conditioned by the cognitive state* $|\psi(j)>$. *The relative direction of the* $|a_{\pm 1}(j)>$ *with respect to the response state cannot be derived but has to be adjusted to the empirical data.*

Remark 9.7 Query experiments *Coarse-grained query experiments are run on ensembles of participants. Their reaction is conveniently described in terms of response statistics* $P(\tilde{A}(j) = \pm 1||\psi(j)>)$ *conditioned by the internal cognitive state* $|\psi(j)>$. *If we allow for more than one query (with respect to the same cognitive state* $|\psi(j)>$*), these could be "compatible" or "incompatible"; incompatible queries are expected to show order effects, similar to the order of quantum mechanical measurements on quantum spins.*

Obviously, a whole bunch of links to other questions and information resources could be of concern; implicitly it has to be understood here that only the question under immediate attention is considered. The validity of such an effective "isolation" would be of much concern even within a quantum scenario in a strict sense.

Such a quantum adaption, like any alternative approach to cognitive modeling, is entirely phenomenological. This means that—contrary to the fundamental spin algebra—a sizable number of important parameters have to be put in by hand. Indeed, there is a long list of such parameters and ad hoc assumptions. The predictive power is thus very limited. And there are always alternative (entirely classical) models available. Nevertheless, for illustrative purposes we close with some pertinent examples [Pothos (2012)]. Whether or not these examples are able to support the ideas of quantum cognition have to be left open. Inconsistencies and unjustified ad hoc assumptions give rise to serious doubts, though.

Example 9.5 Sequential order *Consider the following two coarse-grained queries:*

$\tilde{Q}(j)$*: Is person* X_j *honest (*$X_1 =$ *Gore,* $X_2 =$ *Clinton)? These are part of a former Gallup poll; the concept "honesty" is ill-defined. The queries could be made reasonably well defined via an explicit checklist and input data provided. But this is not done. As a consequence uncontrolled context dependencies must be expected to occur. This is, indeed, the case: There is an empirically found order effect (the answer* A_1 *on question* $\tilde{Q}(1)$*, say, depends on whether or not question* $\tilde{Q}(1)$ *follows question* $\tilde{Q}(2)$*).*

To account for this order effect, quantum cognition assumes that query $\tilde{Q}(1)$ *and query* $\tilde{Q}(2)$ *are "incompatible," that is,* $\theta_{12} \neq 0$*. The order of projections on incompatible questions matters—as known from the quantum analogue in a strict sense:*

$$\hat{P}_{\pm}(1)\hat{P}_{\pm}(2) \neq \hat{P}_{\pm}(2)\hat{P}_{\pm}(1)\,. \tag{9.4}$$

But in order to apply those pairs of projections one has to assume that the pertinent cognitive state $|\psi>$ *is independent of the query* $\tilde{Q}(j)$ *and also independent of the order of the queries; the state only reflects some kind of prior knowledge that Gore would generally be considered more honest than Clinton. This ad hoc prescription is inconsistent: According to Proposition 9.3 each query triggers a pre-dynamics. Why then is it allowed to start from the same* $|\psi(1)>$*, say, if the query* $\tilde{Q}(2)$ *rather than* $\tilde{Q}(1)$ *is to be reacted to? In general: Why not further* *update the response state as the second query comes in—whether compatible or not? Also this updating could (trivially) depend on the sequential order!*

Order effects represent a kind of (short-time) memory. Such a memory could easily be incorporated in the dynamics of the cognitive state itself. No quantum features would be needed.

Example 9.6 Non-symmetrical similarity *Consider the two queries:* $\tilde{Q}(1)$ *"is country X similar to country Y?" and* $\tilde{Q}(2)$*: "is country Y similar to country X?" (X = China, Y = Korea.) Similarity has to be defined with respect to a well-specified subset of properties; the two questions should then lead to identical answers. Again, such specifications are missing, making the queries ill-defined. The participants are deliberately left alone to decide whatever strategy they like. The empirical result (in terms of the response probabilities) is asymmetrical, that is,*

$$P(A_{\pm 1}(1)|\psi>) \neq P(A_{\pm 1}(2)||\psi>). \tag{9.5}$$

The quantum cognition treatment proceeds in three steps:

(i) *Each query is "virtually" decomposed into two sub-queries Q(X) and Q(Y). This decomposition is inconsistent, as the sub-question Q(X), China yes/no, for example, makes no sense, only the full question, is China similar to Korea yes/no.*
(ii) *An additional ad hoc modification is brought in: Q(X) and Q(Y) are assumed to have different dimensionality,* $d_X > d_Y$*—based on the argument that X = China will be characterized by "more information" than Y = Korea. Also this is inconsistent: The effective Hilbert space dimension has been defined by the number of yes/no answers under consideration, rather than by the amount of additional background data not being asked!*
(iii) *The original questions are then re-interpreted to mean:* $\tilde{Q}(1)$*, project first on X and then on Y.* $\tilde{Q}(2)$*, project first on Y and then on X. Queries X and Y are taken to be incompatible, so there could be order effects, as before.*

Not surprisingly, putting in everything as desired, one gets out, what is desired. But to arrive at such a "non-symmetrical similarity" one certainly does not need quantum theory: Classical set theory would be sufficient. Represent China as a big circle, Korea as a small one, let both intersect. Then the common area is certainly larger with respect to Korea than with respect to China. But this ratio has nothing to do with the concept of similarity.

Example 9.7 Conjunction "fallacy" *The participants are conditioned by two different types of information about a person X: An explicit and X-specific one (C_1, represented by a story being told) indicates property A about X should likely be true. From some everyday background experience, C_2, one tends to conclude that property B is unlikely to be found together with property A. It is deliberately left open how to weight the different informational constraints, C_1, C_2.*

The single query $\tilde{Q}$ now reads: Is person X characterized by properties (A and B) or (B)? The participants are empirically found to prefer (A and B) over answer B. Apparently, this result is in conflict with the standard statistical inequality $P(B \cap A|C) \leq P(B|C)$, thus indicating a "violation of the probability sum rule."

Quantum interference is suggested as a cure. For this purpose the query is, again, decomposed into two (incompatible) sub-queries: $\tilde{Q}(1)$: Does X have property A? $\tilde{Q}(2)$: Does X have property B? But now the authors [Pothos (2012)] introduce ad hoc a preferred order: The more likely case, property A, is projected first, and then on B. This implies a higher probability for (A and then B) than for (B) directly. This preference, though, does not follow from the formalism, it is put in by hand.

An alternative simple way out of this conundrum would be to assume that the participants compare $(B \cap A|C_1)$ versus $(B|C_2)$. In this case there is no statistical violation whatsoever.

Example 9.8 Bell's inequality *The authors [Aerts (2011)] start from an artificial binary structure of words and two-word sentences: There are animal pairs A_i, and act pairs, B_j. A typical example: $\{A_i\} = \{horse, bear\} = \{+1, -1\}$; and $\{B_j\} = \{growls, whinnies\} = \{+1, -1\}$. A as well as B is thus mapped on the spin algebra. Out of these inputs we can built $2 \times 2 = 4$ two-word sentences $A - B$ of the form "animal act." These two-word sentences (two-spin systems) are translated into the strict "correlations" $C(A_1, B_1) = C(A_2, B_2) = 1$, $C(A_1, B_2) = C(A_2, B_1) = -1$.*

Each participant is now asked to select one sentence as the best presenting the idea "animal act." On ensemble average one thus obtains four probabilities $P(A_i, B_j)$, $i, j = 1{,}2$. *Eventually, the "expectation value"* $E(A, B)$ *is defined as*

$$E(A, B) = C(A_1, B_1)P(A_1, B_1) + C(A_2, B_2)P(A_2, B_2) + \\ +C(A_1, B_2)P(A_1, B_2) + C(A_2, B_1)P(A_2, B_1)\,. \quad (9.6)$$

This procedure is extended to cover another animal pair, A', *and another action pair,* B', *supposedly in the same Hilbert space. These pairs allow to consider three more sentences,* $\{A', B\}$, $\{A, B'\}$, $\{A', B'\}$, *each with a set of ad hoc correlation numbers* C *as before and a corresponding expectation value* E. *The latter are used as entries in a Bell-type correlation test, cf. Eq. (3.353),*

$$s = E(A', B') + E(A', B) + E(A, B') - E(A, B)\,. \quad (9.7)$$

For the input data derived from the query experiment s *is numerically found to "violate" the Bell inequality.*

This conclusion, though, is seriously flawed. Needless to say that the whole procedure (the various mappings on ± 1*) is entirely arbitrary; other choices would lead to different results. And the four expectation values* E *have been taken to be statistically independent (which could not be the case if there were quantum-like correlations needed to produce Bell violations to begin with).*

Remark 9.8 Non-classical rationality? *Quite often we have to decide, to judge based on limited (insufficient) knowledge (cf. [Jaynes (1957)]). But such situations should not be confused with ill-defined questions. There are various strategies: First choice should be to try to fill the gap, change the ill-defined problem into a well-defined one. If forced to react without this possibility, it seems that humans tend to proceed in a specific way—reflecting the influence of implicit contexts, cf. Remark 9.6. But does this mean that "... considerable evidence has accumulated that naive observers do not typically reason with classical logic?" [Pothos (2012)] And even more controversial: would such apparent quantum features in human thought specify a new way of "rationality"—as advocated by some proponents?*[a]

That is a tall order, indeed. And a category mistake: A response is rational in terms of given logic rules—not in terms of model

[a]This distorted view is best illustrated by the title "Quantum minds: Why we think like quarks" [Buchanan (2011)]. If humans think like quarks, quarks would have to think like humans, cf. Example 9.6. A serious case of sloppy thinking.

properties, whether quantum or classical. Such kinds of insufficiently regulated thinking should hardly be called "reasoning" to begin with. And why should evolution (cf. Section 9.2.3) have come up with the extravaganza of a kind of quantum information processing—just for an undertaking of such limited relevance?

Remark 9.9 Logical paradoxes *Eventually, ill-defined questions should not be confused with logical paradoxes either, like the Barber paradox, often cited in connection with Russel's paradox within set theory [Link (2004)].*

> *The barber is defined to be a man in town who shaves those and only those men who do not shave themselves. Who shaves the barber?*

The answer becomes oscillatory: If the barber does shave himself, he must not shave himself; if he does not shave himself, he must shave himself. The paradox has to do with self-referencing. To avoid the paradox one has to note that there cannot be a barber who cuts the hair of all those who do not cut their own hair.

Quantum physics is fundamental and thus ubiquitous; nevertheless classical phenomena abound, cf. Remark 9.1. Quantum behavior shows up with respect to the appropriate context only. It remains to be seen whether non-classical features may be found also on a higher hierarchical level of description. Up to now evidence is scarce.

Within a different line of thought existing theories (even outside physics) may be found to exhibit analogies with physical scenarios on an entirely formal (mathematical) level. The last item points into this direction:

9.2.6 *Quantum Finance*

Finance is concerned with financial instruments like "futures" and "options" [Baaquie (2002)]. The modeling of financial markets is often based on a top-down (statistical) approach: The use of Langevin type equations combined with noise terms. It has been shown that under appropriate conditions such equations can be recast into the so-called Black–Scholes (BS) equation, a kind of "working horse" in this field. By means of a variable transformation, this dynamical equation takes on the form of a time-dependent Schrödinger equation [Baaquie (2002)].

This quantum formulation presupposes a theoretical understanding in the form of the BS equation. It does not require any justification other than the validity of the mathematical assumptions and approximations involved in its derivation. It is reminiscent of other transformations common within physics, such as, for example, the cubic Schrödinger equation for the propagation of classical electromagnetic fields.

Such kind of mappings allow to transcribe methods from one field to the other—the main potential benefit. Whether there is much gain of insight remains debatable. The name "quantum finance" may easily be misinterpreted to imply a more fundamental theory—which is entirely unjustified.

9.3 Quantum Mechanics, Thermodynamics, and Information

Looking back at the journey we are about to finish now: Quantum thermodynamic processes will continue to present major challenges to theoretical and experimental research; we have not yet reached a complete picture, there is room for surprises. But are there fundamental questions left open?

Remarkably conflicting views are still being promoted: According to [Horodecki (2011)] thermodynamics at the microscopic scale is *fundamentally different* from the macro-scale: it is claimed to be "generically irreversible" in the former case.[a] While according to [Kosloff (2013)], "the laws of thermodynamics are true in any quantum circumstance" (apparent failures derive from flaws in the approximations); this is very close to position expressed in this book, based on the model of "spectral deformation," Definition 4.6. Of course, deviations from this model cannot be excluded, cf.

[a]Such principle differences would be reminiscent of those between classical and quantum computation. Entirely new types of quantum thermodynamic applications might then be expected.

Remark 4.10. And certainly, there are quantum effects in a strict sense, cf., for example, Section 7.4.2 and Section 7.4.3. But by no means are these "essential" for the underlying function. In this sense there is a "peaceful coexistence" between the quantum and the classical approach. Scale invariance would be the dominant experience.

So what is the origin of such opposing views? Is there a way to clarify the debate? It seems the conflict can be traced back (1) to the type of models used (autonomous versus open systems, type of spectral control) and (2) the interpretation of the role played by information.

It has been shown that autonomous machines (Section 7.5.2), indeed, face serious problems; standard thermodynamic control features are typically recovered in an appropriate classical limit only. However, as autonomous systems in a strict sense tend to be exemptions, not the rule (cf. Remark 9.1), these findings hardly ever have "dramatic" consequences.

Information can be seen in different contexts. Here we have subscribed to the idea that information is a kind of short-hand notation within an effective theory, a potentially very powerful substitute, but not independent from the underlying physics. On the other hand, starting from tasks as a kind of "game" (possibly with some failure probability) the information language gains in autonomy (with the implicit danger to miss hidden costs). This point of departure seems to underly investigations like [Horodecki (2011)].

Finally, also the proper definition of entropy in the quantum domain has led to some controversies. The standard definition is the von Neumann form (as widely used, also in this book), but alternatives have been postulated [Horodecki (2011); Dahlsten (2011); Egloff (2012)]. However, these alternatives either apply to non-equilibrium scenarios, cf. Example 4.5, or to certain "games" beyond conventional thermal processing, cf. Section 5.5.5.

A completely different approach to quantum thermodynamics has been suggested by [Gyftopoulos (2005)]. The authors consider entropy a "nonphysical statistical attribute" of matter. (It is formalized as the von Neumann entropy.) In their view thermodynamics does not result from quantum physics, rather there is a generalized evolution equation for both quantum mechanics and thermodynamics (in parallel). This generalized equation comes as a specific non-linear extension of the Liouville–von Neumann equation: Its non-Hamiltonian dissipative term is constructed to pull the momentary state into that with the largest entropy. (In our form of quantum thermodynamics this feature is *derived* for embedded quantum systems, not postulated.)

There are now two independent sources of thermal behavior, so it seems: the one resulting from this "modified quantum dynamics" and the one resulting already from standard quantum dynamics. It is unclear how these might combine in a meaningful way. The second source cannot be avoided, the first source can; Occam's razor would demand to get rid of dispensable complications. We have to wait for future developments.

Bibliography

Aaronson, S. (2013). http://arxiv.org/abs/1306.0159.

Abah, O., et al. (2012). *Phys. Rev. Lett.* **109**, 203006

Abe, S. (2006). *Physica A* **368**, 430.

Abe, S., and Okuyuma, S. (2011). *Phys. Rev. E* **83**, 021121.

Abel, L. (2008). *The Open Cybernetics and Systems Journal* **2**, 252.

Abrams, D. S., and Lloyd, S. (1998). http://arxiv.org/abs/quant-ph/9801041.

Adler, S. L., and Bassi, A. (2009). *Science* **325**, 275.

Aerts, D., Czachor, M., and Sozzo, S. (2011). http://arxiv.org/abs/1104.3345.

Aharonov, D., et al. (2007). *SIAM J. Comput.* **37**, 166.

Alicki, R., et al. (2004). http://arxiv.org/abs/quant-ph/0402012.

Alicki, R., Lidar, D. A., and Zanardi, P. (2006). *Phys. Rev. A* **73**, 052311.

Alicki, R., and Fannes, M. (2012). *Phys. Rev. E* **87**, 042123.

Allahverdyan, A. E., and Nieuwenhuizen, Th. M. (2005). *Phys. Rev. E* **64**, 056117.

Allahverdyan, A. E., Khrennikov, A., and Nieuwenhuizen, Th. M. (2005). *Phys. Rev. A* **72**, 032102.

Allahverdyan, A. E., Johal, R. S., and Mahler, G. (2008). *Phys. Rev. E* **77**, 041118.

Allahverdyan, A. E., and Janzing, D. (2008). *J. Stat. Mechan.: Theory Expt.*, P04001.

Allahverdyan, A. E., Balian, P., and Nieuwenhuizen, Th. M. (2011). http://arxiv.org/abs/1107.2138.

Allahverdyan, A., Hovhnannisyan, K. V. , Janzing, D., and Mahler, G. (2011). *Phys. Rev. E* **84**, 041109.

Allahverdyan, A., Hovhnannisyan, K. V., and Mahler, G. (2012). *Phys. Rev. Lett.* **109**, 248903.

Alvarez, G. A., Rao, D. D. B. , Frydman, L., and Kurizki, G. (2010). *Phys. Rev. Lett.* **105**, 160401.

Amin, M. H. S. (2009). *Phys. Rev. Lett.* **102**, 220401.

Anderson, P. W. (1972). *Science* **177**, 393.

Angulo-Brown, F., et. al. (2012). *Lat. Am. J. Phys. Educ.* **4**, 212.

Antoniou, I., et al. (2001). http://arxiv.org/abs/quant-ph/0012130.

Assis, A. V. D. B. (2011). *Ann.Phys.(Berlin)* **523**, 883.

Aspect, A., Grangier, P., and Roger, G. (1981). *Phys. Rev. Lett.* **47**, 460.

Atmanspacher, H., and Primas, H. (1927). *Foundations of Probability and Physics 3*, ed. by Krennikov, A., pp. 49–61 (AIP Press, New York).

Auletta, G., Ellis, G. F. R., and Jaeger, L. (2008). *J. R. Soc. Interface* **5**, 1159.

Baaquie, B. E., Coriano, C., and Srikant, M. (2002). http://arxiv.org/abs/cond-mat/0208191.

Baez, J. C. (2011). http://www.math.ucr.edu/home/baez/renyi.pdf.

Balatsky, A. V., and Zhu, J. (2003). *Physica E* **18**, 341.

Bandyopadhyay, J. N., et al. (2012). *Phys. Rev. Lett.* **109**, 110502.

Bandyopadhyay, S. http://www.cce.ufes.br/jair/thermo/Teaching_Fallacies.pdf.

Ballentine, L. E. (1998). *Quantum Mechanics* (World Scientific, Singapore).

Barbour, J. (2000). *The End of Time* (Oxford University Press, New York).

Barbour, J. (2011). http://www.fqxi.org/community/essay/winners/2011.1.

Barnum, H., et al. (1996). *Phys. Rev. Lett.* **76**, 2818.

Barnum, H., et al. (2007). *Phys. Rev. Lett.* **99**, 240501.

Bartlett, S. D., Rudolph, T., and Spekkens, R. W. (2006). *Int. J. Quant. Information* **04**, 17.

beim Graben, P., Filk, Th., and Atmanspacher, H. (2012). http://arxiv.org/abs/1204.5454.

Bekenstein, J. D. (1988). *Phys. Rev. A* **37**, 3437.

Bell, J. S. (1966). *Rev. mod. Phys.* **38**, 447.

Belokurov, V. V., et al. (2002). http://arxiv.org/abs/quant-ph/0210149.

Bender, C. M., Brody, D. C., and Meister, B. K. (2000). *J. Phys. A* **33**, 4427.

Bennett, C. H. (1982). *Int. J. Theor. Phys.* **21**, 905.

Bercher, J-F. (2008). http://www.esiee.fr/ bercherj/New/pubs/bercher-iwap08.ppdf

Bertram, G. W. (2011). *Kunst* (Reclam, Stuttgart).

Bérut, A., et al. (2012). *Nature* **483**, 187.

Bian, Z., et al. (2012). http://arxiv.org/abs/1201.1842.

Birjukov, J., Jahnke, Th., and Mahler, G. (2008). *Eur. Phys. J. B* **64**, 105.

Bishop, R. C. (2006). *Analysis* **66**, 44.

Bollinger, J. J., Itano, W. M., Wineland, D. J., and Heinzen, D. J. (1996). *Physs. Rev. A* 54, R4649.

Boole, G. (2009). *An Investigation of the Laws of Thought*, Reprint (Cambridge University Press).

Borowski, P., Gemmer, J., and Mahler, G. (2003). *Eur. Phys. J. B* **35**, 255.

Borowski, P., Gemmer, J., and Mahler, G. (2003). *Europhys. Lett.* **62**, 629.

Bostrom, N. (2002). *Anthropic Bias* (Routledge, New York, London).

Bostrom, N. (2003). *Philosophical Quaterly* **57**, 243.

Bostrom, N. (2008). *MIT Technical Review* **May/June**, 72.

Boukobza, E., and Tannor, D. J. (2006). *Phys. Rev. A* **74**, 063823.

Boykin, P. O., et al. (2002). *PNAS* **99**, 3388.

Braun, S., et al. (2013). *Science* **339**, 52.

Breuer, Th. (1995). *Philosophy of Science* **62**, 197.

Briggs, J. S., and Rost, J. M. (2000). *Eur. Phys. J. D* **10**, 311.

Brillouin, L. (1953). *J. Appl. Phys.* **24**, 1152.

Bridgman, P. W. (1927). *The Logic of Modern Physics* (Macmillan, New York).

Brody, Th. (1993). *The Philosophy Behind Physics* (Springer, Berlin).

Brunner, N., Branciard, C., and Gisin, N. (2008). *Phys. Rev. A* **78**, 052110.

Brunner, N., et al. (2012). *Phys. Rev. E* **85**, 051117.

Brookes, J. C., et al. (2007). *Phys. Rev. Lett.* **98**, 038101.

Bruza, P., Kitto, K., Neslon, D., and McEvoy, C. (2009). *J. Math. Psychology* **53**, 362.

Buchanan, M. (2011). *New Scientist* **2828**, September 5.

Bustos-Martin, R., Refael, G., and von Oppen, F. (2011). *Phys. Rev. Lett.* **111**, 060802.

Cai, J., Popescu, S., and Briegel, H. J. (2010). *Phys. Rev. E* **82**, 021921.

Callen, H. B., and Welton, T. A. (1951). *Phys. Rev.* **83**, 34.

Callender, J. S., and Rost, J. M. (2010). *Sci. Am.* **June 2010**, 58.

Campisi, M. Talkner, P., and Hanggi, P. (2009). *Phys. Rev. Lett.* **102**, 210401.

Campisi, M., Zueco, D., and Talkner, P. (2010). *Chem. Phys.* **375**, 187.

Cassidi, A. C., Clark, C. W., and Rigol, M. (2011). *Phys. Rev. Lett.* **106**, 140405.

Castellani, E. (2000). *Reductionism, emergence, and effective field theories*. http://philsci-archive.pitt.edu/106/.

Caticha, A. (2000). http://arxiv.org/abs/math-ph/0008017.

Cheng, T. (2005). *Relativity, Gravitation and Cosmology* (Oxford University Press, Oxford, New York).

Childs, A. M., Farhi, E., and Preskill, J. (2002). *Phys. Rev. A* **65**, 012322.

Chin, A. W., et al. (2013). *Nature Physics* **9**, 113.

Chiribella, G., et al. (2010). http://arxiv.org/abs/1008.0967.

Chou, C. W., et al. (2010). *Phys. Rev. Lett.* **104**, 070802.

Clune, J., Mouret, J., and Lipson, H. (2013). *Proc. Royal Soc. B* **280**, 20122863.

Cleuren, B., Rutten, B., and Van den Broeck, C. (2012). *Phys. Rev. Lett.* **108**, 120603.

Colbeck, R., and Renner, R. (2011). *Nature Commun.* **2**, 411.

Conrad, M. (1988). In: Herken, R. (Ed.), *The Universal Turing Machine* (Oxford University Press, Oxford).

Conway, J. H., and Kochen, S. (2009). *Notices of the AMS* **56**, 226.

Crespi, A., et al. (2012). http://arxiv.org/abs/1212.2783.

Crenkovic, G. D. (2011). *Information* **2**, 460.

Curzon, F. L., and Ahlborn, B. (1975). *Am. J. Phys.* **43**, 22.

d'Agostini, G. (1999). *Am. J. Phys.* **67**, 1260.

Dahlsten, O. C. O., Renner, R., Rieper, E., and Vedral, V. (2011). *New Journal of Physics* **13**, 053015.

Davies, P. (1990). http://arxiv.org/abs/astro-ph/0403050.

Dalibard, J., Castin, Y., and Molmer, K. (1992). *Phys. Rev. Lett.* **68**, 580.

Dajka, J., Luczka, J., and Hänggi, P. (2011). *Phys. Rev.* **A 84**, 032120.

Davies, P. C. W. (2008). In: *Quantum Aspects of Life* (eds. Abbott, D., Davies, P. C. W., and Pati, A. K.; Imperial College Press, London).

Deutsch, J. M. (1991). *Phys. Rev.* **A 43**, 2046.

Diu, B., Guthmann, C., and Lederer, D., (1989). *Elements de Physique Statistique* (Hermann Editeurs des Sciences et des Arts, Paris).

Deffner, S., Abah, O., and Lutz, E. (2010). *Chem. Phys.* **375**, 200.

del Rio, L., et al. (2011). *Nature* **474**, 61.

Dillenschneider, R., and Lutz, E. (2009). *Eur. Phys. Lett.* **88**, 50003.

Dobler, S., et al. (2012). *PNAS* **109**, 13040.

Donald, M. J. (1987). *J. Stat. Phys.* **49**, 81.

Dowker, F., and Kent, A. (1995). *Phys. Rev. Lett.* **75**, 3038.

Dressel, J., Agrawal, S., and Jordan, A. N. (2010). *Phys. Rev. Lett.* **104**, 240401.

Dürr, J., Nonn, T., and Rempe, G. (1998). *Nature* **395**, 33.

Duncan, T. L. (2005). http://arxiv.org/abs/cond-mat/0501014.

Dunkel, J., and Hilbert, S. (2013). http://arxiv.org/abs/1304.2066.

Egloff, D., et al. (2012) . http://arxiv.org/abs/1207.0434.

Einstein, A., Podolsky, B., and Rosen, N. (1935). *Phys. Rev.* **47**, 777.

Eisert, J., Wilkens, M., and Lewenstein, M. (1999). *Phys. Rev. Lett.* **83**, 3077.

Engel, G. S., et al. (2007). *Nature* **446**, 782.

d'Espagnat, B. (2003). *Veiled Reality* (Westview Press, Boulder, Colorado).

Ducheyne, S. (2006). *Journal of the History of Ideas* **67**, 443.

Esposito, M., et al. (2010). *Phys. Rev. E* **81**, 041106.

Farhi, E., et al. (2001). *Science* **292**, 472.

Feldmann, T., and Kosloff, R. (2003). *Phys. Rev. E* **68**, 016101.

Feynman, R. P., Leigthon, R. B., and Sands, M. (1963). *Feynman Lectures, Vol. I* (Addison Wesley, Reading, Mass.).

Feynman, R. P., Leigthon, R. B., and Sands, M. (1965). *Feynman Lectures, Vol. III* (Addison Wesley, Reading, Mass.).

Feynman, R. (1982). *Int. J. Theoret. Phys.* **21**, 467.

Ford, G. W., and O'Connell, R. F. (2006). *Phys. Rev. Lett.* **96**, 020402.

Franson, J. D. (2013). *Science* **339**, 767.

Fredkin, E. (1990). *Physica D* **45**, 254.

Friston, K. J., and Stephan, K. E. (2007). *Synthese* **159**, 417.

Fuchs, A., and Peres, A (2000). *Phys. Today* **51**, 70.

Garrido, P. L., Marro, J., and de los Santos, F. (eds.) (2011). *Non-equilibrium Statistical Physics Today AIP Conf. Proc.,* **1332** (American Institute of Physics).

Gelbwasser-Klimowsky, D., Alicki, R., and Kurizki, G. (2013). *Phys. Rev. E* **87**, 012140.

Gemmer, J., and Mahler, G. (2001). *Eur. Phys. J. D* **17**, 385.

Gemmer, J., Michel, M., and Mahler, G. (2009). *Quantum Thermodynamics*, 2nd ed. (Springer, Berlin).

Gerlich, S., et al. (2011). *Nature Commun.* **2**, 263.

Ghirardi, G. (2009). http://arxiv.org/abs/0904.0958.

Gisin, N. (1990). *Phys. Lett. A* **113**, 1.

Giulini, D., et al. (1996). *Decoherence and the Appearance of a Classical World in Quantum Theory* (Springer, Berlin).

Gleason, A. M. (2011). *J. Mathematics* **6**, 885.

Glowacki, D. R., Harwey, J. N., and Mulholland, A. J. (2012). *Nature Chemistry* **4**, 169.

Gorin, T., et al. (2006). *Physics Reports* **435**, 33.

Gogolin, C., Müller, M. P., and Eisert, J. (2011). *Phys. Rev. Lett.* **106**, 040401.

Gogolin, C., Kliesch, M., Aolita, L., and Eisert, J. (2013). http://arxiv.org/abs/1306.3995.

Goyal, P. (2012). *Information* **3**, 567.

Granzow, C. M., and Mahler, G. (1998). *Appl. Phys. B* **67**, 733.

Granzow, C. M. (1999). *PhD thesis, Stuttgart* unpublished.

Guerreschi, G. G., et al. (2012). *New J. Phys.* **14**, 053043.

Gyftopoulos, E. P., and Beretta, G. P. (2005). http://arxiv.org/abs/quant-ph/0507187.

Hagen, S., Hameroff, S. R., and Tuszynski, J. A. (2002). *Phys. Rev. E* **65**, 061901.

Haken, H. (1983). *Synergetics. An Introduction.* (Springer, Berlin).

Hameroff, S. R., and Penrose, R. (1996). *J. Conscious. Stud.* **3**, 36.

Harmuth, H. F. (1992). *Information Theory Applied to Space-Time Physics* (Worlds Scientific, Singapore).

Hartmann, M., Mahler,G., and Hess, O. (2004). *Phys. Rev. Lett.* **93**, 080402.

Hartmann, M., Mahler,G., and Hess, O. (2005). *J. Stat. Phys.* **19**, 1139.

Herrmann-Pillath, C. (2010). *Entropy* **12**, 197.

History of Information (Jeremy Norman Inc. 2004–2011), http://www.historyofinformation.com/.

Haake, F. (1991). *Quantum Signatures of Chaos* (Springer, Berlin).

Harrigan, N., and Spekkens, R. W. (2010). *Found Phys.* **40**, 125.

Hatano, N., and Suzuki, M. (2005). http://arxiv.org/abs/math-ph/0506007.

Hilt, S., Shabbir, S, Anders, J., and Lutz, E. (2011). *Phys. Rev. E* **83**, 030102.

Hime, T., et al. (2006). *Science* **314**, 1427.

Hines, A. P., Dawson, C. M., McKenzie, R. M., and Milburn, G. J. (2004). *Phys. Rev. A* **70**, 022303.

Hopkins, A., et al. (2003). *Phys. Rev. B* **68**, 235328.

Holt, J. (2012). *Why Does the World Exist?* (Profile Books Ltd., London).

Hookway, Ch. (1984) (ed.). *Minds, Machines & Evolution* (Cambridge University Press, Cambridge).

Horodecki, M. (2008). *Nature Physics* **4**, 833.

Horodecki, M., and Oppenheimer, J. (2011). *Nature Commun.* http://arxiv.org/abs/1111.3834, 26 June 2013.

Hughston, L. P., Jozsa, R., and Wootters, W. K. (1993). *Phys. Lett. A* **183**, 14.

Husumi, K. (1993). *Prog. Theor. Phys.* **9**, 381.

Hut, P., and von Fraassen, B. (1997). *J. Consciousness Studies* **4**, 2.

Ikeda, T. N., Watanabe, Yu, and Ueda, M. (2011). *Phys. Rev. E* **84**, 021130.

Jacobs, K. (2013). http://arxiv.org/abs/1304.0819.

Jahnke, Th. (2008). *Diploma thesis, University of Stuttgart*, unpublished.

Jahnke, Th., and Mahler, G. (2010). *Eur. Phys. Lett.* **90**, 50008.

Jahnke, Th., Lanery, S., and Mahler, G (2011). *Phys. Rev. E* **83**, 011109.

Jahnke, Th., and Mahler, G. (2011). *Phys. Rev. E* **84**, 011129.

Janotta, P., Gogolin, Ch., Barrett, J., and Brunner, N. (2011). *New J. Phys.* **13**, 063024.

Jarzynski, C. (1997). *Phys. Rev. E* **56**, 5018.

Jarzynski, C. (2011). *Annu. Rev. Condens. Matter Phys.* **2**, 329.

Jaynes, E. (1957). *Phys. Rev.* **106**, 620; **108**, 171.

Jaynes, E. (1965). *Am. J. Phys.* **33**, 391.

Ji, K., and Fine, B. V. (2011). *Phys. Rev. Lett* **106**, 620; **108**, 171.

Johal, R. S. (2010). *Phys. Rev. E* **82**, 061113.

Judson, R. S., and Rabitz, H. (1992). *Phys. Rev. Lett.* **68**, 1500.

Jülicher, F., Ajdari, A., and Prost, J. (1997). *Rev. mod. Phys.* **69**, 1269.

Kampis, G. (1991). *Self-Modifying Systems in Biology and Cognitive Science: A New Framework for Dynamics, Information, and Complexity* (Pergamon, Amsterdam).

Kauffman, S., and Smolin, L. (1997). http://arxiv.org/abs/gr-qc/9703026.

Kieu, T. D. (2004). *Phys. Rev. Lett.* **93**, 140403.

Kim, I., and Mahler, G. (2006). *Eur. Phys. J. B* **54**, 405.

Kim, I., and Mahler, G. (2010). *Phys. Rev. E* **81**, 011101.

Klumpp, S. (2012). *Phys. Scr.* **T151**, 014066.

Kochen, S., and Specker, E. P. (1967). *J. Math. Mech.* **17**, 59.

Kofler, J., and Brukner, C. (2009). http://arxiv.org/abs/1009.2654.

Kofman, A. G., and Kurizki, G. (1996). *Phys. Rev. A* **54**, R3750.

Kosloff, R., Geva, E., and Gordon, J. M. (2000). *J. Appl. Phys.* **87**, 8093.

Kosloff, R. (2013). *Entropy* **15**, 2100.

Krakauer, D. C. (2011). *Chaos* **21**, 037110.

Kolar, M., et al. (2012). *Phys. Rev. Lett.* **109**, 090601.

Kurchan, J. (2005). *Nature* **433**, 222.

Levitin, L. B., and Toffoli, T. (2011). *Int. J. Theor. Phys.* **50**, 3844.

Li, M., and Vitanyi, P. (1993). *An Introduction to Kolmogorov Complexity and its Applications* (Springer, New York).

Liboff, R. (2002). *Introductory Quantum Mechanics* (Addison Wesley).

Lambson, B., Carlton, D, and Bokor, J. (2011). *Phys. Rev. Lett.* **107**, 010604.

Landau, L. D., and Lifschitz, E. M. (1980). *Course of Theoretical Physics, Vol 5. Statistical Physics*, 3rd edition (Pergamon, London).

Landauer, R. (1961). *IBM J. Res. Dev.* **5**, 183.

Landauer, R. (1967). *IEEE Spectrum* **4**, 105.

Lebowitz, J. L. (1993). *Physics Today* **46**(9), 32.

Landauer, R. (1995). *Phil. Trans. R. Soc. London* **A 353**, 367.

Lavanda, B. H., and Dunning-Davies, J. (1991). *Int. J. Theor. Phys.* **30**, 907.

Lebowitz, J. L. (2011). *Non-equilibrium Statistical Physics Today* (eds. Garrido, P. L, Marro, J., and de los Santos, F.) *AIP Conf. Proc.* **1332**, p. 3.

Leggett, A. J. (2001). *J. Phys.: Condens. Matter* **14**, R415.

Leggett, A. J. (2002). *Rev. mod. Phys.* **73**, 307.

Leitsch, A., Schachner, G., and Svozil, K. (2008). *Complex Systems* **18**, 131.

Levine, G., and Muthukumar, V. N. (2011). *Phys. Rev. B* **69**, 113203.

Levy, G., Alicki R., and Kosloff, R. (2012). *Phys. Rev. Lett.* **109**, 248901.

Levy, G., and Kosloff, R. (2012). *Phys. Rev. Lett.* **108**, 070604.

Lieb, E. H., and Simon, B. (1973). *Phys. Rev. Lett.* **31**, 681.

Lieb, E. H., Balian, and Yngvason, J. (2002). http://arxiv.org/abs/math-ph/0204007.

Linden, N., et al. (2009). *Phys. Rev.E* **79**, 061103.

Linden, N., Popescu, S., and Skrzypczyk, P. (2010). *Phys. Rev. Lett.* **105**, 130401.

Lindblad, G. (1976). *Commun. Math. Phys.* **48**, 119.

Link, G. (2004). *One Hundred Years of Russel's Paradox* (Walter de Gruyter, Berlin).

Lloyd, S. (1997). *Phys. Rev. A* **56**, 3374.

Lloyd, S. (2000). *Phys. Rev. A* **62**, 022108.

Lloyd, S. (2002). *Phys. Rev. Lett.* **88**, 237901.

Lloyd, S. (2006). *Programming the Universe* (Alfred A. Knopf, New York).

Loschmidt, J. J. (1876). *Wiener Ber.* **73**, 128.

Lubkin, E. (1978). *J. Math. Phys.* **19**, 1028.

Lundeen, J. S., et al. (2011). *Nature* **474**, 188.

Lyre, H. (1997). In *Time, Temporality, Now* (edited by Atmanspacher, H., and Ruhnau, E. Springer, Berlin), p. 81.

Maccone, L. (2003). *Phys. Rev. Lett.* **103**, 080401.

Mahler, G., and Weberuss, V. A. (1998). *Quantum Networks* (Springer, Berlin).

Mahler, G. (2012). *Physics* **5**, 55.

Margolous, N., and Levitin, L. B. (1998). *Physica* **120 D**, 188.

Mari, A., and Eisert, J. (2012). *Phys. Rev. Lett.* **108**, 120602.

Marshall, W., et al. (2003). *Phys. Rev. Lett.* **91**, 130401.

Martinez, E. A., and Paz, J. P. (2013). *Phys. Rev. Lett.* **110**, 130406.

Matheny, M. H., et al. (2013). *Nano Letters* **13**, 1622.

McFadden, J. (2000). *Quantum Evolution* (Harper Collins, New York).

McKemmish, L. K., et al. (2009). *Phys. Rev. E* **80**, 021912.

Mead, C. A. (1977). *J. Chem. Phys.* **66**, 459.

Mermin, N. D. (2003). *Am. J. Phys.* **66**, 753.

Messiah, A. (2000). *Quantenmechanik, Band 2* (Walter de Gruyter, Berlin).

Misra, B., and Sudarshan, E. C. G. (1977). *J. Math. Phys.* **18**, 756.

Mølmer, K. (2003). *Phys. Rev. A* **55**, 3195.

Mott, N. F. (1931). *Proc. Cambr. Phil. Soc.* **27**, 553.

Moyer, M. (2012). http://blogs.scientificamerican.com/observations/2012/04/12/will-computers-ever-know-everything/

Mukamel, S. (2003). *Phys. Rev. Lett.* **90**, 170604.

Muñoz, E., and Peña, F. J. (2012). *Phys. Rev. E* **86**, 061108.

Myhrvold, N. (2012). http://edge.org/response-detail/2930/.

Nielsen, M. A., and Chuang, I. L. (2000). *Quantum Computation and Quantum Information* (Cambridge University Press, Cambridge).

Oppenheim, J., et al. (2002). *Phys. Rev. Lett.* **89**, 180402.

Otte, A. (2001). *PhD Thesis*, Stuttgart 2001.

Page, D. N. (1993). *Phys. Rev. Lett.* **71**, 1291.

Pais, A. (1982). *Subtle is the Lord: The Science and Life of Albert Einstein* (Oxford University Press, Oxford).

Partovi, M. H. (2008). *Phys. Rev. E* **77**, 021110.

Passon, O. (2006). http://philsci-archive.pitt.edu/3026/1/bohm.pdf.

Pattee, H. H. (2002). *Lecture Notes in Computer Science* **929**, 21.

Pendry, J. B. (1983). *J. Phys. A* **16**, 2161.

Peng, X., et al. (2008). *Phys. Rev. Lett.* **101**, 220405.

Penrose, R. (1989). *The Emperor's New Mind* (Oxford University Press, Oxford New York).

Peres, A. (1990). *Found. Physics* **12**, 1441.

Peres, A. (1996). *Phys. Rev. Lett.* **77**, 1413.

Pike, E. R., and Sarka, S. (1989). *The Quantum Theory of Radiation* (Oxford University Press, Oxford New York).

Pothos, E. M., and Busemeyer, J. R. (2012). *Behavioral Brain Sciences.*, in press.

Plenio, M. B., and Vitelli, V. (2001). *Contemporary Physics* **42**, 25.

Plenio, M. B., and Huelga, S. F. (2002). *Phys. Rev. Lett.* **88**, 197901.

Poon, Ch., and Zhou, K. (2002). *Frontiers in Neuroscience* **5**, 108.

Popescu, S. (2009). http://arxiv.org/abs/1009.2536.

Prior, J., et al. (2010). *Phys. Rev. Lett.* **105**, 050404.

Primas, H. (2007). *Mind & Matter* **5**, 7.

Pusey, M. F., Barrett, J., and Rudolph, T. (2012). *Nature Physics*, **8**, 476.

Quan, H. T., et al. (2007). *Phys. Rev. E* **76**, 031105.

Raizen, M. (2009). *Science* **324**, 1403.

Rapp, A., Mandt, S., and Rosch, A. (2010). *Phys. Rev. Lett.* **105**, 220405.

Rechenberg, P. (2003). *Informatik Spektrum*, 14. Okt. 2003, p. 317.

Renyi, A. (1961). *Proc. 4th Berekly Symposium on Mathematics, Statistics asn Probability*, 1960, p. 547.

Reimann, P. (2008). *Phys. Rev. Lett.* **101**, 190403.

Rempp, F. (2007). *Diploma thesis, University of Stuttgart*, unpublished.

Rezek, Y., et al. (2009). *Europhys. Lett.* **85**, 30008.

Rezek, Y. (2010). *Entropy* **12**, 1885.

Rigol, M., et al. (2007). *Phys. Rev. Lett.* **98**, 050405.

Ritchie, N. W. M., et al. (1991). *Phys. Rev. Lett.* **66**, 1107.

Ritort, F. (2003). *Poincaré Seminar* **2**, 195; http://arxiv.org/abs/cond-mat/0401311.

Ritz, T. Adem, S., and Schulten, K. (2000). *Biophys. J.* **78**, 707.

Roederer, J. G. (2004). *The New Avenues in Bioinformatics* (eds. Seckbach, J., and Rubin, E., Kluwer Academic Publ., Dordrecht).

Roederer, J. G. (2005). *Information and Its Role in Nature* (Springer, New York).

Römer, H., and Filk, Th. (1994). *Statistische Mechanik* (VCH, Weinheim).

Romero, K. M. F., and Franco, R. L. (2012). *Phys. Scr.* **86**, 065004.

Rössler, O. E. (1998). *Endophysics: The world as an Interface* (World Scientific, Singapore).

Rovelli, C. (1996). http://arxiv.org/abs/quant-ph/9609002.

Rovelli, C. (1996). *Int. J. Theoret. Phys.* **35**, 1637.

Rovelli, C. (1998). *Found. Phys.* **28**, 1031.

Rovelli, C. (2009). http://arxiv.org/abs/0903.3832.

Rovelli, C. (2011). *The First Scientist: Anaximander and His Legacy* (Westholme Publishing, Chicago).

Sagawa, T., and Ueda, M (2008). *Phys. Rev. Lett.* **100**, 080403.

Saunders, S. (2004). *Proc. Royal Soc. London A* **460**, 1.

Salthe, S. N. (2011). *Information* **2**, 417.

Sarowar, M., et al. (2010). *Nature Physics* **6**, 462.

Sarandy, M. S., Wu, L. A., and Lidar, D. A. (2004). *Quantum Information Processing* **3**, 331.

Sasaki, K., Kanada, R., and Amari, S. (2007). *J. Phys. Soc. Japan* **76**, 023003.

Sayrin, C., et al. (2011). *Nature* **477**, 73.

Schilpp, P. A. (ed.) (1969). *Albert Einstein: Philosopher-Scientist* Vol. 1 (Open Court, Illinois).

Schmidt, H., and Mahler, G. (2005). *Phys. Rev. E* **72**, 016117.

Schreiber, D. J., and Wipplinger, H. P. (eds.) (2011). *Jonas Burgert. Experimentation In Vivo* (Verlag der Buchhandlung Walther König, Köln).

Schröder, H., and Mahler, G. (2010). *Phys. Rev. E* **81**, 021118.

Schweber, S. S. (1993). *Physics Today* **Nov. 1993**, 34.

Scovil, H. E. D., and Schulz-DuBois, E. O. (1959). *Phys. Tev. Lett.* **2**, 262.

Scully, M. O., et al. (2003). *Science* **299**, 862.

Seifert, U. (2012). *Rep. Prog. Phys.* **75**, 16001.

SETI Institute. http://www.seti.org

Shannon, C. E. (1948). *Bell Sys. Tech. J.* **27**, 379; 623.

Sher, S., and McKenzie, C. R. M. (2011). In: Keren, G. (Ed.), *Perspectives on Framing*, pp. 35 (Psychology Press, Talor & Francis Group., UK).

Sinha, U., et al. (2010). http://arxiv.org/abs/1007.4193.

Smolin, L. (2001). http://arxiv.org/gr-qc/0104097.

Smolin, L., and Smith, G. (2013). http://arxiv.org/abs/1305.4904.

Srikhirin, P., Aphornratana, S., and Chungpaibulpatana, S. (2011). *Renewable and Sustainable Energy Reviews* **5**, 343.

Stapp, H. (2007). *Mindful Universe* (Springer, The Frontiers Collection, Berlin, Heidelberg).

Strauss, Y., et al. (2008). http://arxiv.org/abs/0802.2448.

Susskind, L. (2008). *The Black Hole War* (Little, Brown and Comp., New York).

Symul, T., Assad, M., and Lam, P. K. (2011). *Appl. Phys. Lett.* **98**, 231103.

Szilard, L. (1929). *Zeitschr. f. Physik* **53**, 840.

Talkner, P., Lutz, E., and Hänggi, P. (2007).*Phys. Rev. E* **75**, 050102.

Teifel, J., and Mahler, G. (2010). *Eur. Phys. J. B* **75**, 275.

Thirring, W. (2008). *Quantum Mathematical Physics: Atoms, Molecules and Large Systems* (Springer, Berlin, Heidelberg, New York).

Thomas, G., and Johal, R. S. (2011). *Phys. Rev. E* **83**, 031135.

Titulaer, U. M., and Glauber, R. J. (1965). *Phys. Rev.* **140**, B 676.

Toffoli, T. (1982). *Int. J. Theoret. Phys.* **21**, 165.

Tonner, F., and Mahler, G. (2006). *Fortschr. Phys.* **54**, 939.

Toyabe, S., et al. (2010). *Nature Physics* **6**, 988.

Tsallis, C. (2000). http://arxiv.org/abs/cond-mat/0010150.

Turin, L. (1982). *Chem. Senses* **21**, 773.

Unruh, W. G. (1994). *Phys. Rev. A* **50**, 882.

van Kampen, N. G. (2008). *Stochastic Processes in Physics and Chemistry* (North Holland, Amsterdam).

Volokitin, A. I. (2011). http://arxiv.org/abs/1112.4912.

Vutha, A. C. (2010). http://arxiv.org/abs/1001.3322.

Waldherr, G., and Mahler, G. (2010). *Eur. Phys. Lett.* **89**, 40012.

Waldherr, G., and Mahler, G. (2010). *Phys. Rev. E* **81**, 061122.

Weimer, H., Henrich, M. J., Rempp, F., Schröder, H., and Mahler, G. (2008). *Eur. Phys. Lett.* **83**, 30008.

von Weizsäcker, C. F (1980). *The Unity of Nature* (Farrar Straus & Giroux, New York).

Wharton, K. (2011). http://arxiv.org/abs/1106.1254.

Wheeler, J. A., and Ford, K. (2002). *Geons, Black Holes & Quantum Foam* (W. W. Norton & Comp., New York).

Wiseman, H. M., and Eisert, J. (2007). http://arxiv.org/abs/0705.1232.

Wiseman, H. M., and Gambetta, J. M. (2011). http://arxiv.org/abs/1110.0069.

Wissner-Gross, A. D., and Freer, C. E. (2013). *Phys. Rev. Lett.* **110**, 168702.

Wittmann, B., et al. (2011). http://arxiv.org/abs/1111.0760.

Wright, L. (1973). *The Philosophical Review* **82**(2), 139.

Xu, N., et al. (2012). *Phys. Rev. Lett.* **108**, 130501.

Yan, Z., and Chen, J. (1990). *J. Phys. D* **23**, 136.

Yan, Z. (2000). *Eur. J. Phys.* **21**, 625.

Yirka, B. (2012). http://phys.org/news/2012-11-physicists-theorize-entangled-quantum-batteries.html.

Yockey, H. P. (2005). *Information Theory, Evolution, and the Origin of Life* (Cambridge University Press, Cambridge).

Youssef, M., Mahler, G., and Obada, A. S. F. (2009). *Phys. Rev. E* **80**, 061129.

Zanardi, P. (2001). *Phys. Rev. Lett.* **87**, 077901.

Zhen, Y., et al. (2012). *Science* **337**, 1634.

Zeilinger, A., et al. (2005). *Nature* **433**, 230.

Zukowski, M., et al. (1993). *Phys. Rev. Lett.* **71**, 4287.

Zurek, W. H., and Paz, J. P. (1996). http://arxiv.org/abs/quant-ph/9612037.

Zurek, W. H. (2003). *Rev. mod. Phys.* **75**, 715

Zurek, W. H. (2003). *Phys. Rev. A* **67**, 012320.

Zyczkowski, K. (2003). *Open Sys. & Information Dyn.* **10**, 297.

Index